AF560509

Global Warming and Forest:
An Overview of Current Knowledge

Global Warming and Forest: An Overview of Current Knowledge

Pawan Pattanaik

RANDOM PUBLICATIONS
NEW DELHI (INDIA)

Global Warming and Forest: An Overview of Current Knowledge

Reprinted - 2018

ISBN 978-93-5111-833-6

Published in 2016 in India by

RANDOM PUBLICATIONS

4376-A/4B, Gali Murari Lal, Ansari Road
New Delhi-110 002
Phone : +9111-43580356, 011-23289044, 011-43142548
e-mail: sales@randompublications.com,
info@randompublications.com, randomexports@gmail.com

Type Setting by : Friends Media, Delhi-110089
Printed at : Mehra Printers, Delhi-110 092

Preface

Global warming is one of the most controversial issues facing the world today. It is controversial not only because it involves very serious, far-reaching—global—issues that can permanently and negatively affect the world, but also because it involves the lifestyles and personal choices of everyone on Earth. While it is true that some people and ecosystems may suffer the negative effects more than others and some people may have to be willing to make larger lifestyle changes, everyone will have a responsibility and a stake in the outcome. If ecologically sound decisions are made, then life will be much better for not only today's population but also the generations of the future. If people choose not to make wise environmental choices today, then the future of upcoming generations will pay the price.

The teeming life of forests, and the physical structures containing them, are in continuous flux with incoming solar energy, the atmosphere, the water cycle and the carbon cycle—in addition to the influences of human activities. The complex relationships both add to and subtract from the equations that dictate the warming of the planet.

Forests play an important role in regulating the earth's temperature and weather patterns by storing large quantities of carbon and water. This regulatory function has a profound effect on both the local and the global climate. Locally, trees provide shade, which in turn lowers summer temperatures and prevents the soil from drying out, they reduce heat loss from the ground in winter and reduce storm damage by providing shelter from wind. Globally, forests regulate the global carbon cycle, having a profound effect on the climate. As well as this, deforestation is also contributing to climate change. Indeed the CO2 released each year from forest loss is higher than that released by our yearly transport emissions.

This book alerts readers to another global threat that could be worse than carbon emissions, namely, the problem of nitrogen overfertilization.

–Author

Contents

1

Global Warming

Global warming is one of the most controversial issues facing the world today. It is controversial not only because it involves very serious, far-reaching—global—issues that can permanently and negatively affect the world, but also because it involves the lifestyles and personal choices of everyone on Earth. While it is true that some people and ecosystems may suffer the negative effects more than others and some people may have to be willing to make larger lifestyle changes, everyone will have a responsibility and a stake in the outcome. If ecologically sound decisions are made, then life will be much better for not only today's population but also the generations of the future. If people choose not to make wise environmental choices today, then the future of upcoming generations will pay the price.

This stage presents the current effects of global warming on ecosystems and what that means for the future, as well as the scientific findings of more than 2,500 scientists worldwide working together to better understand the pertinent aspects of global warming and what they mean to every individual on Earth. It then discusses global warming's present and potential effects on ecosystems if global warming continues unabated and why the Earth will become a very different place to try to adapt to in the future. In order to understand what certain parts of the world will be like in the face of global warming, it is first necessary to understand the concept of ecosystems and the related signs and effects of global warming.

GLOBAL WARMING SCENARIOS

Although the science remains uncertain, 1990 marked the year when many governments accepted the impending reality of global climate change. Some governments began to announce targets for reductions in carbon dioxide in order to prompt the international negotiation process. Some governments remain unconvinced that the greenhouse effect is real. Others accept the science, but not the need to act dramatically. This volume is concerned not with the scientific debate as such, but with the basis for climate change policy in OECD countries. Crucial to the whole process of policy development is an assessment of the impacts of climate change. These impacts will appear first as physical changes in the environment, but will eventually translate into social and economic changes as well. Impacts are important, essentially because they represent the potential benefits that might be realised from taking political action against the climate change problem.

If impacts are projected to be small, or if it is likely that the forecasted impacts could be easily managed, then the benefits of measures to contain global warming will be small, and the balance of costs and benefits will militate against major policy action. On the other hand, if the impacts are projected to be large, or if there is a real risk of some of them taking on "catastrophic" proportions, then early and firm action would be prudent. Put another way, the cost-benefit balance, however imperfectly understood, is fundamental to the policy process.

Because the costs of environmental policy are typically easier to quantify than the benefits, policy discussions often begin with an exogenously-determined "target", and seek to minimise the costs of achieving that target. This is more cost-effectiveness analysis than it is cost-benefit analysis. It has the advantage of being conceptually simpler, but it runs the risk that the initial target will be inappropriately set. If this target is too stringent (or not stringent enough), there is a likelihood that both economic and environmental resources will be "wasted". Given the long time horizons involved in the climate change issue, and given the world-wide economic and environmental implications of that issue, the potential for such wastage is very high indeed. It is a basic premise of this book that economic resources should be allocated to resolving the climate change problem (i.e. costs should be incurred) only to the extent that the benefits (i.e. reduced damages potential) warrant these costs. In order to define this "tradeoff" point, an effort has to be made to measure the potential impacts of global warming.

The Intergovernmental Panel on Climate Change (IPCC) was set up by the United Nations Environment Programme (UNEP) and the World Meteorological Organisation (WMO) in 1988 to investigate global warming. It established three working groups to help it fulfil its mandate. Working Group 1 (Chaired by the United Kingdom) has looked at the scientific evidence in support of global warming. Working Group 2 (chaired by the Soviet Union) has investigated potential impacts. Working Group 3 (chaired by the United States) has evaluated various policy response options.

Working Group 1 has forecast that global warming will indeed occur, and has assessed four possible scenarios:

Scenario A: "Business as usual" i.e. a continuation of current trends in greenhouse gas emissions, which will produce a rate of warming in terms of global mean surface temperatures of 0.2°C-0.5°C per decade in the 21st century, with a "best guess" of 0.3°C. This would produce a 1 degree rise in temperature by 2025 and 3 degrees by 2100, compared to 1990 levels.

Scenario B: Deforestation is halted, natural gas is increasingly substituted for coal, energy conservation occurs, but temperatures still rise by 2°C by 2100 compared to today, i.e. a rise of 0.2°C per decade.

Scenario C: A greater switch to renewable energy sources in the second half of the 21st century holds the temperature rise to a little above 0.1°C per decade.

Scenario D: Assumes that the switch to renewables occurs in the first half of the 21st century, which stabilises gas concentrations in the atmosphere.

The temperatures in question are global mean temperatures. For policy purposes, these estimates are not very useful. What matters more is the regional variation which will exist around about the mean. For example, in the "business as usual" scenario, temperatures are expected to be higher than the mean in Southern Europe and central North America, with reduced summer rainfall and soil moisture. By 2030, winter temperatures could rise in these regions by 2°C (perhaps more in Central North America), and summer temperatures by 2-3°C. In the Sahel, warming will increase by 1-3°C. South East Asia might experience temperature rises of 1-2°C, and Australia some 1-2°C in summer and 2°C in winter. Sea levels rise in the "business as usual" scenario by some 6 cms per decade due to thermal expansion of the oceans, and the melting of some land ice, producing a rise of some 20 cms by 2030, and 65 cms by 2100.

Regional Shares of Emissions

The importance of the USA and USSR/Eastern Europe are clearly seen, as is the role of European Community countries (which make up the bulk of the "rest of OECD"). The rapid growth of the developing countries' shares is also seen - note the very rapid rise in the share of Centrally Planned Asia (China mainly), and other developing countries. Any agreement on emission controls must, therefore, secure the co-operation of both developed and developing countries, if it is to be successful.

Impacts

The Working Group 2 report emphasises the fact that uncertainty is a major problem to be overcome in assessing the socio-economic impacts of climate change. Given the large uncertainties that are involved, it is understandable that most of the impacts described in their report are outlined in qualitative terms only. Nevertheless, Working Group 2 foresees significant impacts in virtually all economic sectors, assuming that temperatures and sea levels rise as predicted. More specifically, Working Group 2 recommends that efforts should be intensified to better understand the linkages between physical and socio-economic impacts, and that better methodologies for quantifying these impacts need to be developed.

Table. Regional contributions to Greenhouse Gas Emissions

CO_2 Anthropogenic Emissions Shares (%)				
Source	1985	2000	2015	2050
United States	21	19	16	12
Rest of OECD	22	19	16	12
USSR and Eastern Europe	22	22	19	18
Centrally Planned Asia	10	13	16	21
Other Developing Countries	25	28	32	37
Commercial Energy	86	87	89	92
Tropical Deforestation	12	11	9	6
Other	2	2	2	2
Total (10^9 tonnes Carbon)	5.99	8.05	10.27	16.95
Average Annual Growth Rate	1.6%			
Methane Anthropogenic Emissions Shares (%)				
Source	1985	2000	2015	2050
United States	12	11	9	8
Rest of OECD	13	12	12	10
USSR and Eastern Europe	13	14	14	15
Other Developing Countries	17	16	17	19
Other Developing Countries	46	47	49	48
Fuel Production	18	22	26	32
Enteric Fermentation	23	24	23	22
Rice Cultivation	34	31	29	24
Landfills	9	10	10	14
Tropical Deforestation	6	6	5	4
Other	9	7	7	
Total (10^1 tonnes CH_4)	320.1	399.5	476.8	710.5
Average Annual Growth Rate	1.2%			

To some extent, this conclusion is based on the three methodological studies contained in this book. Working Group 2 recognised early in its mandate that hard data on the impacts of climate change simply does not exist. Some studies had been carried out on potential *physical* impacts, but virtually no guidance existed on the possible "downstream" social, and economic response functions. For that reason, Working Group 2 asked OECD to work on developing "more robust" methodologies for assessing these types of impact. The papers presented in this book are a direct response to this request.

Policy Options

How should the international community respond to the issue of global warming? Working Group 3 points out that there are basically three options:

Do Nothing

The arguments for doing nothing are:

- That global warming is all very uncertain anyway;
- That doing something may be expensive and that costs borne today will benefit generations yet to come (an inequitable situation);

- That delaying present action will improve the level of scientific information about global warming; and
- That the world will adapt "automatically" through migration, changing agricultural production, switching expenditures to sea level defences, and so on.

The arguments against doing nothing are:

- That global warming is irreversible. The longer the delay, the more warming the world is committed to. If the worst fears are realised, it will be too late to correct it;
- That scientific information can be improved even while further action is being considered;
- That delay is "intergenerationally unfair" — current generations have an obligation not to impose heavy costs on future generations; and
- That action now could be relatively cheap (e.g. energy conservation), and will likely yield other environmental benefits (e.g. reduced acid rain; reduced congestion, if the contribution of road travel to CO_2 is properly tackled, etc.).

Reduce Emissions and Invest in Carbon "Sinks"

The arguments for reducing emissions are very much the same as the arguments against doing nothing. The additional advantage of acting to reduce the problem at source is that it gives greater certainty compared to adaptation policies. Most importantly, global warming may have a large potential for "surprise" events, e.g. major climatic incidents, simply because we know so little about how climatic systems function. These surprises may be very damaging. By reducing emissions, we lower the potential for unpleasant surprises.

The obvious argument against reducing greenhouse gas (GHG) emissions is that, while it may be fairly cheap in the short run, further reductions in emissions will get more and more expensive. A number of studies have estimated costs of adjustment, and some have suggested that major reductions in GHGs could be very costly in economic terms.

Afforestation offers the potential for "fixing" CO_2, although due allowance has to be made for carbon release once the forest rotation has ended and the wood decays.

Much depends on how the wood is used. For example, wood for paper pulp would release carbon fairly quickly, or would contribute to methane releases in landfill sites. Wood used for furniture and building could store carbon for much longer. Afforestation in the tropics might fix carbon on an even bigger scale.

Afforestation also introduces the idea of "carbon offsets". An electric power utility might be allowed to build a new power station, provided it "offset" the projected emissions with carbon-fixing trees. Since it does not matter where the emissions and fixing take place (in terms of global warming), the prospect is opened up for tropical afforestation as an offset to increased carbon emissions in the rich world. This is already happening. One US utility has agreed to conserve tropical forest in Guatemala. Dutch utilities may well plant new forests in Colombia, Ecuador and Peru. The arguments against afforestation as a "cure" are:

- It could be very expensive;
- Much modern afforestation is regarded as being environmentally unattractive anyway (e.g. conifer plantations); and
- In order to overcome the problem of merely postponing carbon releases (because of wood decay), afforestation would have to proceed at a faster and faster pace over time.

Invest in Adaptation

Under the "do nothing" option, adaptation is "natural": it simply comes about because people respond to climatic change. But governments could also invest in adaptation by e.g. raising existing sea defences and building new ones; investing in new research and development (e.g. more climate resistant crops); protecting groundwater systems from satinisation; relocating communities; etc. The argument for adaptation is that it may be a lot cheaper than emission reduction, especially if costs could be spread out over very long periods of time.

The argument against adaptation is that it does nothing to reduce further warming, and simply imposes bigger and bigger cost burdens on future generations. Although it is illustrative to fix our attention on certain threshold points in time (e.g. the year in which CO_2 emission levels are expected to reach twice their current magnitude), it should not be forgotten that warming will continue long beyond this point, if no action is taken.

Choosing a Policy Mix

In the end, international global warming policy will probably consist of some measure of all three of these options. In particular:

- Some warming will be allowed to take place, simply because the economic dislocation caused by aiming for stabilisation of temperature is likely to be too great;
- Because some warming will take place, both "natural" and "engineered" adaptations will occur;
- The international community is already focusing on emission reduction (coupled with afforestation) as its primary policy response. From the discussion above, it is clear that, so long as the welfare of future generations matters, this is a sensible approach.

EFFECTS OF GLOBAL WARMING

Although not every scientist worldwide may look at global warming in the same way, they do overwhelmingly agree that the Earth's atmosphere is getting warmer. Worldwide temperatures have risen more than 1°F over the past century, and 17 of the past 20 years have been the hottest ever recorded. A special report issued by Time magazine on April 3, 2006, the Intergovernmental Panel on Climate Change in their third report, released in 2001, had analysed data from the past two decades representing properties such as air and ocean temperatures and the habitat characteristics and patterns of wildlife. Examples of observed changes included"shrinkage of glaciers, thawing of permafrost, later freezing and earlier breakup of ice on rivers and lakes, lengthening of mid- to high-latitude growing seasons, poleward and altitudinal shifts of plant and animal ranges, declines of some plant and animal populations, and earlier flowering of trees, emergence of insects, and egg-laying in birds. Associations between changes in regional temperatures and observed changes in physical and biological systems have been documented in many aquatic, terrestrial, and marine environments."

The IPCC is an organized group of more than 2,500 climate experts from around the world that consolidates their most recent scientific findings every five to seven years into a single report, which is then presented to the world's political leaders. The IPCC was established in 1988 by the World Meteorological Organization and the United Nations Environment Programme to specifically address the issue of global warming. As a result of their comprehensive analysis, they have determined that this steady warming has had a significant impact on at least 420 animal and plant species and also on natural processes. Furthermore, this has not just occurred in one geographical location but worldwide. In the IPCC's fourth report, released in February 2007, they concluded that it is"very likely" that heat-trapping emissions from human activities have caused"most of the observed increase in globally averaged temperatures since the mid-20th century."

Also, in the February 2007 report, they concluded the following:

- "Human induced warming over recent decades is already affecting many physical and biological processes on every continent. Nearly 90 per cent of the 29,000 observational data series examined revealed changes consistent with the expected response to global warming, and the observed physical and biological responses have been the greatest in the regions that have warmed the most."

In these studies, scientists have been able to break down the natural and human-caused components in order to see how much of an effect humans have had. Human effects can include activities such as burning fossil fuels, agricultural practices, deforestation, industrial processes, the introduction of invasive plant or animal species, and various types of land-use change. In many cases, scientists do not need to look very far to see the effects a warming world is having on the environment and the Earth's ecosystems. Glaciers worldwide are melting at an accelerated rate never seen before. The cap of ice on top of Kilimanjaro is rapidly disappearing, the glaciers of world-renowned Glacier National Park in the United States and Canada are melting and projected to be gone in the next few decades, and the glaciers in the European Alps are experiencing a similar fate. In the world's tropical oceans, vast expanses of beautiful, brilliantly coloured coral reefs are dying off as oceans slowly become too warm.

Unable to survive the higher temperatures, the corals are undergoing a process called bleaching and are turning white and dying. In the Arctic, as temperatures climb, ice is melting at accelerated rates, leaving polar bears stranded, destroying their feeding and breeding grounds, and causing them to starve and drown. Permafrost is melting at accelerated rates. As the ground thaws, it is disrupting the physical and chemical components of the ecosystem by causing the ground to shift and settle, toppling buildings and twisting roads and railroad tracks, as well as releasing methane gas into the atmosphere. Weather patterns are also changing. El Niño events are triggering destructive weather in the eastern Pacific. There has been an increase in extreme weather events, such as hurricanes. Droughts have become more prevalent in some geographical areas, such as parts of Asia, Africa, Australia, and the American Southwest.

Animal and plant habitats have been disrupted, and, as temperatures continue to climb, there have been several documented migrations of individual species moving northward or to higher elevations on individual mountain ranges. Migration patterns are also being affected, such as those already documented of beluga whales, butterflies, and polar bears. Spring is also arriving earlier in some areas, which is now influencing the timing of bird and fish migration, egg laying, leaf unfolding, and spring planting for agriculture. In fact, based on satellite imagery documentation of the Northern Hemisphere, growing seasons have steadily become longer since 1980. While species have been faced with changing environments in the past and have been able to adapt in many cases, the IPCC climate change scientists view this current rate of change with alarm. They fully expect the magnitude of these changes to increase with the temperatures over the next century and beyond.

The concern is that many species and ecosystems will not be able to adapt as rapidly as the effects of global warming will cause the environment to change. In addition, there will also be other disturbances, such as floods, insect infestations, and the spread of disease, wildfire, and drought. Any of these additional challenges can destroy a species or habitat. In particular, alpine and polar species are especially vulnerable to the effects of climate change because as species move northward or higher on mountains, these species' habitats will shrink, leaving them with nowhere to go. With so much evidence, most scientists no longer doubt that global warming is real, nor do they question the fact that humans are to blame.

All it takes is a look at the air quality over significant population and industrial centres to begin to grasp the effect that humans can have on the environment. Based on temperature records kept before the beginning of the Industrial Revolution, carbon dioxide in the atmosphere has increased 30 per cent above those earlier levels. Not only are the levels higher, but they increase annually. The IPCC, at current conditions, by the year 2100, the average temperature is expected to increase between 2° and 11.5°F- an amount more than 50 per cent higher than what was predicted only 50 years ago. Within the IPCC's predicted temperature range, at the lower end, storms would become more frequent and intense, droughts would be more severe, and coastal areas would be flooded by rising sea levels from melting glaciers and ice caps. There would be enough of a disruption that ecosystems worldwide would be thrown out of balance and altered. If, however, the temperature rise falls towards the higher end of the estimate, the results on ecosystems worldwide would be disastrous. Sea levels could rise so much that entire islands of low elevation, such as the Maldives, could completely disappear. Other areas, such as the Nile Delta and much of the United States' coastal southeast could become completely uninhabitable. Climate zones could shift, completely disrupting land-use practices. For instance, the current agricultural region of the Great Plains in the United States could be shifted to Canada. The southern portion of the United States could become more like Central or South America. Siberia would no longer be a frozen, desolate landscape. Parts of Africa could become dry, desolate wastelands. If this were to happen, it would have a severe impact on the production of agriculture.

Areas currently equipped to produce agriculture would no longer be able to, and areas that were able, based on climate, may not have the financial resources or the proper soils. The ripple effect of these disruptions would be felt worldwide. Millions of people would be forced to migrate from newly uninhabitable regions to new areas where they could survive. This would also affect public health. Rising seas would contaminate freshwater with salt water; there would be more heat-related illnesses and deaths; and disease-carrying rodents and insects, such as mice, rats, mosquitoes, and ticks, would spread diseases such as malaria, encephalitis, Lyme disease, and dengue fever. Scientists of the IPCC agree that one of the most serious aspects of all this drastic change is that it is happening so fast. These changes are happening at a faster pace than the Earth has seen in the last 100 million years. While humans may be able to pick up and move to a new location, animals and their associated ecosystems cannot. The choices people make and the actions they take today will determine the fate of other life and their ecosystems tomorrow.

THE TRUTH ABOUT GLOBAL WARMING

On both sides of this multi-faceted global warming debate, data is often misrepresented or misused, scientific facts and processes are contorted, and conclusions from both supporters and opponents are hotly contested. While the politicians and mainstream media generally provide a case to support man made global warming, it is the job of individuals with no political or financial bias to sort through the fact and fiction. In an age where media propaganda is rife, that may be no easy task.

To begin with, climate is not just the 'weather' conditions of a particular town or city, it is a pattern closely calculated within a period of every 30 years. It is comparative record and analysis of precipitation, radiations, temperatures, winds, and other meteorological conditions. In the 21st century, our climatic conditions are undergoing a sea change. The term 'global warming' has been an often-discussed term in the UN councils and world environment meetings. Both nature and humans have been equally responsible in bringing about the climatic changes. Though we cannot blame the nature as it has to take its course invariably, but we can surely do our bit in saving the planet.

Let us begin by analyzing how nature and man-made causes were responsible for global warming. Natural Causes: It has been said that nature completes its cycle every 40,000 years and in between, we have undergone Ice Age and evolution of humankind to its present form. What changes have occurred that is destroying our planet right now?

- Continent Shifts- 200 million years ago, the 7 continents was a large piece of landmass known as 'Pangaea'. When the continents broke up, it affected the physical features of the landmass, the climate, and oceanic currents. Till today, the Himalayan range constantly grows and is heading towards Asian landmass.
- Volcanic Eruption: When volcanoes erupt, they leave behind a large trail of gases, dust, vapor, and ash. The elements remain in the stratosphere for innumerable years. These create a cloud in the atmosphere and block the sunrays. When the solar radiations do not reach the atmosphere, cooling sets in. This situation occurred in the year 1816 when the Tambora volcano erupted in 1815. The cooling effects lead to frosty summers over Western Europe, Canada and US.
- Oceanic Currents: 71% of the earth consists of water. The oceans soak up most of the sunlight reaching earth. The ocean currents do not remain static; they govern the heat and cold balance throughout the Earth. The flow of the currents determines the coldness of countries like North Atlantic and Alaska. So, when the heat absorbed by water rise up as water vapor (a greenhouse gas), it leads to cloud formation and gives a cooling effect. This phenomenon was exactly what took place during the last Ice Age (14,000 years ago).

Man-made Causes: Difficulty lies in actually pointing the exact cause when man became the destroyer of nature, and indirectly of climate. Even a 0.1% of change in natural patterns disrupts the ecological system. Man has been using nature for his benefits since the times we leant agriculture. One definite period of drastic climatic changes would be the beginning of Industrial Revolution. Much as it had benefited us in terms of technological advancement, it robbed us of nature and we are still paying for its consequences. How does man contribute towards climatic changes?

THE ANSWER IS SIMPLE- GLOBAL WARMING.

Global Warming is the "increase in the average temperature of the Earth's near-surface air and oceans since the mid-20th century and its projected continuation". Greenhouse gases, like carbon dioxide, contribute to global warming heavily. Man contributes to global warming in the form of deforestation, pollution, unchecked population growth, and by harming the ecological system. These are explained briefly below:

- *Deforestation:* Trees, or greenery, in a much broader term, are a prerequisite condition to protect the earth. Greenery covers the earth and maintains the balance in atmosphere through the cycle of inhaling carbon dioxide and exhaling oxygen for us to breathe. Felling of tree has been prevalent since time immemorial but with time, man has become reckless. Deforestation disrupts the ecological cycle of the natural world. Without the green cover, how is the carbon dioxide going to process itself?
- *Pollution:* Deforestation meant more land, and with technological advancements, more land meant urbanization of cities. Remember-industrial revolution? With factory productions and uninhibited burning of fossil fuels, man only contributed woe to nature. Climate began to change. Of course, climate changes are not instant, it took and 100s of years and an equal amount of time for man to understand the repercussions of his handiwork. Carbon emissions continue today through the medium of cars, burning electricity, and others.
- *Population:* Two things strike here- transportation and food production. Food production means increase in the level of methane. Why? It is because animals are put into use for producing food. A large population means increase in production of cars and therefore, more burning of fossil fuels like petroleum.
- *Ecological System:* Do you have any explanation as to why daily 100s of species are on the verge of extinction? Humans and animals alike have particular ecological systems founded upon million of years of evolutionary processes. However, through deforestation, decrease in forest lands, basically, the lack of green cover, we have been ravishing

the habitat of the animal kingdom in the name of civilization. Similarly, excessive fishing is making them extinct. Deposit of factory spoils on the sea or ocean bed has been poisoning the water habitat. It will not be a surprise if within few years, the plant kingdom and the water bodies die out permanently.

All the factors explained above are contributing to the increase of global warming. You will notice that some areas are getting warmer and some are equally getting colder, which means the seasonal balance have been tilting. The marauding effects of humankind on nature would take another hundreds of years to set it back on track, which would be just the beginning. In addition, we have to begin at the earliest. We need to give nature back what we have taken away from her unflinchingly.

CONTROL THE HARM ALREADY DONE

First, nature cannot be controlled. What we can do now is to set in motion the reverse steps that can rectify the harm done to climatic conditions. Nations and countries should be standing together to restore the much cherished and desired 'greenery' all over the earth. The main problem as to why we least worried about nature and its climatic changes is because of the lackadaisical, callous, and least bothered attitude of our own.

There is a lack of motivation to protect our environment, our lovable planet. Undeniably, man has been putting off taking note of the depleting green cover until the last moment. Deducting quite a handful of volunteers and government working organizations, who has seriously though upon this matter? WE think that saving nature and reversing the climate is not our prerogative. Instead, we leave it upon governmental organizations and we blame them for any wrongdoings. How tragic could it get? The keyword here is global cooperation. There are some basic things man can take course too, which will help in restoring the lost glory of our precious environment. So, let's see what we can do to benefit nature?

- Begin to recycle home waste. Separate the renewable and non-renewable wastes. Use the organic material as compost for plants. Use electricity when required only. More than saving you few dollars, it will reduce environmental pollution. Grow plants and shrubs at your home to contribute to the green cover.
- Take to cycles or battery mode bikes as a way of transportation. This will not only reduce carbon emissions, and thereby pollution, it will clear the environment, which will in effect have an impact on the climate. A cleaner environment means a safer atmosphere. If you cannot do without cars, then use alternate sources of energy.
- It is necessary to preserve the fossil fuels. It had taken million of years to form the fossils, which we are using now and so, it will take

another few millions to form fossil fuels. In addition, reduction in carbon emission will balance the increasing world average temperature. Reduce dependence upon non-biodegradable sources of energy like plastics.

- Imbalanced climate affects the health too. Respiratory problems, waterborne diseases, mosquito- borne diseases would soon become a common phenomenon if not taken care of. Try to maintain a pure and clean environment at home.
- Encourage your company to contribute in carbon footprint assessment. Only an assessment would help to devise strategies to overcome increase carbon exposure.

These steps and many more, actually innumerable ways, are there to reverse the climatic degradation only if man becomes more aware and understands the repercussions of environmental hazards seriously. The climate reversal techniques will not only reduce the increasing global warming but also reduce the temperature, thus balancing the heat with the cold. Overall, the precautionary measures upheld now will lay a precedent for the future generations to take care of the greenery of nature and its bounty. Humans, plant kingdom, and animal kingdom are an organic species. We need to learn to share the ecological system with other life forms.

Global warming is (presently) demonstrated in increases in global average air and ocean temperatures and widespread melting of snow and ice. This is most likely caused by the observed increase in anthropogenic greenhouse gas concentrations in the atmosphere. Global warming will almost certainly continue due to the time scales associated with climate processes, even if greenhouse gas concentrations were to be stabilised at present levels. The increasing temperatures and the melting of snow and ice are leading to rising global average sea levels. The uptake of anthropogenic carbon is leading to increasing acidification of the oceans. The frequencies and intensities of extreme weather are changing.

Weather extremes – such as hot and cold temperatures, storms and cyclones, and heavy rainfalls with risks of flooding will increase in frequency in many regions. Seasons are changing, with early spring and late autumn. Plant and animal ranges are likely to be shifting and some plant and animal populations are likely to decline. A warmer climate is favourable for the survival of pathogenic microorganisms and parasites, and vector habitats are expanding geographically, leading to new human and animal disease scenarios. Even today, we see impacts of climate change in many natural and man made systems, and these impacts will probably become more and more visible in the coming decades. Examples of systems and sectors most likely to be affected are ecosystems, food, fibre and forest products, coastal systems and low-lying areas, polar and boreal areas, settlements and industries linked to climate-sensitive resources, handling of water (drinking, waste, urban runoff) in cities. Human and animal health aspects are also critical.

Climate change will almost certainly affect many vital functions in society as well as in the natural environment, and will pose a major threat to global sustainable development. Responses to climate change are therefore imperative. This entails a continuous risk management process, where combinations of adaptation and mitigation measures can reduce the risks associated with climate change. Research funded by Formas shall contribute to an increased capacity in society and ecosystems to respond to climate change. Many questions regarding the functions and interactions of various systems related to climate change are still to be answered in order to improve our capacity to respond to climate change.

Research is needed on the interactions between natural systems and man made infrastructure, for instance in urban planning and construction. Moreover, research is essential in the development of effective measures for mitigation and adaptation. Analyses of interactions between natural and human systems are necessary, as well as analyses of interactions between different societal interests and processes. Analyses of equity and burden sharing between people and countries, generations, and sectors are also needed.

GLOBAL WARMING ON ECOSYSTEMS

In a study on the effects of global warming on the Earth's ecosystems conducted by Chris Thomas, a conservation biologist at the University of Leeds in the United Kingdom, he states"Climate change now represents at least as great a threat to the number of species surviving on Earth as habitat destruction and modification." Thomas worked with a group of 18 scientists worldwide in the largest study of its type ever accomplished. The end result of their study came down to this conclusion:"By 2050, rising temperatures, made more severe through human-induced input, such as the burning of fossil fuels, could send more than a million of Earth's land-dwelling plants and animals down the road to extinction." The research team worked by themselves in six biodiversity-rich areas around the world, ranging from Australia to South Africa. As they gathered field data about species distribution and regional climate, they programmed the information into computer climate models.

The purpose of the computer models was to simulate the direction and distance individual species would migrate in response to temperature and climate changes. Once all team members had collected their specific data, they combined it into a single model in order to understand the global concept. Once the model had been carefully evaluated, it was determined that by the year 2050, at predicted global warming rates, 15 to 37 per cent of the 1,103 species studied could be at risk of extinction. When the study area was expanded to cover the entire Earth, the researchers estimated that worldwide more than a million species could begin to face extinction by 2050."This study makes clear that climate change is the biggest new extinction threat," said coauthor Lee Hannah of Conservation International in Washington, D.C."In some cases we found that there will no longer be anywhere climatically suitable for these species to live; in other cases they may be unable to reach distant regions where the climate will be suitable.

Other species are expected to survive in much reduced areas, where they may then be at risk from other threats," said coauthor Guy Midgley of the National Botanical Institute in Cape Town, South Africa."Seeing the range of responses across all 1,103 species, it becomes obvious that we have a lot of work left to do before we can accurately predict what types of animals and plants are most at risk. This range of responses shows that species will not be able to move as whole biological communities, and that the typical natural communities we recognize today will probably not exist under future conditions. Figuring out what will replace them requires a lot of imagination," said coauthor Alison Cameron of the University of Leeds. Chris Thomas, almost all future climate projections expect more warming and even more extinction between 2050 and 2100, and, even though projections are only made to 2100, temperatures will still keep going up and more warming will occur after that.

This group of researchers say that taking action now to slow global warming is important to make sure that climate change ends up on the low end of the prediction in order to avoid"catastrophic extinctions." Thomas also stated that because there may be a time lag between the climate changing and the last individual of a species dying off, the rapid reduction of greenhouse gas emissions may enable some species to survive. It is also important to keep in mind that although some species may be able to migrate successfully to a new location, some plant and animal species that live in high mountain or polar ecosystems cannot move farther to escape warming temperatures. Similarly, coral reef systems cannot just pick up and move to a new location.

Long-established communities have remained where they are in order to survive. Robert Puschendorf, a biologist at the University of Costa Rica, believes these estimates"might be optimistic." As global warming interacts with other factors such as habitat destruction, increase in invasive species, and the buildup of carbon dioxide in the environment, there may be more than a million species that face extinction. In connection with Chris Thomas's study, Richard J. Ladle, *et al.,* from Oxford University"Dangers of Crying Wolf Over Risk of Extinctions," bringing up an issue important to all scientific investigations. It cautioned that the scientific world needs to"learn how to deal with increasingly sensationalist mass media."

They warn that policy-makers must be informed by a"balanced assessment of scientific knowledge and not popular perception created by commercially driven media. Departure from rational objectivity risks undermining public trust in the natural sciences and could play into the hands of antienvironmentalists. This places responsibilities on both scientists and journalists to ensure fair and accurate reporting of their work." Ladle's report clarified that a large portion of the media incorrectly focused on the idea that more than a million species would definitely go extinct by 2050 when the study actually concluded that the extinctions will occur eventually and not in the next 50 years.

They reported that 21 out of 29 substances quoting Thomas's study misinformed the public by inferring the species would be extinct by 2050. Their chief concern is that the media, or anyone else, through unclear presentation, could inadvertently increase public cynicism and complacency about climate change and biodiversity loss. The resulting message is that when valuable studies, such as Chris Thomas's, are conducted and the results are released, in order to receive the best results from the scientific community, policy-makers, and the public in general, it is critical not to sensationalize or misrepresent the facts. Otherwise, it puts the real message in jeopardy and may even do more harm than good. Both plant and animal species are at risk due to the effects of global warming.

The World Wildlife Fund, the golden toad and the harlequin frog of Costa Rica have already disappeared as a direct result of global warming. As different components of an ecosystem change, it can upset the natural balance in many ways. For instance, it can disrupt a species by having spring show up a week or two earlier. Over time, delicate balances have been set up between animals and the food they eat. If a certain animal relies on a specific food, but global warming has already caused the food to grow through its life cycle before the animal is ready to use it, then it will have a direct negative effect on that animal-the food will not be available when the animal needs it. That animal's health and existence may be threatened. Furthermore, this could cause a ripple effect in the food chain, having an impact on more than one species.

One example of this is when spring comes earlier than it has in the past. The timing of feeding for newly hatched birds may not correspond to the availability of worms or insects, impacting the fledglings' chances of survival. The WWF, climate records compared with long-term records of flowering and nesting times show a noticeable shift away from each other, depicting global warming trends. In Britain, flowering time and leaf-on records, which date back to 1736, have provided concrete evidence of climate-related seasonal shifts. Long-term trends towards earlier bird breeding, earlier spring migrant arrival, and later autumn departure dates have also been recorded in North America. Changes in migratory patterns have also been documented in Europe.

Once again, the WWF documents that climate change and global warming are currently affecting species in many ways. Animals and plants that require cooler temperatures in order to survive, which need to either migrate northward or higher in elevation in a mountain ecosystem, are already being documented in several places worldwide. This is occurring in the European Alps, in Queensland in Australia, and in the rain forests of Costa Rica. Fish in the North Sea have been documented migrating northward. Fish populations that used to inhabit areas around Cornwall, England, have migrated as far north as the Shetland and Orkney Islands. WWF global warming experts believe, based on this evidence, that"the impacts on species are becoming so significant that their movements can be used as an indicator of a warming world.

They are silent witnesses of the rapid changes being inflicted on the Earth." In fact, these same scientists believe that global warming could begin causing extinctions of animal species in the near future because the heating caused by accelerated global warming has a severe impact on the Earth's many delicate ecosystems-both on the land and the species that live on it. Worldwide, there are species and habitats that have now been identified as being threatened and endangered due to the effects of global warming. Because ecosystems can be altered to the point where the damage becomes irreversible and species must either adapt to survive or face extinction, it is critical that the issue of global warming be addressed and acted upon now before it is too late.

THE EFFECTS OF GLOBAL WARMING ON COASTAL LOCATIONS

The Pew Centre on Global Climate Change is a nonprofit organization that brings together business leaders, policy makers, scientists, and other experts worldwide to create a new approach to managing the problem of global warming-what they refer to as an extremely"controversial issue." Not slanted in any particular way politically or economically, they"approach the issue objectively and base their research and conclusions on sound science, straight talk, and a belief that experts worldwide can work together to protect the climate while sustaining economic growth." The Pew Centre was established in 1998 and has"issued more than 100 reports from top-tier researchers on key climate topics such as economic and environmental impacts and practical domestic and international policy solutions."

The centre's leaders and staff hold briefings with members of Congress and administration officials in the U.S. government, as well as with leaders of international governments. They also work with 43 major corporations in business roles to encourage them to help promote practical solutions to solving the global warming crisis. In January 2007, the Centre was one of the inaugural members of the U.S. Climate Action Partnership-an alliance of major businesses and environmental groups that calls on the federal government to enact legislation requiring significant reductions of greenhouse gas emissions. The Pew Centre on Global Climate Change, in terms of global warming, the biggest impact on estuarine and marine systems will be temperature change, sea-level rise, the availability of water from precipitation and run-off, wind patterns, and storminess. In these oftenfragile systems, temperature has a direct and serious effect. For the sea life living within the ocean, temperature directly affects an organism's biology, such as birth, reproduction, growth, behaviour, and death. The remainder of this part will further explore the specific effects global warming will have on temperature, sea-level rise, wind circulation, aquaculture, algae blooms and disease, and the long-term effects of coastal building.

Temperature

Temperature extremes (both high and low) can be deadly to living organisms. Many species are vulnerable to temperatures just a few degrees higher than what they are accustomed to. Even increases in temperature as little as 1.7°F (1°C) can seriously harm certain species. For example, from 1976-77, the species of reef fish off Los Angeles diminished by 15-25 per cent when temperatures abruptly warmed 1.7°F (1°C). Even if the temperature is not high enough to kill, it can be high enough to negatively influence the organism's life functions such as metabolism, growth, behaviour, and physiological factors such as the timing of reproduction rates of egg and larval development. Temperature also influences where organisms can survive and controls how large a given population can become.

One area where this has long been documented is along the West Coast of North America. The nearshore sea surface is predominantly warm and the offshore area is predominantly cool, creating a region that has long been famous as a rich, diverse fishery. Temperature differences can also influence interaction between species, such as predator-prey, parasite-host, and other relationships that may develop over the struggle for limited resources. If temperatures change the distribution patterns of organisms, it could also change the balance of predators, prey, parasites, and competitors in an ecosystem, completely readjusting balances, food chains, behaviours, and the equilibrium of the ecosystem. Global warming can also change the way that species interact by changing the timing of physiological events. One of the key changes that it could alter is the timing of reproduction for many species. Rising temperature could interfere with the timing of birth being correlated with food availability of that species.

This can be a problem, for example, for bird species that migrate and depend on a specific food source to be available when they reach their breeding grounds. If warmer temperatures have changed the timing and it is a few weeks off of when the food will be available and no longer synchronized with migrating birds, it could leave the birds without available food, threatening their survival. Temperature also plays an important role with oxygen because it directly influences the amount that water can hold. The warmer the water, the less oxygen it holds. As global warming continues to force temperatures to rise, less oxygen will be available for marine species, which could then threaten their survival. A significant amount of data has already been collected on these variables. The Pew Centre, areas that still need much more data collected and research applied concern the ocean temperatures influence on the interactions among organisms, such as predator-prey relationships, parasite-host relationships, and the interplay of species concerning competition for resources.

It is also possible to model the effects of sea-level rise in shallow continental margins, such as flooding wetlands and shoreline erosion. What is more difficult to predict and needs more research are the global warming effects on precipitation amounts, wind patterns, and intense weather. Precipitation directly affects estuaries because it affects the run-off into estuaries, which influences estuarine circulation.

Sea-Level Rise

Sea-level rise will mount from the melting of glacial and polar land ice. The Pew Centre, the effects of sea-level rise will vary by location, how fast the sea level rises, and the biogeochemical responses of the individual ecosystems involved. One of the areas identified as being the most susceptible to damage from sea-level rise is the low-lying, flat wetlands located in the middle and south Atlantic and the Gulf of Mexico (involving states such as Alabama, Florida, Mississippi, Georgia, and Texas). A wetland is an area on shore that has a wet, spongy soil. These areas are also referred to as swamps, marshes, and bogs. As sea levels rise, ocean water will submerge and erode the shorelines. In the natural areas covered with marshes and mangroves (common to Florida and the Gulf states), sea levels will flood the wetlands and waterlog the soils.

Because the plants that live in the wetlands, which do not contain salty water, are not accustomed to salt, the salty ocean water will kill them. Because wetlands provide habitat for wildlife, including several migrating birds, this would also destroy their habitat. Based on research by the Pew Centre, if the wetlands located along the Gulf of Mexico are not bordered by human development (such as homes) and the inland slope is relatively flat, and the sea-level rise is gradual, they would have a chance of surviving by migrating further inland to escape the inundation of salt water. The Pew Centre states that"these areas would be safe at least at the rates of rise currently projected for the next 50 to 100 years." The areas that would be hard hit are those that cannot migrate inland because urbanization has grown right up to their shoreline, effectively removing any potential wetland habitat. This is very detrimental to the environment because wetlands are an important part of the biological productivity of coastal systems. Marshes provide many critical services; they function not only as habitats for wildlife, but as nurseries for breeding and raising young and as refuges from predators. Wetlands function as part of an integrated system.

If they are jeopardized, their loss will affect the availability and transfer of nutrients, the flow of energy, and the availability of natural habitat needed by multitudes of organisms already living there. One of the most unfortunate losses will be those areas where rare, threatened, or endangered plant and animal species live, such as the American alligator, Florida black bear, West Indian manatee, Florida panther, southern bald eagle, snowy egret, and roseate spoonbill. If invading salt water destroys these habitats, all these species could become extinct.

Wind Circulation

Another consideration in the marine environment is wind circulation. Winds are created by the uneven heating at the Earth's surface, specifically the high thermal gradients between the equator and the poles. Under global warming conditions, however, the polar regions will be subjected to higher temperatures, thereby reducing the temperature gradient between the equator and the poles. This could cause a weakening in the overall wind circulation around the Earth. The winds drive the surface currents of the Earth's oceans. A slowdown of the ocean surface circulation could negatively affect the structure and function of both the open ocean and the near shore ecosystems. Weak currents would make it difficult for any currents to transport nutrients where they need to be deposited. If wind speed and direction are altered, which is what experts at the Pew Centre have predicted will occur with global warming, the productivity of estuarine and marine systems will also be affected.

One of the reasons why some coastal areas are so rich and productive in fishery resources, such as the West Coast of the United States, is because of the coldwater upwelling that occurs along the coastal margins. The thermal properties and circulation patterns function to bring the cold water from the deep ocean waters to the surface along the coastlines, supplying the coastal fisheries with an abundance of fish. Nutrients that are delivered from these deeper waters to the surface help provide for the phytoplankton along the coasts. The upwelling process is controlled by the winds that blow along shore. Upwelling can be minimized, however, if the water column is stratified by differences in temperature or salinity, which act as a barrier to upward movement through the water column. The effects of this type of scenario were played out off the coast of California from 1951 to 1993. The sea surface warmed during this period 2.5°F (1.5°C), resulting in a decreased upwelling of nutrientrich cold waters. This caused a 70 per cent decline in the abundance of zooplankton, which in turn harmed the coastal food webs, negatively affecting fish and birds. There are different opinions on this matter, however. Andrew Bakum of the National Marine Fisheries Service of the National Oceanic and Atmospheric Administration (NOAA), his research leads him to believe that global warming may actually increase coastal upwelling in some areas.

He believes this is possible because most greenhouse gas models project more warming over land than over open oceans, which would increase the temperature contrast between the land and the ocean. This resultant contrast would strengthen the low-pressure cells that usually form over land that are adjacent to offshore high-pressure cells. The relationship of forces between these two pressure cells would create alongshore winds that would create a strong upwelling. On the East Coast of the United States, circulation patterns could slow down transport of nutrients and fish species, such as blue crabs and bluefish, which would lower the abundances of these particular species along the coasts and within estuaries. When their populations are decreased, the coastal ecosystem becomes threatened.

Aquaculture

Coastal aquaculture, the farming of fish from the ocean in special facilities, has become a key food source in recent years. The Food and Agriculture Organization (FAO), global aquaculture production has been increasing since the 1950s at a growth rate of about 10 per cent per year since 1990. By 2030, the FAO projects that aquaculture harvests will be greater than capture harvests. Marine aquaculture is a steadily growing industry. The effect that global warming will have on aquaculture is mixed. On one hand, the higher temperatures could enhance growth rates of aquaculture species and could make it possible to run aquaculture operations in locations that are currently too cold. On the other hand, areas that are suitable now for aquaculture may become too warm for existing species to survive.

The Pew Centre's research suggests that as conditions change, if the temperatures rise slowly enough, aquaculture operations should be able to keep up with the changes and keep negative effects to a minimum.

Algae Blooms and Disease

One major drawback of aquaculture facilities is that they affect local ecosystems when their concentrated wastes pollute the surrounding environment, encouraging algae blooms. If global warming becomes an issue, algae blooms will become more common. Warm water holds less oxygen and accelerates the microbial decomposition of the aquaculture wastes. Oxygen concentrations are further lowered, exacerbating the problem. As global warming continues and temperatures climb upward, disease will become more prevalent, distributing the pathogens to more areas. Not only does this affect ocean life, it also affects humans. An example of this has been showed with oysters.

The protozoan Perkinsus marinus (Dermo) is a pathogen that threatens the health of oysters. Cool ocean temperatures keep infection by this pathogen to a minimum. As water temperatures have risen in recent years, the spread of this pathogen has been documented spreading to the northeastern states of the United States. Based on this evidence, Dr. T. Cook of Rutgers University's Institute of Marine and Coastal Sciences says longterm climate changes may produce shifts in salinity and temperature that enable pathogens to spread. Climate change could also affect the distribution of other aquatic diseases in a similar manner. Warmer coastal waters along with eutrophication (an increase in chemical nutrients—usually compounds containing nitrogen or phosphorus—in an ecosystem) can also increase the intensity of harmful algal blooms, which can destroy habitat and shellfish nurseries and are also toxic to both marine species and humans. Unfortunately, these blooms have been recently increasing worldwide, possibly because they are correlated to global warming.

One of the concerns is that consumption of shellfish that have ingested harmful algae can cause neurotoxic poisoning in humans. The Pew Centre, there are two schools of thought concerning species adaptation and survival in marine ecosystems. One group thinks that marine ecosystems will have far fewer extinctions than terrestrial ecosystems because marine species have wider geographic temperature ranges and the ability and opportunity to migrate to new habitats. The other group thinks that the lack of evidence of recent marine extinctions is simply because the ocean systems are so vast that scientists are not aware of them, and there is not enough data collected about the oceans.

Long-term Effects of Coastal Building

The New York Times on July 25, 2006, entitled "Climate Experts Warn of More Coastal Building," 10 climate experts from around the United States say that the unchecked coastal development in geographic locations that are vulnerable to hurricanes and other forms of severe weather, along with the lack of government regulation on development, is the biggest reason why extreme, violent weather is currently causing so much human suffering, loss of life, and property damage. The substance states that the opinion of the experts is that: "Whatever the relationship between hurricanes and climate, hurricanes are hitting the coasts and houses should not be built in their path." One of the main problems is social/political in nature.

Because coastal areas are popular places to live, pressures are put on states and Congress to obtain discounted insurance for property known to be in harm's way. The climate experts claim that reimbursement for property loss wrongly encourages victims of disastrous weather to build again in the wrong place. "Federal disaster policies, while providing obvious humanitarian benefits, also serve to promote risky behaviour in the long run." The climate scientists also stressed that storms like Katrina "are inevitable even in a stable climate." One of the scientists, Philip J. Klotzback, a hurricane researcher at Colourado State University who does not support the notion that global warming is spurring stronger or more frequent hurricanes, stated, "The social and economic trends are completely clear.

There is likely to be destructiveness in tropical storms anyway because more people now live in vulnerable coastal areas." Kerry A. Emanuel, a climatologist at the Massachusetts Institute of Technology who does believe that the building energy of hurricanes in recent decades is related to human-driven warming of the seas, also warns people from building in these areas.

DETERMINES GLOBAL TEMPERATURE AND CLIMATE

Global temperature and climate are largely determined by the balance of incoming energy from the sun, minus outgoing radiation. Incoming light radiation from the sun has short-wavelengths and can readily pass through the atmosphere, but after being absorbed and re-radiated from Earth's surfaces the out-going infrared radiation has longer wave-lengths and is less able to pass through the atmosphere. The so-called "greenhouse gases" absorb and re-radiate a portion of the outgoing long-wave radiation back towards earth, acting like a heat-trapping blanket. Even slight changes in the ratio of incoming and outgoing solar energy have significant influence on our global climate system. Even though greenhouse gasses make up less than 1 per cent of Earth's atmosphere, our global climate is quite sensitive to changes in their concentration. Ice-core data from Greenland and Antarctica tells us that atmospheric levels of carbon dioxide ($CO_{2)}$ vary somewhat predictably with cycles of ice ages and warm interglacial periods. The ice cores also show that atmospheric CO_2 is increasing almost 100 times faster today than during past climate cycles, and that current concentrations of CO_2 are higher than at any time in at least the last 800,000 years. Given the difficulty of rapidly changing our resource-intensive lifestyles, we'll be lucky if global atmospheric CO_2 concentration merely doubles. More likely it will go much higher before we control our appetite for fossil fuels and land exploitation.

While CO_2 is of primary concern among greenhouse gasses, there are others such as methane (CH_4) that contribute to global warming. CO_2 is unique in that is has a very long, approximately 100 year, "residence time" in the atmosphere. Concentrations of CO_2 in the atmosphere will likely remain far above "normal" for centuries, because the millions of tons of CO_2 released to the atmosphere during the agricultural revolution, the industrial revolution, and the automobile revolution will not reach a new equilibrium until biological and geophysical processes (in the oceans and on land) have a chance to capture and store most of the "extra" carbon.

We have a moral obligation to leave future generations with choices and opportunities for survival. We must avoid irreversible harm to the planet's life support systems including a livable climate and function ecosystems that sustain life.

CARBON MOVE IN AND OUT OF THE ATMOSPHERE

There is a fixed amount of carbon on planet earth which is distributed among several carbon reservoirs or pools in the atmosphere, biosphere, hydrosphere, and lithosphere. In the grand scheme, carbon is neither created nor destroyed but continually moves between these various pools owing to the operation of natural and human-induced processes. The root cause of global climate change is that human activity has shifted massive quantities of carbon to the atmosphere from forests, soil, and fossil deposits.

In the atmosphere carbon is stored as CO_2, methane (CH_4), and other organic compounds. Carbon moves *into* the atmosphere from decomposition of organic matter, respiration by living organisms, combustion, volcanic activity, burning fossil fuels, degassing of waterbodies, etc. Carbon moves *out of* the atmosphere via photosynthesis, rock weathering, dissolution in water, etc. All plants, including forests and many micro-organisms, use photosynthesis which takes CO_2 out of the air to build sugars that can be used by the cell to build cellulose or other complex carbon molecules that comprise plant biomass.

This process is called "primary production" and it feeds the bottom of the global food chain. Virtually all life on earth, including humans, relies directly or indirectly on photosynthesis. Most terrestrial plants share a significant portion of their photosynthate with soil organisms, a cooperative relationship that builds a large and complex underground ecosystem that also stores carbon. Plants shed dead leaves and wood which also builds carbon stores in the soil.

In the hydrosphere (e.g. the oceans) carbon is stored mostly as dissolved $CO_2$and other dissolved organic compounds that originated in some photosynthetic life form. Carbon moves *into* the ocean from the atmosphere and biosphere via dissolving of gaseous CO_2 in cold seas, leaching from soil, and input of organic matter from river systems and the biosphere. Carbon moves *out of* the ocean primarily via photosynthesis (*e.g.* phytoplankton and cyanobacteria), degassing of warm seas, and deposition in marine sediments.

In the biosphere carbon is stored as live or recently dead plants, animals, and micro-organisms both in the ocean and on land (*e.g.*, forests and soils). Forests dominate the terrestrial carbon cycle, harbouring 86 per cent of the planet's above-ground carbon and 73 per cent of the planet's soil carbon. Carbon enters *into* the biomass pool via photosynthesis, then becomes entrained and cycled through the entire global food chain. Carbon moves *out of* the biomass pool through decomposition and respiration or through deposition in long-term storage in soil or geologic and fossil deposits.

In fossil deposits, the carbon from long-dead plants and animals are stored as coal, oil, "natural gas," or kerogen. These can be thought of as both "ancient sunlight" and "ancient atmosphere." Carbon moves *into* the fossil pool via deposition and storage in low-oxygen conditions. Carbon moves *out of* fossil pool mostly via industrial exploitation and combustion.

In the non-fossil lithosphere carbon is stored in carbonate rocks such as limestone and chalk. Carbon moves *into* these geologic structures mostly through ocean deposition. A portion of the oceanic carbon is taken up to make the shells of marine organisms that fall to the deep ocean floor where they may be subducted beneath the earth's crust and end up in long-term geologic storage, *e.g.* the Cliffs of Dover. Carbon moves *out of* the lithosphere mostly via volcanic activity and human industry such as the manufacture of cement which heats limestone and releases significant quantities of CO_2.

The advent and diversity of life on earth has had a profound impact on the global carbon cycle and now plays a fundamental role in determining whether or not we have a livable climate. The abiotic carbon cycle that existed before the proliferation of life was less stable than the carbon cycle that developed after marine organisms started to make calcium carbonate shells and deposit carbon in deep storage which has helped buffer CO_2 extremes over long time scales. Scientists have found a correlation between biodiversity and levels of atmospheric CO_2 over the last 370 million years.

Human activity, mostly in just the recent era, has dramatically reallocated global carbon stores from the other carbon reservoirs into the atmosphere where it can influence our climate. For example, burning fossil fuels and heating limestone to make cement move carbon from long-term fossil and geologic storage into the atmosphere. Logging kills trees - stops carbon-uptake via photosynthesis, and moves carbon from living forests and soil into the atmosphere. Land uses such as agriculture, livestock grazing, and draining swamps move carbon from the soil to the atmosphere.

Climate Change Affect the Pacific Northwest

While predicting the local *weather* is an uncertain science, *climate* prediction is actually more accurate because the focus is on large-scale trends rather than local details. We know that the planet as a whole is almost certain to become warmer on average, and scientists expect an acceleration of the hydrologic cycle as warmer temperatures lead to increased evaporation from the oceans and more transpiration from plants. However, the effects of climate change will not be uniform around the globe. Significant uncertainty remains about how global trends will express themselves regionally. Future climate in the Pacific Northwest is even more uncertain because of complex topography and uncertain changes in precipitation, but our close proximity to the moderating influence of the Pacific Ocean likely offers a slight buffer from climate extremes.

The Pacific Northwest should expect continued climate variability. Existing cycles of cool-wet winters and warm-dry summers will likely continue, though they will be superimposed on a warmer average climate. Both floods and droughts have been part of our past and will almost certainly be part of our future, and both will likely get worse, but we don't know if these climate extremes will be expressed with more frequency or more intensity, or both.

It is reasonable to expect more precipitation, mostly during our existing wet seasons. More of our winter precipitation will fall as rain instead of snow, so storage of water in snowpacks will likely decrease (on average). We should expect milder winters, earlier melting of snow packs, earlier spring run-off, longer periods of summer low stream flow, and more drought.

Importantly, earth's biogeochemical systems are complex and not at equilibrium. There are many feedbacks that lead to non-linear behaviour, so we should NOT expect climate changes to be slow and predictable. Small changes in CO_2 and global temperature can lead to large and/or rapid changes in climate and ecosystems. Accordingly, the rate of current and future global changes may be unprecedented, chaotic, and highly disruptive.

PROTECT FORESTS FROM THE PERILS OF CLIMATE CHANGE

Jerry Franklin points out that "forest management can either exacerbate or reduce the effects of climatic change on the productivity and biological diversity of northwest forestscapes." To increase the chances that we will continue to enjoy the diverse benefits we receive from northwest forests, we must maintain and enhance their ability to respond to change. The key components of such a strategy are:

- Maintain biodiversity in all its dimensions. This will be critical, because genetic diversity is like a library of possibilities that have worked well during past climate variability, representing the sum of "tools" available for the future.
- Protect intact native ecosystems where species relations have stood the test of time and remain robust;
- Provide refugia and allow species to migrate. Buffer and expand protected areas to provide connectivity along climatic gradients. Manage the entire landscape to be amenable to dispersal of native species.
- Protect streams. Cold water fish are particularly vulnerable to climate change because of increased winter flooding, reduced summer stream flow, and increased stream temperature. To mitigate expected effects on fish we should provide generous riparian buffers to help shade streams and maintain lower stream temperatures. To render streams more resilient to hydrologic extremes, such as flooding, we should manage whole watersheds to improve their ability to absorb, store, and slowly release water. This can be accomplished in part by reducing disturbance of vegetation and soils, reducing road densities, and retaining abundant woody debris.

LOGGING RELEASES SIGNIFICANT AMOUNTS OF CARBON

Not surprisingly, logging accelerates the transfer of carbon to the atmosphere by killing trees that would otherwise continue to capture and store carbon through photosynthesis and growth. Killing trees also stops them from pumping carbon into the soil where much of the carbon in forests is stored. Logging actually accelerates the rate of decomposition of wood via several mechanisms. By removing the forest canopy and exposing the soil to more sunlight, logging raises soil temperature which increases the rate of decay. Logging also breaks up woody material in the forest thereby decreasing the average piece size and increasing the surface area exposed to microbial decomposition. Finally, logging debris is often burned on site or as part of an industrial process.

Traditional logging also increases the risk of disturbances. Logging increases wind damage by creating exposed edges and increasing wind speeds within forest stands. Logging often increases the wildfire hazard by making the stand hotter, dryer, and windier; by moving the most flammable small fuels from the forest canopy to the forest floor (*i.e.*, logging slash) where they are more available for combustion; and by initiating the establishment of dense stands of young trees with interlocking branches (resinous fuels) close to the ground.

Logging roads also increase the risk of human-caused fire ignitions and spread tree diseases like Port Orford cedar root disease that kill trees and release carbon. Scientists estimate that a large fraction of all the carbon transferred to the atmosphere by humans has been released due to forest exploitation. In recent decades CO_2 emissions resulting from human-induced changes to forests exceed CO_2 emissions from all motor vehicle sources combined, but forest releases are less than total emissions from all uses of fossil fuels. After logging an old-growth forest, the site remains a net source of carbon for more than 20 years, and depending on the conditions, the site does not rebuild pre-logging carbon stores for a century or more. As a result of widespread clearcutting and aggressive slash burning, the Pacific Northwest has contributed huge quantities of carbon to the atmosphere.

Increase Carbon Storage in Forests

Here in the Pacific Northwest we live in the midst of a globally significant carbon pool that should be nurtured and conserved to help keep carbon out of the atmosphere. Temperate old-growth forests of the Pacific Northwest contain some of the highest amounts of biomass per acre measured anywhere in the world.

About half of the dry weight of forest biomass is comprised of carbon. The latest IPCC Mitigation Report notes that "Forest-related mitigation activities can considerably reduce emissions from sources and increase CO_2 removals by sinks at low costs..." The IPCC also states that more than 1/3 of the potential mitigation available from forests is located outside the tropics and half of the forest mitigation will come from changes in forest practices, rather than simply preventing deforestation.

The objectives of forest management with respect to mitigating climate change should be two-fold effort to *protect* and *restore* forests —

- Minimize the release of additional forest carbon into the atmosphere. The best way to *retain* carbon in existing forests is to protect mature and old-growth forests and roadless areas.
- Rebuild depleted carbon stores within forested landscapes. Probably the best way to *rebuild* forest carbon stores in forests is to allow forests that were previously logged or burned to regrow and become mature and old-growth forests.

There are significant complementary benefits of managing forests for carbon storage to ameliorate global climate change. If done carefully, forests managed to provide public services such as clean water, habitat for fish and wildlife, soil conservation, and an enhanced amenity-based economy will also store large amounts of carbon over time.

Forests exhibit a quality known as "ecological inertia" which recognizes that established forests are generally long-lived, resilient to disturbance, and help create conditions suitable for their own survival. This means that our northwest forests may be able to persist through some climate changes and continue to store carbon and provide other benefits, as long as they are not clearcut or severely disturbed. This implies that if we want continued carbon storage in forests that are at the edges of their suitable range we should avoid stand-replacing logging methods (such as clearcutting) and, where ecologically appropriate, we may need to strategically reduce fuels to reduce the risk of stand-replacing fire.

Such fuel reduction must be done carefully however, because excessive removal of vegetation not only compromises carbon storage in both plants and soil, but can also increase fuel loads and fire hazard. Recent fire/fuel models indicate that forest fire hazard can be managed reasonably well by treating about 20-30 per cent of the landscape in strategic locations.

Treating fuel on every acre is neither needed or desired. Logging need not be the primary tool for accomplishing fuel reduction, because non-commercial techniques, such as low-intensity prescribed fire, are available and effective.

Forest Management Recommendations

Private forestlands: Short-rotation clearcutting typically practiced by private industrial forest land-owners is probably the worst possible way to manage forests for carbon storage, because the young forests never develop large carbon stores; significant soil carbon is lost during and after clearcutting, slash disposal, and site preparation; and the resulting wood products produced have limited longevity. Where logging is expected to continue, scientists recommend that carbon release can be mitigated if forest managers:

- Allow trees to grow much longer before harvest (*i.e.*, longer rotations),
- Retain more live trees on every acre during harvest (*i.e.*, thin instead of clearcut),
- Retain more dead wood after harvest (*e.g.* protect snags, practice less intensive slash disposal and site preparation), and
- Take steps to reduce road systems and prevent soil erosion, which would help store more carbon in forest soils.

Public lands: Federal forests can help mitigate climate change if they are restored to their natural-sustainable level of biomass and biodiversity. Large stores of carbon exist within roadless areas and mature and old-growth forests on federal lands. These should be protected from harvest, while previously logged younger forests should be carefully restored to a mature and old-growth condition that has optimal biomass storage. This management approach luckily complements other highly sought-after forest values that are currently under-represented in our forests. Careful management of forests for carbon storage can help resolve ongoing controversies over forestry's impact on water quality, old-growth, roadless areas, fish and wildlife habitat, and scenic values.

Market Solutions: Given humanity's slow response to the growing evidence of human-induced climate change and its consequences, aggressive approaches such as market intervention are now needed. The debate continues on whether a carbon tax or cap-and-trade system is better, but either is better than nothing. A carbon tax system establishes the price of carbon and the market determines how much is sequestered and not emitted. In a cap-and-trade carbon market, government would determine how much total carbon can be emitted from all sources and the market would determine who is allowed to emit the carbon and at what price.

Under current international climate protocols it is possible that forest owners of the Pacific Northwest might seek compensation for storing "extra" carbon. This would reward forest managers for storing carbon that would otherwise be transferred to the atmosphere and help off-set some of the economic costs of managing forests for carbon storage. However, there are unresolved issues about how to account for the full carbon consequences of proposed forest management activities.

For instance, the Kyoto Protocol has some "perverse incentives" that could reward carbon-poor young forests at the expense of carbon-rich old forests, though this is not scientifically supported.

In contrast to the sink management proposed in the Kyoto protocol, which favours young forest stands, we argue that preservation of natural old-growth forests may have a larger effect on the carbon cycle than promotion of regrowth.... [I]ncreasing life-span of the stand, proportionally more carbon can be transferred into a permanent pool of soil carbon (passive soil organic matter or black carbon)... [R]eplacing unmanaged old-growth forest by young Kyoto stands... will lead to massive carbon losses to the atmosphere mainly by replacing a large pool with a minute pool of regrowth and by reducing the flux into a permanent pool of soil organic matter.

Carbon stored in wood products generally do not last as long as they would if left safe inside a mature tree, but we can improve the carbon storage equation by using less wood and by increasing the lifespan of wood products. It's not just American's big cars and SUVs that are a problem; it's also their increasingly large houses. We should consider policies to help reverse the national trend towards larger houses, and we should build houses that last for centuries instead of just decades.

What about Forest Fires?

We cannot avoid the fundamentally dynamic nature of forests. Fire is an unavoidable part of life in western forests and we must stop fighting a losing battle against the inevitable. Most western forests are in some ways *dependent* upon disturbances such as fire, and past fire suppression has exacerbated rather than solved the problem of fire. Our goal should not be to prevent all damage from fires, insects, etc. Fire should be allowed to operate within natural bounds, as long as it doesn't threaten public safety. Communities and property owners in forest settings must take responsibility for becoming fire resilient or fire permeable. We should maintain healthy forest habitat by allowing natural disturbance processes to operate and expect forest carbon stores to ebb and flow, while also allowing forests to grow for long periods (and store lots of carbon) in between these natural disturbances. We must take a long-term and landscape view, so that we *optimize* carbon storage at any given point in space and time in order to *maximize* carbon storage over large landscapes and long time frames.

Fuels could be reduced in forests that are significantly outside the natural range of variability, but this must be done in a strategic and limited way that protects all large fire resilient trees and spatially disconnects large expanses of excessive fuels, while retaining as much biomass as sustainably possible. Current enthusiasm for fuel reduction must be tempered with a realization that removing too much fuel makes forests hotter, dryer, and windier which increases fire hazard and increases decomposition rates, both of which counter carbon storage and other objectives. After fire, the goal should be to retain carbon on site and allow the recovering forest to grow into a mature and old-growth condition. Aggressive replanting as recommended by the timber industry is unsupported because it establishes a dense fuel-laden condition that is susceptible to drought and is soon ripe for another fire. Natural regeneration of forests leads to more diverse and less dense forests, which is preferable from a climate change perspective because the resulting habitat diversity and spatial discontinuity are more resilient to future hazards.

2

Greenhouse Effect and Global Warming

The "greenhouse effect" is widely discussed in the media, and although its details are complicated, its principles are not difficult to understand. Without a greenhouse effect, radiation from the Sun (mostly in the form of visible light) would travel to Earth and be changed into heat, only to be lost to space. This scenario can be sketched as follows:

Sun's radiation → absorbed by Earth → Re-radiated to space as heat →

The greenhouse effect is a process where energy from the sun readily penetrates into the lower atmosphere and onto the surface of Earth and is converted to heat, but then cannot freely leave the planet. This can be sketched as follows: Sun's Radiation ! absorbed by Earth ! some re-radiated to space as heat ! some trapped by the atmosphere. Due to the presence of certain "greenhouse gases" that trap heat, like carbon dioxide, methane, water vapour, and CFC's, the atmosphere retains the sun's radiation and warms up the planet. By increasing the abundance of these gases in the atmosphere, humankind is increasing the overall warming of the Earth's surface and lower atmosphere, a process called "global warming."

THE RADIATION BALANCE

Another way to think about the greenhouse effect is to consider that according to physics the radiation we receive from the Sun must be equally balanced by the heat Earth radiates out to space. If we were to give back less energy than we receive, our planet would soon be too hot for life. Likewise, if we were to give back more energy that we receive, our planet would soon be too cold for life. This can be written as a balanced equation of radiation:

If we were to measure the temperature of the Earth from space, the Earth's "surface" would show a temperature appropriate for this requirement of energy balance: a measurement of roughly -18 degrees Celsius (about 0 °F). At this temperature, our planet radiates a quantity of heat into space that is equivalent to the amount of energy received from the Sun.

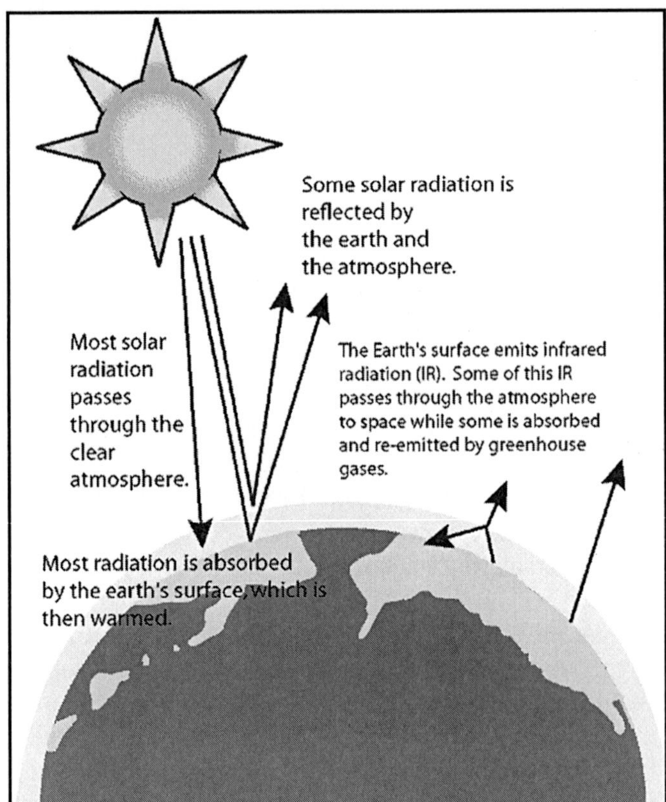

Fig. Solar Radiation input to Earth = Earth's output of Re-radiated Heat

At this point you may be asking how we can speak of "global warming" when we have just stated that the Earth (as seen from space) MUST stay at the same temperature? And how is it that the temperature of the Earth's surface is only a chilly 0°F? The key to understanding this apparent contradiction is to remember that we live at the bottom of the atmosphere. As far as the radiation balance is concerned, the lower atmosphere and the surface of Earth form part of a "warm interior" of the planet.

The apparent temperature "surface" that we would see from space is located well above the real surface of the Earth where we live. This apparent temperature "surface" is about 5000 metres up (17,000 feet) within the atmosphere. To get a better handle on this concept consider the following: the difference in elevation between 0 metres and 5,000 metres corresponds to a difference in temperature of about 60°F. In other words, at sea level it is 60°F warmer than it would be without the atmosphere. For the last 100 years or so this apparent temperature "surface" has been moving upward in the atmosphere as a result of global warming. As the apparent "surface" rises, the bottom of the atmosphere gets warmer, a fact that can be seen in the positions of the snow line (the elevation where snow begins to form) and tree line (the elevation where it becomes to cold for trees to grow). However, despite all these changes happening in the lower atmosphere, the overall temperature of the planet as seen from space stays the same.

How is it possible that the Earth exactly balances the incoming sunlight with the outgoing heat radiation?. The answer is simple: the amount of heat radiation from Earth is precisely tied to the temperature of the atmosphere. If the temperature of the apparent "surface" is too low and Earth radiates too little heat to keep the balance, Earth will warm up and radiate more heat into space. If the temperature of the apparent "surface" is too high and Earth radiates more heat than it receives, the planet will become colder and radiate less energy back to space. Overall, this "negative feedback" stabilises the radiation balance despite all the variations of temperature from one place to another and within the vertical column of the atmosphere. It sets the temperature so that the incoming and outgoing energy is balanced.

AVERAGE TEMPERATURES ON THE MOON

We can get another idea about what the temperature on Earth would be like without a greenhouse atmosphere by contemplating the Moon. The Earth's satellite has no atmosphere because its gravitational force is not strong enough to retain gas for long. It has the same distance from the Sun as the Earth, but its temperature varies enormously: where the Sun is shining, the Moon's temperature rises to 230°F and where it is dark falls to negative 290°F. The average surface temperature of the moon, about the same distance as the Earth from the Sun, is also near 0°F, but of course, the moon has no atmosphere. By contrast, the average surface temperature of the Earth is 60°F at sea level. On Earth, the contrast between maximum and minimum temperatures would not be as great as on the Moon, even without an atmosphere, because the Earth rotates once in a day, while the Moon only rotates once in a month. However, without an atmosphere the Earth's contrast between day and night and the contrast between summer and winter would be very large indeed.

Not all the gases in the atmosphere are equally active in keeping Earth warm. In fact, the atmosphere's most abundant gas, molecular nitrogen, does very little in this regard, and the same is true for the second most abundant gas, molecular oxygen. The most important ingredient of the air for producing the greenhouse effect is water vapour. However, its abundance depends on the air's temperature. The warmer the air, the more water vapour it can hold. (As air cools, the vapour condenses into rain or snow.) It is carbon dioxide that moves the air towards higher temperature, so that water vapour can take over and warm it some more.

Carbon dioxide molecules intercept infrared radiation, warming the air and increasing water vapour through evaporation from the sea surface and from plants and soil moisture. Water vapour then increases the temperature even more. The process is checked by a rise in infrared radiation to space and by formation of clouds. Unfortunately, the role of clouds in the radiation balance is as yet poorly understood. Different types of clouds have different effects, and this makes the calculations complicated and the results uncertain.

SOLAR VARIATIONS

One last point to consider when discussing the greenhouse effect is the amount of sunlight coming in to Earth. The quantity of sunlight we receive depends on the size and the brightness of the Sun and the distance between it and the Earth. As far as we know, the size of the Sun does not change much over the time spans we are considering; we can assume it is constant. The sun's brightness varies only a little, about one-thousandth over an eleven-year sunspot cycle, but perhaps more over longer time spans. We can take that as constant too, calling the incoming energy flux "the solar constant." One of the contentious issues in the discussions about the global warming of the last 100 years that has not been fully resolved is the question of whether a brighter Sun may have contributed to the recently observed temperature rise.

THE EARTH'S GREENHOUSE EFFECT

The realisation that Earth's climate might be sensitive to the atmospheric concentrations of gases that create a greenhouse effect is more than a century old. Fleming and Weart provided an overview of the emerging science. In terms of the energy balance of the climate system, Edme Mariotte noted in 1681 that although the Sun's light and heat easily pass through glass and other transparent materials, heat from other sources does not. The ability to generate an artificial warming of the Earth's surface was demonstrated in simple greenhouse experiments such as Horace Benedict de Saussure's experiments in the 1760s using a 'heliothermometre' to provide an early analogy to the greenhouse effect.

It was a conceptual leap to recognise that the air itself could also trap thermal radiation. In 1824, Joseph Fourier, citing Saussure, argued 'the temperature [of the Earth] can be augmented by the interposition of the atmosphere, because heat in the state of light finds less resistance in penetrating the air, than in repassing into the air when converted into non-luminous heat'. In 1836, Pouillit followed up on Fourier's ideas and argued 'the atmospheric stratum...exercises a greater absorption upon the terrestrial than on the solar rays'. There was still no understanding of exactly what substance in the atmosphere was responsible for this absorption. In 1859, John Tyndall identified through laboratory experiments the absorption of thermal radiation by complex molecules. He noted that changes in the amount of any of the radiatively active constituents of the atmosphere such as water or CO_2 could have produced 'all the mutations of climate which the researches of geologists reveal'. In 1895,

Svante Arrhenius followed with a climate prediction based on greenhouse gases, suggesting that a 40% increase or decrease in the atmospheric abundance of the trace gas CO_2 might trigger the glacial advances and retreats. One hundred years later, it would be found that CO_2 did indeed vary by this amount between glacial and interglacial periods. However, it now appears that the initial climatic change preceded the change in CO_2 but was enhanced by it.

G. S. Callendar solved a set of equations linking greenhouse gases and climate change. He found that a doubling of atmospheric CO_2 concentration resulted in an increase in the mean global temperature of 2°C, with considerably more warming at the poles, and linked increasing fossil fuel combustion with a rise in CO_2 and its greenhouse effects: 'As man is now changing the composition of the atmosphere at a rate which must be very exceptional on the geological time scale, it is natural to seek for the probable effects of such a change. From the best laboratory observations it appears that the principal result of increasing atmospheric carbon dioxide... would be a gradual increase in the mean temperature of the colder regions of the Earth.' In 1947, Ahlmann reported a 1.3°C warming in the North Atlantic sector of the Arctic since the 19th century and mistakenly believed this climate variation could be explained entirely by greenhouse gas warming.

Similar model predictions were echoed by Plass in 1956: 'If at the end of this century, measurements show that the carbon dioxide content of the atmosphere has risen appreciably and at the same time the temperature has continued to rise throughout the world, it will be firmly established that carbon dioxide is an important factor in causing climatic change'. In trying to understand the carbon cycle, and specifically how fossil fuel emissions would change atmospheric CO_2, the interdisciplinary field of carbon cycle science began. One of the first problems addressed was the atmosphere-ocean exchange of CO_2. Revelle and Suess explained why part of the emitted CO_2 was observed to accumulate in the atmosphere rather than being completely absorbed by the oceans. While CO_2 can be mixed rapidly into the upper layers of the ocean, the time to mix with the deep ocean is many centuries. By the time of the TAR, the interaction of climate change with the oceanic circulation and biogeochemistry was projected to reduce the fraction of anthropogenic CO_2 emissions taken up by the oceans in the future, leaving a greater fraction in the atmosphere.

In the 1950s, the greenhouse gases of concern remained CO_2 and H_2O, the same two identified by Tyndall a century earlier. It was not until the 1970s that other greenhouse gases–CH_4, N_2O and CFCs–were widely recognised. By the 1970s, the importance of aerosol-cloud effects in reflecting sunlight was known and atmospheric aerosols were being proposed as climate-forcing constituents. Charlson and others built a consensus that sulphate aerosols were, by themselves, cooling the Earth's surface by directly reflecting sunlight.

Moreover, the increases in sulphate aerosols were anthropogenic and linked with the main source of CO_2, burning of fossil fuels. Thus, the current picture of the atmospheric constituents driving climate change contains a much more diverse mix of greenhouse agents.

PAST CLIMATE OBSERVATIONS, ASTRONOMICAL THEORY AND ABRUPT CLIMATE CHANGES

Throughout the 19th and 20th centuries, a wide range of geomorphology and palaeontology studies has provided new insight into the Earth's past climates, covering periods of hundreds of millions of years. The Palaeozoic Era, beginning 600 Ma, displayed evidence of both warmer and colder climatic conditions than the present; the Tertiary Period was generally warmer; and the Quaternary Period showed oscillations between glacial and interglacial conditions. Louis Agassiz developed the hypothesis that Europe had experienced past glacial ages, and there has since been a growing awareness that long-term climate observations can advance the understanding of the physical mechanisms affecting climate change. The scientific study of one such mechanism–modifications in the geographical and temporal patterns of solar energy reaching the Earth's surface due to changes in the Earth's orbital parametres–has a long history. The pioneering contributions of Milankovitch to this astronomical theory of climate change are widely known, and the historical review of Imbrie and Imbrie calls attention to much earlier contributions, such as those of James Croll, originating in 1864.

The pace of palaeoclimatic research has accelerated over recent decades. Quantitative and well-dated records of climate fluctuations over the last 100 kyr have brought a more comprehensive view of how climate changes occur, as well as the means to test elements of the astronomical theory. By the 1950s, studies of deep-sea cores suggested that the ocean temperatures may have been different during glacial times. Ewing and Donn proposed that changes in ocean circulation actually could initiate an ice age. In the 1960s, the works of Emiliani and Shackleton showed the potential of isotopic measurements in deepsea sediments to help explain Quaternary changes. In the 1970s, it became possible to analyse a deep-sea core time series of more than 700 kyr, thereby using the last reversal of the Earth's magnetic field to establish a dated chronology. This deep-sea observational record clearly showed the same periodicities found in the astronomical forcing, immediately providing strong support to Milankovitch's theory.

Ice cores provide key information about past climates, including surface temperatures and atmospheric chemical composition. The bubbles sealed in the ice are the only available samples of these past atmospheres. The first deep ice cores from Vostok in Antarctica provided additional evidence of the role of astronomical forcing. They also revealed a highly correlated evolution of temperature changes and atmospheric composition, which was subsequently confirmed over the past 400 kyr and now extends to almost 1 Myr. This discovery drove research to understand the causal links between greenhouse gases and climate change.

The same data that confirmed the astronomical theory also revealed its limits: a linear response of the climate system to astronomical forcing could not explain entirely the observed fluctuations of rapid ice-age terminations preceded by longer cycles of glaciations. The importance of other sources of climate variability was heightened by the discovery of abrupt climate changes. In this context, 'abrupt' designates regional events of large amplitude, typically a few degrees celsius, which occurred within several decades–much shorter than the thousand-year time scales that characterise changes in astronomical forcing. Abrupt temperature changes were first revealed by the analysis of deep ice cores from Greenland.

Oeschger et al. recognised that the abrupt changes during the termination of the last ice age correlated with cooling in Gerzensee and suggested that regime shifts in the Atlantic Ocean circulation were causing these widespread changes. The synthesis of palaeoclimatic observations by Broecker and Denton invigourated the community over the next decade. By the end of the 1990s, it became clear that the abrupt climate changes during the last ice age, particularly in the North Atlantic regions as found in the Greenland ice cores, were numerous, indeed abrupt and of large amplitude. They are now referred to as Dansgaard-Oeschger events. A similar variability is seen in the North Atlantic Ocean, with north-south oscillations of the polar front and associated changes in ocean temperature and salinity. With no obvious external forcing, these changes are thought to be manifestations of the internal variability of the climate system.

The importance of internal variability and processes was reinforced in the early 1990s with analysis of records with high temporal resolution. New ice cores new ocean cores from regions with high sedimentation rates, as well as lacustrine sediments and cave stalagmites produced additional evidence for unforced climate changes, and revealed a large number of abrupt changes in many regions throughout the last glacial cycle. Long sediment cores from the deep ocean were used to reconstruct the thermohaline circulation connecting deep and surface waters and to demonstrate the participation of the ocean in these abrupt climate changes during glacial periods.

By the end of the 1990s, palaeoclimate proxies for a range of climate observations had expanded greatly. The analysis of deep corals provided indicators for nutrient content and mass exchange from the surface to deep water, showing abrupt variations characterised by synchronous changes in surface and deep-water properties. Precise measurements of the CH_4 abundances in polar ice cores showed that they changed in concert with the

Dansgaard-Oeschger events and thus allowed for synchronisation of the dating across ice cores. The characteristics of the antarctic temperature variations and their relation to the Dansgaard-Oeschger events in Greenland were consistent with the simple concept of a bipolar seesaw caused by changes in the thermohaline circulation of the Atlantic Ocean. This work underlined the role of the ocean in transmitting the signals of abrupt climate change.

Abrupt changes are often regional, for example, severe droughts lasting for many years have changed civilizations, and have occurred during the last 10 kyr of stable warm climate. This result has altered the notion of a stable climate during warm epochs, as previously suggested by the polar ice cores. The emerging picture of an unstable oceanatmosphere system has opened the debate of whether human interference through greenhouse gases and aerosols could trigger such events. Palaeoclimate reconstructions cited in the FAR were based on various data, including pollen records, insect and animal remains, oxygen isotopes and other geological data from lake varves, loess, ocean sediments, ice cores and glacier termini.

These records provided estimates of climate variability on time scales up to millions of years. A climate proxy is a local quantitative record that is interpreted as a climate variable using a transfer function that is based on physical principles and recently observed correlations between the two records. The combination of instrumental and proxy data began in the 1960s with the investigation of the influence of climate on the proxy data, including tree rings, corals and ice cores.

Phenological and historical data are also a valuable source of climatic reconstruction for the period before instrumental records became available. Such documentary data also need calibration against instrumental data to extend and reconstruct the instrumental record. With the development of multi-proxy reconstructions, the climate data were extended not only from local to global, but also from instrumental data to patterns of climate variability. Most of these reconstructions were at single sites and only loose efforts had been made to consolidate records.

Mann et al. made a notable advance in the use of proxy data by ensuring that the dating of different records lined up. Thus, the true spatial patterns of temperature variability and change could be derived, and estimates of NH average surface temperatures were obtained. The Working Group I WGI FAR noted that past climates could provide analogues. Fifteen years of research since that assessment has identified a range of variations and instabilities in the climate system that occurred during the last 2 Myr of glacial-interglacial cycles and in the super-warm period of 50 Ma. These past climates do not appear to be analogues of the immediate future, yet they do reveal a wide range of climate processes that need to be understood when projecting 21stcentury climate change.

SOLAR VARIABILITY AND THE TOTAL SOLAR IRRADIANCE

Measurement of the absolute value of total solar irradiance is difficult from the Earth's surface because of the need to correct for the influence of the atmosphere. Langley attempted to minimise the atmospheric effects by taking measurements from high on Mt. Whitney in California, and to estimate the correction for atmospheric effects by taking measurements at several times of day, for example, with the solar radiation having passed through different atmospheric pathlengths. Between 1902 and 1957, Charles Abbot and a number of other scientists around the globe made thousands of measurements of TSI from mountain sites. Values ranged from 1,322 to 1,465 W m-2, which encompasses the current estimate of 1,365 W m-2. Foukal et al. deduced from Abbot's daily observations that higher values of TSI were associated with more solar faculae.

In 1978, the Nimbus-7 satellite was launched with a cavity radiometre and provided evidence of variations in TSI. Additional observations were made with an active cavity radiometre on the Solar Maximum Mission, launched in 1980. Both of these missions showed that the passage of sunspots and faculae across the Sun's disk influenced TSI. At the maximum of the 11-year solar activity cycle, the TSI is larger by about 0.1% than at the minimum. The observation that TSI is highest when sunspots are at their maximum is the opposite of Langley's hypothesis. As early as 1910, Abbot believed that he had detected a downward trend in TSI that coincided with a general cooling of climate. The solar cycle variation in irradiance corresponds to an 11-year cycle in radiative forcing which varies by about 0.2 W m-2.

There is increasingly reliable evidence of its influence on atmospheric temperatures and circulations, particularly in the higher atmosphere. Calculations with three-dimensional models suggest that the changes in solar radiation could cause surface temperature changes of the order of a few tenths of a degree celsius. For the time before satellite measurements became available, the solar radiation variations can be inferred from cosmogenic isotopes and from the sunspot number. Naked-eye observations of sunspots date back to ancient times, but it was only after the invention of the telescope in 1607 that it became possible to routinely monitor the number, size and position of these 'stains' on the surface of the Sun. Throughout the 17th and 18th centuries, numerous observers noted the variable concentrations and ephemeral nature of sunspots, but very few sightings were reported between 1672 and 1699.

This period of low solar activity, now known as the Maunder Minimum, occurred during the climate period now commonly referred to as the Little Ice Age. There is no exact agreement as to which dates mark the beginning and end of the Little Ice Age, but from about 1350 to about 1850 is one reasonable estimate. During the latter part of the 18th century, Wilhelm Herschel noted the presence not only of sunspots but of bright patches, now referred to as faculae, and of granulations on the solar surface. He believed that when these indicators of activity were more numerous, solar emissions of light and heat were greater and could affect the weather on Earth. Heinrich Schwabe published his discovery of a '10-year cycle' in sunspot numbers. Samuel Langley compared the brightness of sunspots with that of the surrounding photosphere. He concluded that they would block the emission of radiation and estimated that at sunspot cycle maximum the Sun would be about 0.1% less bright than at the minimum of the cycle, and that the Earth would be 0.1°C to 0.3°C cooler.

These satellite data have been used in combination with the historically recorded sunspot number, records of cosmogenic isotopes, and the characteristics of other Sun-like stars to estimate the solar radiation over the last 1,000 years. These data sets indicated quasi-periodic changes in solar radiation of 0.24 to 0.30% on the centennial time scale. These values have recently been re-assessed. The TAR states that the changes in solar irradiance are not the major cause of the temperature changes in the second half of the 20th century unless those changes can induce unknown large feedbacks in the climate system. The effects of galactic cosmic rays on the atmosphere and those due to shifts in the solar spectrum towards the ultraviolet range, at times of high solar activity, are largely unknown. The latter may produce changes in tropospheric circulation via changes in static stability resulting from the interaction of the increased UV radiation with stratospheric ozone. More research to investigate the effects of solar behaviour on climate is needed before the magnitude of solar effects on climate can be stated with certainty.

BIOGEOCHEMISTRY AND RADIATIVE FORCING

The modern scientific understanding of the complex and interconnected roles of greenhouse gases and aerosols in climate change has undergone rapid evolution over the last two decades. While the concepts were recognised and outlined in the 1970s, the publication of generally accepted quantitative results coincides with, and was driven in part by, the questions asked by the IPCC beginning in 1988. Thus, it is instructive to view the evolution of this topic as it has been treated in the successive IPCC reports. The WGI FAR codified the key physical and biogeochemical processes in the Earth system that relate a changing climate to atmospheric composition, chemistry, the carbon cycle and natural ecosystems. The science of the time, as summarised in the FAR, made a clear case for anthropogenic interference with the climate system.

In terms of greenhouse agents, the main conclusions from the WGI FAR Policymakers Summary are still valid today:

- 'Emissions resulting from human activities are substantially increasing the atmospheric concentrations of the greenhouse gases: CO_2, CH_4, CFCs, N_2O';
- 'Some gases are potentially more effective';

- Feedbacks between the carbon cycle, ecosystems and atmospheric greenhouse gases in a warmer world will affect CO_2 abundances; and
- GWPs provide a metric for comparing the climatic impact of different greenhouse gases, one that integrates both the radiative influence and biogeochemical cycles.

The climatic importance of tropospheric ozone, sulphate aerosols and atmospheric chemical feedbacks were proposed by scientists at the time and noted in the assessment. For example, early global chemical modelling results argued that global tropospheric ozone, a greenhouse gas, was controlled by emissions of the highly reactive gases nitrogen oxides, carbon monoxide and non-methane hydrocarbons. In terms of sulphate aerosols, both the direct radiative effects and the indirect effects on clouds were acknowledged, but the importance of carbonaceous aerosols from fossil fuel and biomass combustion was not recognised.

The concept of radiative forcing as the radiative imbalance in the climate system at the top of the atmosphere caused by the addition of a greenhouse gas was established at the time and summarised in the WGI FAR. Agents of RF included the direct greenhouse gases, solar radiation, aerosols and the Earth's surface albedo. What was new and only briefly mentioned was that 'many gases produce indirect effects on the global radiative forcing'. The innovative global modelling work of Derwent showed that emissions of the reactive but nongreenhouse gases–NOx, CO and NMHCs–altered atmospheric chemistry and thus changed the abundance of other greenhouse gases. Indirect GWPs for NOx, CO and VOCs were proposed.

The projected chemical feedbacks were limited to short-lived increases in tropospheric ozone. By 1990, it was clear that the RF from tropospheric ozone had increased over the 20th century and stratospheric ozone had decreased since 1980 but the associated RFs were not evaluated in the assessments. Neither was the effect of anthropogenic sulphate aerosols, except to note in the FAR that 'it is conceivable that this radiative forcing has been of a comparable magnitude, but of opposite sign, to the greenhouse forcing earlier in the century'. Reflecting in general the community's concerns about this relatively new measure of climate forcing, RF bar charts appear only in the underlying FAR stages, but not in the FAR Summary. Only the long-lived greenhouse gases are shown, although sulphate aerosols direct effect in the future is noted with a question mark.

The cases for more complex chemical and aerosol effects were becoming clear, but the scientific community was unable at the time to reach general agreement on the existence, scale and magnitude of these indirect effects. Nevertheless, these early discoveries drove the research agendas in the early 1990s. The widespread development and application of global chemistrytransport models had just begun with international workshops. In the

Supplementary Report to the FAR, the indirect chemical effects of CO, NOx and VOC were reaffirmed, and the feedback effect of CH_4 on the tropospheric hydroxyl radical was noted. Aerosolclimate interactions still focused on sulphates, and the assessment of their direct RF for the NH was now somewhat quantitative as compared to the FAR. Stratospheric ozone depletion was noted as causing a significant and negative RF, but not quantified. Ecosystems research at this time was identifying the responses to climate change and CO_2 increases, as well as altered CH_4 and N_2O fluxes from natural systems; however, in terms of a community assessment it remained qualitative.

By 1994, with work on SAR progressing, the Special Report on Radiative Forcing reported significant breakthroughs in a set of stages limited to assessment of the carbon cycle, atmospheric chemistry, aerosols and RF. The carbon budget for the 1980s was analysed not only from bottomup emissions estimates, but also from a top-down approach including carbon isotopes. A first carbon cycle assessment was performed through an international model and analysis workshop examining terrestrial and oceanic uptake to better quantify the relationship between CO_2 emissions and the resulting increase in atmospheric abundance. Similarly, expanded analyses of the global budgets of trace gases and aerosols from both natural and anthropogenic sources highlighted the rapid expansion of biogeochemical research.

The first RF bar chart appears, comparing all the major components of RF change from the pre-industrial period to the present. Anthropogenic soot aerosol, with a positive RF, was not in the 1995 Special Report but was added to the SAR. In terms of atmospheric chemistry, the first open-invitation modelling study for the IPCC recruited 21 atmospheric chemistry models to participate in a controlled study of photochemistry and chemical feedbacks. These studies demonstrated a robust consensus about some indirect effects, such as the CH_4 impact on atmospheric chemistry, but great uncertainty about others, such as the prediction of tropospheric ozone changes.

The model studies plus the theory of chemical feedbacks in the CH_4-CO-OH system firmly established that the atmospheric lifetime of a perturbation of CH_4 emissions was about 50% greater than reported in the FAR. There was still no consensus on quantifying the past or future changes in tropospheric ozone or OH. In the early 1990s, research on aerosols as climate forcing agents expanded. Based on new research, the range of climaterelevant aerosols was extended for the first time beyond sulphates to include nitrates, organics, soot, mineral dust and sea salt.

Quantitative estimates of sulphate aerosol indirect effects on cloud properties and hence RF were sufficiently well established to be included in assessments, and carbonaceous aerosols from biomass burning were recognised as being comparable in importance to sulphate. Ranges are given in the special report for direct sulphate RF and biomass-burning aerosols.

The aerosol indirect RF was estimated to be about equal to the direct RF, but with larger uncertainty. The injection of stratospheric aerosols from the eruption of Mt. Pinatubo was noted as the first modern test of a known radiative forcing, and indeed one climate model accurately predicted the temperature response. In the one-year interval between the special report and the SAR, the scientific understanding of aerosols grew. The direct anthropogenic aerosol forcing was reduced to -0.5 W m^{-2}. The RF bar chart was now broken into aerosol components with a separate range for indirect effects. Throughout the 1990s, there were concerted research programmes in the USA and EU to evaluate the global environmental impacts of aviation. Several national assessments culminated in the IPCC Special Report on Aviation and the Global Atmosphere which assessed the impacts on climate and global air quality.

An open invitation for atmospheric model participation resulted in community participation and a consensus on many of the environmental impacts of aviation. The direct RF of sulphate and of soot aerosols was likewise quantified along with that of contrails, but the impact on cirrus clouds that are sometimes generated downwind of contrails was not. The assessment re-affirmed that RF was a first-order metric for the global mean surface temperature response, but noted that it was inadequate for regional climate change, especially in view of the largely regional forcing from aerosols and tropospheric ozone. By the end of the 1990s, research on atmospheric composition and climate forcing had made many important advances. The TAR was able to provide a more quantitative evaluation in some areas. For example, a large, open-invitation modelling workshop was held for both aerosols and tropospheric ozone-OH chemistry.

This workshop brought together as collaborating authors most of the international scientific community involved in developing and testing global models of atmospheric composition. In terms of atmospheric chemistry, a strong consensus was reached for the first time that science could predict the changes in tropospheric ozone in response to scenarios for CH_4 and the indirect greenhouse gases and that a quantitative GWP for CO could be reported. Further, combining these models with observational analysis, an estimate of the change in tropospheric ozone since the pre-industrial era–with uncertainties–was reported. The aerosol workshop made similar advances in evaluating the impact of different aerosol types. There were many different representations of uncertainty in the TAR, and the consensus RF bar chart did not generate a total RF or uncertainties for use in the subsequent IPCC Synthesis Report.

CRYOSPHERIC TOPICS

The cryosphere, which includes the ice sheets of Greenland and Antarctica, continental glaciers, snow, sea ice, river and lake ice, permafrost and seasonally frozen ground, is an important component of the climate system. The cryosphere derives its importance to the climate system from a variety of effects, including its high reflectivity for solar radiation, its low thermal conductivity, its large thermal inertia, its potential for affecting ocean circulation and atmospheric circulation, its large potential for affecting sea level, and its potential for affecting greenhouse gases. Studies of the cryospheric albedo feedback have a long history. The albedo is the fraction of solar energy reflected back to space. Over snow and ice, the albedo is large compared to that over the oceans.

In a warming climate, it is anticipated that the cryosphere would shrink, the Earth's overall albedo would decrease and more solar energy would be absorbed to warm the Earth still further. This powerful feedback loop was recognised in the 19th century by Croll and was first introduced in climate models by Budyko and Sellers. But although the principle of the albedo feedback is simple, a quantitative understanding of the effect is still far from complete. For instance, it is not clear whether this mechanism is the main reason for the highlatitude amplification of the warming signal. The potential cryospheric impact on ocean circulation and sea level are of particular importance. There may be 'large-scale discontinuities' resulting from both the shutdown of the large-scale meridional circulation of the world oceans and the disintegration of large continental ice sheets. Mercer proposed that atmospheric warming could cause the ice shelves of western Antarctica to disintegrate and that as a consequence the entire West Antarctic Ice Sheet would lose its land connection and come afloat, causing a sea level rise of about five metres.

The importance of permafrost-climate feedbacks came to be realised widely only in the 1990s, starting with the works of Kvenvolden, MacDonald and Harriss et al. As permafrost thaws due to a warmer climate, CO_2 and CH_4 trapped in permafrost are released to the atmosphere. Since CO_2 and CH_4 are greenhouse gases, atmospheric temperature is likely to increase in turn, resulting in a feedback loop with more permafrost thawing. The permafrost and seasonally thawed soil layers at high latitudes contain a significant amount of the global total amount of soil carbon. Because global warming signals are amplified in high-latitude regions, the potential for permafrost thawing and consequent greenhouse gas releases is thus large. In situ monitoring of the cryosphere has a long tradition. For instance, it is important for fisheries and agriculture. Seagoing communities have documented sea ice extent for centuries. Records of thaw and freeze dates for lake and river ice start with Lake Suwa in Japan in 1444, and extensive records of snowfall in China were made during the Qing Dynasty. Records of glacial length go back to the mid-1500s.

Internationally coordinated, long-term glacier observations started in 1894 with the establishment of the International Glacier Commission in Zurich, Switzerland. The longest time series of a glacial mass balance was started in 1946 at the Storglaciären in northern Sweden, followed by Storbreen in Norway. Today a global network of mass balance monitoring for some 60 glaciers is coordinated through the World Glacier Monitoring Service. Systematic measurements of permafrost began in earnest around 1950 and were coordinated under the Global Terrestrial Network for Permafrost.

The main climate variables of the cryosphere are in principle observable from space, given proper calibration and validation through in situ observing efforts. Indeed, satellite data are required in order to have full global coverage. The polar-orbiting Nimbus 5 satellite, launched in 1972, yielded the earliest all-weather, allseason imagery of global sea ice, using microwave instruments and enabled a major advance in the scientific understanding of the dynamics of the cryosphere. Launched in 1978, the Television Infrared Observation Satellite yielded the first monitoring from space of snow on land surfaces. The number of cryospheric elements now routinely monitored from space is growing, and current satellites are now addressing one of the more challenging elements, variability of ice volume. Climate modelling results have pointed to high-latitude regions as areas of particular importance and ecological vulnerability to global climate change.

It might seem logical to expect that the cryosphere overall would shrink in a warming climate or expand in a cooling climate. However, potential changes in precipitation, for instance due to an altered hydrological cycle, may counter this effect both regionally and globally. By the time of the TAR, several climate models incorporated physically based treatments of ice dynamics, although the land ice processes were only rudimentary. Improving representation of the cryosphere in climate models is still an area of intense research and continuing progress.

OCEAN AND COUPLED OCEAN-ATMOSPHERE DYNAMICS

Developments in the understanding of the oceanic and atmospheric circulations, as well as their interactions, constitute a striking example of the continuous interplay among theory, observations and, more recently, model simulations. The atmosphere and ocean surface circulations were observed and analysed globally as early as the 16th and 17th centuries, in close association with the development of worldwide trade based on sailing. These efforts led to a number of important conceptual and theoretical works. For example, Edmund Halley first published a description of the tropical atmospheric cells in 1686, and George Hadley proposed a theory linking the existence of the trade winds with those cells in 1735. These early studies helped to forge concepts that are still useful in analysing and understanding both the atmospheric general circulation itself and model simulations.

A comprehensive description of these circulations was delayed by the lack of necessary observations in the higher atmosphere or deeper ocean. The balloon record of Gay-Lussac, who reached an altitude of 7,016 m in 1804, remained unbroken for more than 50 years. The stratosphere was independently discovered near the turn of the 20th century by Aßmann and Teisserenc de Bort, and the first manned balloon flight into the stratosphere was made in 1901. Even though it was recognised over 200 years ago that the oceans' cold subsurface waters must originate at high latitudes, it was not appreciated until the 20th century that the strength of the deep circulation might vary over time, or that the ocean's Meridional Overturning Circulation may be very important for Earth's climate. By the 1950s, studies of deep-sea cores suggested that the deep ocean temperatures had varied in the distant past. Technology also evolved to enable measurements that could confirm that the deep ocean is not only not static, but in fact quite dynamic. By the late 1970s, current metres could monitor deep currents for substantial amounts of time, and the first ocean observing satellite revealed that significant information about subsurface ocean variability is imprinted on the sea surface.

At the same time, the first estimates of the strength of the meridional transport of heat and mass were made, using a combination of models and data. Since then the technological developments have accelerated, but monitoring the MOC directly remains a substantial challenge, and routine observations of the subsurface ocean remain scarce compared to that of the atmosphere. In parallel with the technological developments yielding new insights through observations, theoretical and numerical explorations of multiple equilibria began. Chamberlain suggested that deep ocean currents could reverse in direction, and might affect climate. The idea did not gain momentum until fifty years later, when Stommel presented a mechanism, based on the opposing effects that temperature and salinity have on density, by which ocean circulation can fluctuate between states. Numerical climate models incorporating models of the ocean circulation were developed during this period, including the pioneering work of Bryan and Manabe and Bryan.

The idea that the ocean circulation could change radically, and might perhaps even feel the attraction of different equilibrium states, gained further support through the simulations of coupled climate models. Model simulations using a hierarchy of models showed that the ocean circulation system appeared to be particularly vulnerable to changes in the freshwater balance, either by direct addition of freshwater or by changes in the hydrological cycle. A strong case emerged for the hypothesis that rapid changes in the Atlantic meridional circulation were responsible for the abrupt Dansgaard-Oeschger climate change events.

Although scientists now better appreciate the strength and variability of the global-scale ocean circulation, its roles in climate are still hotly debated. Is it a passive recipient of atmospheric forcing and so merely a diagnostic consequence of climate change, or is it an active contributor? Observational evidence for the latter proposition was presented by Sutton and Allen, who noticed SST anomalies propagating along the Gulf Stream/North Atlantic

Current system for years, and therefore implicated internal oceanic time scales. Is a radical change in the MOC likely in the near future? Brewer et al. and Lazier showed that the water masses of the North Atlantic were indeed changing in the modern observational record, a phenomenon that at least raises the possibility that ocean conditions may be approaching the point where the circulation might shift into Stommel's other stable regime. Studying the interactions between atmosphere and ocean circulations was also facilitated through continuous interactions between observations, theories and simulations, as is dramatically showed by the century-long history of the advances in understanding the El Nino-Southern Oscillation phenomenon.

This coupled air-sea phenomenon originates in the Pacific but affects climate globally, and has raised concern since at least the 19th century. Sir Gilbert Walker describes how H. H. Hildebrandsson noted largescale relationships between interannual trends in pressure data from a worldwide network of 68 weather stations, and how Lockyer and Lockyer confirmed Hildebrandsson's discovery of an apparent 'seesaw' in pressure between South America and the Indonesian region. Walker named this seesaw pattern the 'Southern Oscillation' and related it to occurrences of drought and heavy rains in India, Australia, Indonesia and Africa. He also proposed that there must be a certain level of predictive skill in that system. El Nino is the name given to the rather unusual oceanic conditions involving anomalously warm waters occurring in the eastern tropical Pacific off the coast of Peru every few years. The 1957–1958 International Geophysical Year coincided with a large El Nino, allowing a remarkable set of observations of the phenomenon.

A decade later, a mechanism was presented that connected Walker's observations to El Nino. This mechanism involved the interaction, through the SST field, between the east-west atmospheric circulation of which Walker's Southern Oscillation was an indicator and variability in the pool of equatorial warm water of the Pacific Ocean. Observations made in the 1970s showed that prior to ENSO warm phases, the sea level in the western Pacific often rises significantly.

By the mid-1980s, after an unusually disruptive El Nino struck in 1982 and 1983, an observing system had been put in place to monitor ENSO. The resulting data confirmed the idea that the phenomenon was inherently one involving coupled atmosphere-ocean interactions and yielded much-needed detailed observational insights.

By 1986, the first experimental ENSO forecasts were made. The mechanisms and predictive skill of ENSO are still under discussion. In particular, it is not clear how ENSO changes with, and perhaps interacts with, a changing climate. The TAR states '...increasing evidence suggests the ENSO plays a fundamental role in global climate and its interannual variability, and increased credibility in both regional and global climate projections will be gained once realistic ENSOs and their changes are simulated'. Just as the phenomenon of El Nino has been familiar to the people of tropical South America for centuries, a spatial pattern affecting climate variability in the North Atlantic has similarly been known by the people of Northern Europe for a long time.

The Danish missionary Hans Egede made the following wellknown diary entry in the mid-18th century: 'In Greenland, all winters are severe, yet they are not alike. The Danes have noticed that when the winter in Denmark was severe, as we perceive it, the winter in Greenland in its manner was mild, and conversely'. Teisserenc de Bort, Hann, Exner, Defant and Walker all contributed to the discovery of the underlying dynamic structure. Walker, in his studies in the Indian Ocean, actually studied global maps of sea level pressure correlations, and named not only the Southern Oscillation, but also a Northern Oscillation, which he subsequently divided into a North Pacific and a North Atlantic Oscillation. However, it was Exner who made the first correlation maps showing the spatial structure in the NH, where the North Atlantic Oscillation pattern stands out clearly as a north-south oscillation in atmospheric mass with centres of action near Iceland and Portugal.

The NAO significantly affects weather and climate, ecosystems and human activities of the North Atlantic sector. But what is the underlying mechanism? The recognition that the NAO is associated with variability and latitudinal shifts in the westerly flow of the jet stream originates with the works of Willett, Namias, Lorenz, Rossby and others in the 1930s, 1940s and 1950s. Because atmospheric planetary waves are hemispheric in nature, changes in one region are often connected with changes in other regions, a phenomenon dubbed 'teleconnection'. The NAO may be partly described as a high-frequency stochastic process internal to the atmosphere. This understanding is evidenced by numerous atmosphere-only model simulations. It is also considered an expression of one of Earth's 'annular modes'.

It is, however, the low-frequency variability of this phenomenon that fuels continued investigations by climate scientists. The long time scales are the indication of potential predictive skill in the NAO. The mechanisms responsible for the correspondingly long 'memory' are still debated, although they are likely to have a local or remote oceanic origin. Bjerknes recognised the connection between the NAO index and sea surface conditions. He speculated that ocean heat advection could play a role on longer time scales. The circulation of the Atlantic Ocean is radically different from that of the Indian and Pacific Oceans, in that the MOC is strongest in the Atlantic with warm water flowing northwards, even south of the equator, and cold water returning at depth. It would therefore not be surprising if the oceanic contributions to the NAO and to the Southern Oscillation were different.

Earth's climate is characterised by many modes of variability, involving both the atmosphere and ocean, and also the cryosphere and biosphere.

Understanding the physical processes involved in producing low-frequency variability is crucial for improving scientists' ability to accurately predict climate change and for allowing the separation of anthropogenic and natural variability, thereby improving the ability to detect and attribute anthropogenic climate change. One central question for climate scientists is to determine how human activities influence the dynamic nature of Earth's climate, and to identify what would have happened without any human influence at all.

THE HUMAN FINGERPRINT ON GREENHOUSE GASES

The high-accuracy measurements of atmospheric CO_2 concentration, initiated by Charles David Keeling in 1958, constitute the master time series documenting the changing composition of the atmosphere. These data have iconic status in climate change science as evidence of the effect of human activities on the chemical composition of the global atmosphere. Keeling's measurements on Mauna Loa in Hawaii provide a true measure of the global carbon cycle, an effectively continuous record of the burning of fossil fuel. They also maintain an accuracy and precision that allow scientists to separate fossil fuel emissions from those due to the natural annual cycle of the biosphere, demonstrating a long-term change in the seasonal exchange of CO_2 between the atmosphere, biosphere and ocean. Later observations of parallel trends in the atmospheric abundances of the $13CO_2$ isotope and molecular oxygen uniquely identified this rise in CO_2 with fossil fuel burning.

To place the increase in CO_2 abundance since the late 1950s in perspective, and to compare the magnitude of the anthropogenic increase with natural cycles in the past, a longerterm record of CO_2 and other natural greenhouse gases is needed. These data came from analysis of the composition of air enclosed in bubbles in ice cores from Greenland and Antarctica. The initial measurements demonstrated that CO_2 abundances were significantly lower during the last ice age than over the last 10 kyr of the Holocene. From 10 kyr before present up to the year 1750, CO_2 abundances stayed within the range 280 ± 20 ppm. During the industrial era, CO_2 abundance rose roughly exponentially to 367 ppm in 1999 and to 379 ppm in 2005.

Direct atmospheric measurements since 1970 have also detected the increasing atmospheric abundances of two other major greenhouse gases, methane and nitrous oxide. Methane abundances were initially increasing at a rate of about 1% yr–1 but then slowed to an average increase of 0.4% yr–1 over the 1990s with the possible stabilisation of CH4 abundance. The increase in N2O abundance is smaller, about 0.25% yr–1, and more difficult to detect. To go back in time, measurements were made from firn air trapped in snowpack dating back over 200 years, and these data show an accelerating rise in both CH4 and N_2O into the 20th century. When ice core measurements extended the CH4 abundance back 1 kyr, they showed a stable, relatively constant abundance of 700 ppb until the 19th century when a steady increase brought CH4 abundances to 1,745 ppb in 1998 and 1,774 ppb in 2005. This peak abundance is much higher than the range of 400 to 700 ppb seen over the last half-million years of glacial-interglacial cycles, and the increase can be readily explained by anthropogenic emissions. For N_2O the results are similar: the relative increase over the industrial era is smaller, yet the 1998 abundance of 314 ppb, rising to 319 ppb in 2005 is also well above the 180- to-260 ppb range of glacial-interglacial cycles

Several synthetic halocarbons are greenhouse gases with large global warming potentials. The chemical industry has been producing these gases and they have been leaking into the atmosphere since about 1930. Lovelock first measured CFC-11 in the atmosphere, noting that it could serve as an artificial tracer, with its north-south gradient reflecting the latitudinal distribution of anthropogenic emissions. Atmospheric abundances of all the synthetic halocarbons were increasing until the 1990s, when the abundance of halocarbons phased out under the Montreal Protocol began to fall. In the case of synthetic halocarbons, ice core research has shown that these compounds did not exist in ancient air and thus confirms their industrial human origin.

At the time of the TAR scientists could say that the abundances of all the well-mixed greenhouse gases during the 1990s were greater than at any time during the last half-million years and this record now extends back nearly one million years. Given this daunting picture of increasing greenhouse gas abundances in the atmosphere, it is noteworthy that, for simpler challenges but still on a hemispheric or even global scale, humans have shown the ability to undo what they have done. Sulphate pollution in Greenland was reversed in the 1980s with the control of acid rain in North America and Europe and CFC abundances are declining globally because of their phase-out undertaken to protect the ozone layer.

GLOBAL SURFACE TEMPERATURE

Shortly after the invention of the thermometre in the early 1600s, efforts began to quantify and record the weather. The first meteorological network was formed in northern Italy in 1653 and reports of temperature observations were published in the earliest scientific journals. By the latter part of the 19th century, systematic observations of the weather were being made in almost all inhabited areas of the world. Formal international coordination of meteorological observations from ships commenced in 1853.

Inspired by the document Suggestions on a Uniform System of Meteorological Observations, the International Meteorological Organization was formed in 1873. Its successor, the World Meteorological Organization, still works to promote and exchange standardised meteorological observations.

Yet even with uniform observations, there are still four major obstacles to turning instrumental observations into accurate global time series:

1. Access to the data in usable form;
2. Quality control to remove or edit erroneous data points;
3. Homogeneity assessments and adjustments where necessary to ensure the fidelity of the data; and
4. Area-averaging in the presence of substantial gaps.

Koppen was the first scientist to overcome most of these obstacles in his quest to study the effect of changes in sunspots. Much of his data came from Dove but wherever possible he used data directly from the original source, because Dove often lacked information about the observing methods. Koppen considered examination of the annual mean temperature to be an adequate technique for quality control of far distant stations. Using data from more than 100 stations, Koppen averaged annual observations into several major latitude belts and then area-averaged these into a near-global time series shown in Figure.

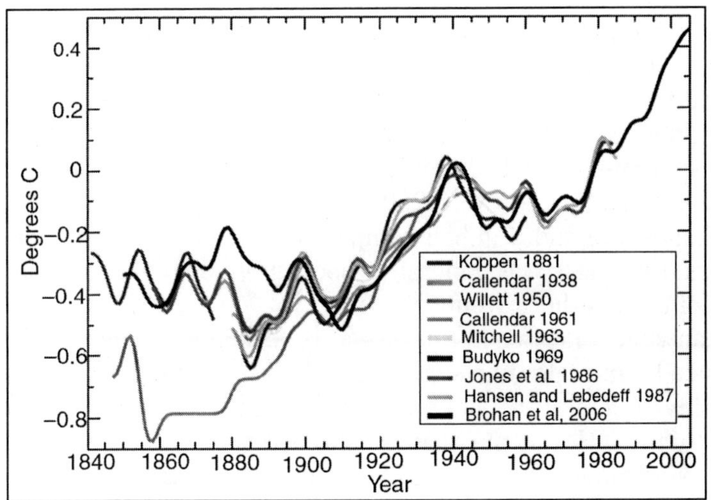

Fig. Published Records of Surface Temperature Change Over Large Regions.

Callendar produced the next global temperature time series expressly to investigate the influence of CO_2 on temperature. Callendar examined about 200 station records. Only a small portion of them were deemed defective, based on quality concerns determined by comparing differences with neighbouring stations or on homogeneity concerns based on station changes documented in the recorded metadata. After further removing two arctic stations because he had no compensating stations from the antarctic region, he created a global average using data from 147 stations. Most of Callendar's data came from World Weather Records. Initiated by a resolution at the 1923 IMO Conference, WWR was a monumental international undertaking producing a 1,196-page volume of monthly temperature, precipitation and pressure data from hundreds of stations around the world, some with data starting in the early 1800s. In the early 1960s, J. Wolbach had these data digitised. The WWR project continues today under the auspices of the WMO with the digital publication of decadal updates to the climate records for thousands of stations worldwide. Willett also used WWR as the main source of data for 129 stations that he used to create a global temperature time series going back to 1845. While the resolution that initiated WWR called for the publication of long and homogeneous records, Willett took this mandate one step further by carefully selecting a subset of stations with as continuous and homogeneous a record as possible from the most recent update of WWR, which included data through 1940. To avoid over-weighting certain areas such as Europe, only one record, the best available, was included from each 10° latitude and longitude square. Station monthly data were averaged into five-year periods and then converted to anomalies with respect to the five-year period 1935 to 1939. Each station's anomaly was given equal weight to create the global time series.

Callendar in turn created a new near-global temperature time series in 1961 and cited Willett as a guide for some of his improvements. Callendar evaluated 600 stations with about three-quarters of them passing his quality checks. Unbeknownst to Callendar, a former student of Willett, Mitchell in work first presented in 1961, had created his own updated global temperature time series using slightly fewer than 200 stations and averaging the data into latitude bands. Landsberg and Mitchell compared Callendar's results with Mitchell's and stated that there was generally good agreement except in the data-sparse regions of the Southern Hemisphere.

Meanwhile, research in Russia was proceeding on a very different method to produce large-scale time series. Budyko used smoothed, hand-drawn maps of monthly temperature anomalies as a starting point. While restricted to analysis of the NH, this map-based approach not only allowed the inclusion of an increasing number of stations over time but also the utilisation of data over the oceans.

Increasing the number of stations utilised has been a continuing theme over the last several decades with considerable effort being spent digitising historical station data as well as addressing the continuing problem of acquiring up-to-date data, as there can be a long lag between making an observation and the data getting into global data sets. During the 1970s and 1980s, several teams produced global temperature time series. Advances especially worth noting during this period include the extended spatial interpolation and station averaging technique of Hansen and Lebedeff and the Jones et al. painstaking assessment of homogeneity and adjustments to account for discontinuities in the record of each of the thousands of stations in a global data set. Since then, global and national data sets have been rigorously adjusted for homogeneity using a variety of statistical and metadata-based approaches.

One recurring homogeneity concern is potential urban heat island contamination in global temperature time series. This concern has been addressed in two ways. The first is by adjusting the temperature of urban stations to account for assessed urban heat island effects. The second is by performing analyses that, like Callendar, indicate that the bias induced by urban heat islands in the global temperature time series is either minor or nonexistent.

As the importance of ocean data became increasingly recognised, a major effort was initiated to seek out, digitise and quality-control historical archives of ocean data. This work has since grown into the International Comprehensive Ocean- Atmosphere Data Set which has coordinated the acquisition, digitisation and synthesis of data ranging from transmissions by Japanese merchant ships to the logbooks of South African whaling boats. The amount of sea surface temperature and related data acquired continues to grow.

As fundamental as the basic data work of ICOADS was, there have been two other major advances in SST data. The first was adjusting the early observations to make them comparable to current observations. Prior to 1940, the majority of SST observations were made from ships by hauling a bucket on deck filled with surface water and placing a thermometre in it. This ancient method eventually gave way to thermometres placed in engine cooling water inlets, which are typically located several metres below the ocean surface. Folland and Parker developed an adjustment model that accounted for heat loss from the buckets and that varied with bucket size and type, exposure to solar radiation, ambient wind speed and ship speed.

They verified their results using time series of night marine air temperature. This adjusted the early bucket observations upwards by a few tenths of a degree celsius. Most of the ship observations are taken in narrow shipping lanes, so the second advance has been increasing global coverage in a variety of ways. Direct improvement of coverage has been achieved by the internationally coordinated placement of drifting and moored buoys. The buoys began to be numerous enough to make significant contributions to SST analyses in the mid-1980s and have subsequently increased to more than 1,000 buoys transmitting data at any one time. Since 1982, satellite data, anchored to in situ observations, have contributed to near-global coverage. In addition, several different approaches have been used to interpolate and combine land and ocean observations into the current global temperature time series. To place the current instrumental observations into a longer historical context requires the use of proxy data.

Figure depicts several historical 'global' temperature time series, together with the longest of the current global temperature time series, that of Brohan et al.. While the data and the analysis techniques have changed over time, all the time series show a high degree of consistency since 1900. The differences caused by using alternate data sources and interpolation techniques increase when the data are sparser. This phenomenon is especially showed by the pre-1880 values of Willett's time series. Willett noted that his data coverage remained fairly constant after 1885 but dropped off dramatically before that time to only 11 stations before 1850. The high degree of agreement between the time series resulting from these many different analyses increases the confidence that the changes they are indicating are real.

Despite the fact that many recent observations are automatic, the vast majority of data that go into global surface temperature calculations–over 400 million individual readings of thermometres at land stations and over 140 million individual in situ SST observations–have depended on the dedication of tens of thousands of individuals for well over a century. Climate science owes a great debt to the work of these individual weather observers as well as to international organisations such as the IMO, WMO and the Global Climate Observing System, which encourage the taking and sharing of high-quality meteorological observations. While modern researchers and their institutions put a great deal of time and effort into acquiring and adjusting the data to account for all known problems and biases, century-scale global temperature time series would not have been possible without the conscientious work of individuals and organisations worldwide dedicated to quantifying and documenting their local environment.

DETECTION AND ATTRIBUTION

Using knowledge of past climates to qualify the nature of ongoing changes has become a concern of growing importance during the last decades, as reflected in the successive IPCC reports. While linked together at a technical level, detection and attribution have separate objectives. Detection of climate change is the process of demonstrating that climate has changed in some defined statistical sense, without providing a reason for that change. Attribution of causes of climate change is the process of establishing the most likely causes for the detected change with some defined level of confidence. Using traditional approaches, unequivocal attribution would require controlled experimentation with our climate system.

However, with no spare Earth with which to experiment, attribution of anthropogenic climate change must be pursued by:

- Detecting that the climate has changed;
- Demonstrating that the detected change is consistent with computer model simulations of the climate change 'signal' that is calculated to occur in response to anthropogenic forcing; and
- Demonstrating that the detected change is not consistent with alternative, physically plausible explanations of recent climate change that exclude important anthropogenic forcings

Both detection and attribution rely on observational data and model output. Estimates of century-scale natural climate fluctuations remain difficult to obtain directly from observations due to the relatively short length of most observational records and a lack of understanding of the full range and effects of the various and ongoing external influences. Model simulations with no changes in external forcing provide valuable information on the natural internal variability of the climate system on time scales of years to centuries. Attribution, on the other hand, requires output from model runs that incorporate historical estimates of changes in key anthropogenic and natural forcings, such as well-mixed greenhouse gases, volcanic aerosols and solar irradiance. These simulations can be performed with changes in a single forcing only or with simultaneous changes in a whole suite of forcings.

In the early years of detection and attribution research, the focus was on a single time series - the estimated global-mean changes in the Earth's surface temperature. While it was not possible to detect anthropogenic warming in 1980, Madden and Ramanathan and Hansen et al. predicted it would be evident at least within the next two decades. A decade later, Wigley and Raper used a simple energy-balance climate model to show that the observed change in global-mean surface temperature from 1867 to 1982 could not be explained by natural internal variability. This finding was later confirmed using variability estimates from more complex coupled oceanatmosphere general circulation models. As the science of climate change progressed, detection and attribution research ventured into more sophisticated statistical analyses that examined complex patterns of climate change. Climate change patterns or 'fingerprints' were no longer limited to a single variable or to the Earth's surface. More recent detection and attribution work has made use of precipitation and global pressure patterns, and analysis of vertical profiles of temperature change in the ocean and atmosphere. Studies with multiple variables make it easier to address attribution issues. While two different climate forcings may yield similar changes in global mean temperature, it is highly unlikely that they will produce exactly the same 'fingerprint'. Such model-predicted fingerprints of anthropogenic climate change are clearly statistically identifiable in observed data. The common conclusion of a wide range of fingerprint studies conducted over the past 15 years is that observed climate changes cannot be explained by natural factors alone. A substantial anthropogenic influence is required in order to best explain the observed changes. The evidence from this body of work strengthens the scientific case for a discernible human influence on global climate.

FACTORS AFFECTING GREENHOUSE RADIATIVE FORCING

A number of basic factors affect the behaviour of different greenhouse gases as forcing agents within the climate system. First, the absorption strength and wavelength of the absorption in the thermal infrared are of fundamental importance in dictating whether a molecule can be an important greenhouse forcing agent; this effect is modified by the overlap between the absorption bands and those of other gases present in the atmosphere.

For example, the natural quantities of CO_2 are so large (compared to other trace gases) that the atmosphere is very opaque over short distance at the centre of its 15μm absorption band. The addition of a small amount of gas capable of absorbing at this wavelength has negligible effect on the net radiative flux at the tropopause.

Other greenhouse gases with absorption bands in the more transparent regions of the infrared spectrum, in particular between 10 and 12μm, will have a far greater radiative forcing effect. Second, the atmospheric residence time (lifetime) of a greenhouse gas can greatly influence its potential as a radiative forcing agent. Gases that remain in the atmosphere for considerable periods of time, before being removed via their sinks, will have a greater forcing potential over longer time horizons.

Third, the existing quantities of a greenhouse gas in the atmosphere can dictate the effect that additional molecules of that gas can have. For gases such as halocarbons, where the naturally occurring concentrations are zero or very small, their forcing is close to linear for present-day concentrations.

Gases such as methane and nitrous oxide are present in such quantities that significant absorption is already occurring, and it is found that their forcing is approximately proportional to the square root of their concentration.

For carbon dioxide, parts of the spectrum are already so opaque that additional molecules are almost ineffective; the forcing is found to be only logarithmic in concentration.

As well as the direct effects on radiative forcing, many greenhouse gases also have indirect radiative effects on the climate through their interactions with atmospheric chemical processes.

For example, the oxidation of methane in the atmosphere leads to additional production of CO_2. Certain halocarbons such as the CFCs significantly affect the distribution of ozone, another greenhouse gas, in the atmosphere.

The hydroxyl (OH) radical, itself not a greenhouse gas, is extremely important in the troposphere as a chemical scavenger. Reactions with OH largely control the atmospheric lifetime, and, therefore, the concentrations of a number of greenhouse gases, in particular methane and many of the halocarbons.

GREENHOUSE WARMING POTENTIALS

Although there are a number of ways of measuring and contrasting the radiative forcing potential of different greenhouse gases, the Global Warming Potential (GWP) is perhaps the most useful, particularly as a policy instrument.

GWPs take account of the various factors influencing the radiative forcing potential of greenhouse gases. Such measures combine the calculations of the absorption strength of a molecule with assessments of its atmospheric lifetime; it can also include the indirect greenhouse effects due to chemical changes in the atmosphere caused by the gas. A number of GWPs are listed in Table.

Table. Global Warming Potentials of the Major Greenhouse Gases

Trace Gas	Global Warming Potential (relative to CO_2)		
	Integration	**Time Horizon,**	**Years**
	20	**100**	**500**
CO_2	1	1	1
CH_4 (incl. indirect)	62	24.5	7.5
N_2O	290	320	180
CFC-12	7900	8500	4200
HCFC-22	4300	1700	520

ΔF-ΔC RELATIONSHIPS

The change in net radiative flux (Wm^{-2}) at the tropopause, ΔF, associated with a particular greenhouse gas, is usually expressed as some function of the change in atmospheric concentration of that gas. Direct-effect ΔF-ΔC relationships are calculated using detailed radiative-convective models.

The form of the ΔF-ΔC relationship depends primarily on the existing gas concentration, as explained in. For low/moderate/high concentrations, the form is well approximated by a linear/square-root/logarithmic dependence of ΔF on concentration.

For example, the ΔF-ΔC relationship for CO_2 is given by:

$$\Delta F = 6.3 \ln (C/C_0)$$

where C_0 is the initial CO_2 concentration, C is the final concentration and ln is the natural logarithm. This relationship is valid for concentrations up to 1000ppmv. Alternatively, the ΔF-ΔC relationship for CFC-12 is given by:

$$\Delta F = 0.22 (X-X_0)$$

where X_0 is the initial CFC-12 concentration and X is the final concentration. The relationship holds for X less than 2ppbv (2000pptv). Nevertheless, it should be appreciated that such relationships are empirical in nature and are therefore subject to uncertainties.

- First, there are uncertainties in the basic spectroscopic data for many gases.
- Second, uncertainties arise through details in the radiative-convective modelling.
- Third, the assumptions used to model DF-DC relationships are subject to uncertainties, for example the assumed vertical profile of

concentration, temperature and moisture changes, and the indirect effects on radiative forcing due to chemical interactions.

GREENHOUSE RADIATIVE FORCING

From the modelled ΔF-ΔC relationships, the increase in radiative forcing due to the enhanced greenhouse gas concentrations can be calculated. Instrumental records exist for the most recent decades, whilst proxy data are used to calculate greenhouse gas concentrations earlier in the study period.

Table gives the concentration changes for five of the commonly known greenhouse gases, whilst Table details their contribution to radiative forcing for a number of time intervals. Here, 1765 has been regarded as the onset of the Industrial Revolution.

Table. Changes in Atmospheric Concentration of the Major Greenhouse Gases Since 1750

Year	CO_2(ppmv)	CH_4(ppbv)	N_2O (ppbv)	CFC11 (pptv)	CFC12 (pptv)
1765	279	790	275	0	0
1900	296	974	292	0	0
1960	316	1272	297	18	30
1970	325	1421	299	70	121
1980	337	1569	303	158	273
1992	355	1714	311	270	504

Table. Change in Radiative Forcing (Wm^{-2}) due to Concentration Changes in Greenhouse Gases

Time period	CO_2	CH_4	N_2O	CFC11	CFC12	Total[†]
1765-1900	0.37	0.1	0.027	0.0	0.0	0.53
1765-1960	0.79	0.24	0.045	0.004	0.008	1.17
1765-1970	0.96	0.30	0.054	0.014	0.034	1.48
1765-1980	1.20	0.36	0.068	0.035	0.076	1.91
1765-1990	1.50	0.42	0.10	0.062	0.14	2.45

Changes in CO_2 concentration over the last two centuries have contributed most to the greenhouse radiative forcing. Over the period 1765 to 1990, CO_2 forcing has accounted for 61% of the total enhanced greenhouse forcing. Nevertheless, other greenhouse gases, in particular the halocarbons, are now accounting for an increasing proportion of the total greenhouse forcing, due to their relatively larger GWPs. The total increase in radiative forcing since 1765 is shown in Figure.

The total increase in direct radiative greenhouse forcing is approximately 2.5Wm-2. This value should be compared to the solar constant, 1368Wm-2, the amount of total solar radiation intercepted by the Earth. The average amount of radiation arriving at the top of the troposphere is about 270Wm-2. This figure is lower than the solar constant, for it takes into account the latitudinal and temporal variations in insulation.

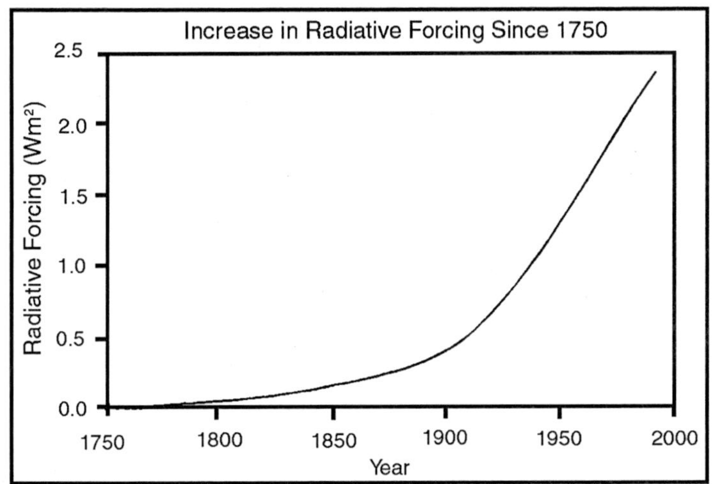

Fig. Increase in Radiative Forcing since 1750

RADIATIVE FORCING OF OZONE

The ΔF-ΔC relationship for atmospheric ozone is more complex than for other trace gases because of its marked vertical variations in absorption and concentration. Changes in ozone can cause greenhouse forcing by influencing both solar and infrared radiation. The net change in radiative forcing is strongly dependent on the vertical distribution of ozone concentration changes, and is particularly sensitive to variations around the tropopause.

Decreases in stratospheric ozone, principally over the Antarctic at altitudes between about 14 and 24km, have been occurring since the 1970s due to the anthropogenic release of CFCs and halons. These changes in ozone substantially perturb both solar and long-wave radiation.

While the solar effects due to ozone loss are determined by the total column ozone amounts, the long-wave effects are determined both by the amount and its vertical location (Lacis *et al.*, 1990). In general, stratospheric ozone loss will tend to increase the solar forcing, resulting in surface-troposphere warming, whilst decreasing greenhouse forcing, with consequent surface-troposphere cooling. Most models unambiguously demonstrate that changes in greenhouse forcing are dominant compute a global mean forcing of -0.2 ?0.1Wm^{-2} between 1970 and 1990. Such a value represents a significant offset to the positive greenhouse forcing from changes in halocarbons over the same period, estimated at 0.22Wm^{-2}.

The build up of tropospheric ozone due to chemical reactions involving precursors produced in various industrial processes has potentially important consequences for radiative forcing. Hauglustine *et al.* (1994) used a 2-D radiative-convective model to estimate that changes in tropospheric ozone since pre-industrial times have contributed a global mean forcing of $0.55Wm^{-2}$.

This agrees well with other studies, both modelling and observational. Nevertheless, the increases in tropospheric ozone will be highly regional and so will the positive greenhouse forcing associated with them.

ATMOSPHERIC AEROSOLS

Atmospheric aerosol particles are conventionally defined as those particles suspended in air having diametres in the region of 0.001 to 10 μm. They are formed by the reaction of gases in the atmosphere, or by the dispersal of material at the surface. Although making up only 1 part in 10^9 of the mass of the atmosphere, they have the potential to significantly influence the short-wave radiative transfer. The recent addition of anthropogenic aerosols to the atmosphere has introduced a negative change in radiative forcing which partially offsets the positive greenhouse forcing.

SOURCES AND SINKS OF AEROSOLS

Atmospheric aerosol particles may be emitted as particles or formed in the atmosphere from gaseous precursors (secondary sources).

Summarises the estimated recent annual emissions into the troposphere or stratosphere from the major sources of atmospheric aerosol, both natural and anthropogenic, primary and secondary. These include sulphates from the oxidation of sulphur-containing gases, nitrates from gaseous nitrogen species, organic materials from biomass combustion and oxidation of VOCs†, soot from combustion, and mineral dust from aeolian (wind-blown) processes. It should be noted that each aerosol flux estimate is a best guess, and uncertainty ranges of ±100% are not untypical.

Removal of aerosol mass is mainly achieved by transfer to the Earth's surface or by volatilisation. Such transfer is brought about by precipitation (wet deposition) and by direct uptake at the surface (dry deposition). The efficiency of both these deposition processes, and hence the time spent in the atmosphere by an aerosol particle, is a complex function of the aerosol's physical and chemical characteristics (e.g. particle size), and the time and location of its release.

For fine sulphate aerosols (0.01 to 0.1μm) released into or formed near the Earth's surface, an average lifetime is typically of the order of several days. This time scale is dominated mainly by the frequency of recurrence or precipitation. Conversely, particles transported into or formed in the upper troposphere are likely to remain there for weeks or months because of the less efficient precipitation scavenging. Stratospheric aerosols, formed as a consequence of large volcanic eruptions can remain there for up to one or two years.

Table. Recent Global Annual Emissions Estimates from Major Aerosols, Mt

Source	Flux Estimate	Particle Size†
Natural		
Primary		
Mineral aerosol	1500	mainly coarse
Sea salt	1300	coarse
Volcanic dust	33	coarse
Organic aerosols	50	coarse
Secondary		
Sulphates from biogenic gases	90	fine
Sulphates from volcanic SO_2	12	fine
Organic aerosols from VOCs	55	fine
Nitrates from NOx	22	mainly coarse
Total	3062	
Anthropogenic		
Primary		
Industrial dust	100	coarse and fine
Soot	10	mainly fine
Biomass burning	80	fine
Secondary		
Sulphates from SO_2	140	fine
Organic aerosols from VOCs	10	fine
Nitrates from NOx	40	mainly coarse
Total	380	

Owing to the short lifetime of aerosol particles in the atmosphere, particularly in the troposphere, and the non-uniform distribution of sources, their geographical distribution is highly non-uniform. As a consequence, the relative importance of the numerous sources shown in Table varies considerably over the globe. In certain areas of the Northern Hemisphere, for example Europe and North America, industrial sources are relatively much more important.

RADIATIVE FORCING BY AEROSOLS

Atmospheric aerosols influence climate in two ways, directly through the reflection and absorption of solar radiation (in both the troposphere and stratosphere), and indirectly through modifying the optical properties and lifetimes of clouds (mainly in the troposphere). Estimation of aerosol radiative forcing is more complex and hence more uncertain than radiative forcing due to the well-mixed greenhouse gases for several reasons. First, both the direct and indirect radiative effect of aerosol particles are strongly dependent on the particle size and chemical composition and cannot be related to mass source strengths in a simple manner, For this reason, anthropogenic aerosols contribute between 10 and 20% of the atmospheric mass burden, yet 50% to the global mean aerosol optical depth +. Second, the indirect radiative effects of aerosols depend on complex processes involving aerosol particles and the nucleation and growth of cloud droplets. Third, most aerosols have short lifetimes (days to weeks) and therefore their spatial distribution is highly inhomogeneous and strongly correlated with their sources.

DIRECT RADIATIVE FORCING

Aerosol particles in the 0.1 to 1.0μm diametre range have the highest efficiency per unit mass for optical interactions with incoming solar radiation due to the similarity of particle size and radiation wavelength. Sulphate aerosols and organic matter are therefore most effective at scattering and absorbing the short-wave radiation, with the effect of negative radiative forcing on the climate system The negative radiative forcing due to large injections of sulphate aerosol precursors into the stratosphere from volcanic eruptions is well known.

The major contributions to the anthropogenic component of the aerosol optical depth arise from sulphates produced from sulphur dioxide released during fossil fuel combustion and from organics released by biomass burning. To date, estimations of global negative forcing associated with anthropogenic aerosols are based on a number of modelling studies, due primarily to a lack of observational data.

Using a aerosol radiative-convective model, obtained a figure of -0.6 Wm^{-2} for direct global radiative forcing due to anthropogenic sulphate alone, whilst Kiehl and Brigleb calculated a value of -0.3Wm^{-2} for the Earth, and -0.43Wm^{-2} for the Northern Hemisphere. Figure shows the geographic distribution of annual mean direct radiative forcing (Wm^{-2}) from anthropogenic sulphate aerosols. The regionality of the forcing, due to localised emission sources and the short lifetimes of tropospheric sulphate aerosols, is clearly noticeable.

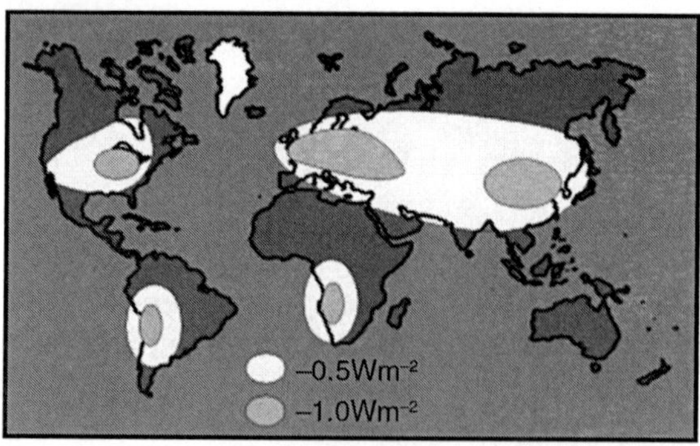

Fig. Global Aerosol Forcing

Other model-calculated values for direct global radiative forcing due to sulphate aerosols include -0.25Wm^{-2} and -0.9Wm^{-2}. Such a large sensitivity of the results of these simulations clearly indicate the need for more observational data on the chemical and physical properties of aerosols as well as more refined sulphate distribution calculations.

The effects of aerosols emitted as a result of biomass burning has received much less attention and global estimates of radiative forcing due to this source are subject to considerable uncertainty. The direct global-mean radiative forcing since pre-industrial times may lie in the range -0.05 to -0.6Wm^{-2}

Indirect Radiative Forcing

The global energy balance is sensitive to cloud albedo, and most particularly towards marine stratus (low-level) clouds which cover about 25% of the Earth.

Cloud albedo is itself sensitive to changes in the cloud droplet number concentration.

This droplet number depends, in a complex manner, on the concentration of cloud condensation nuclei (CCN), which, in turn, depends on aerosol concentration. Through this indirect effect, the negative radiative forcing caused by anthropogenic aerosol emissions might be further increased.

Particles of sizes around 0.1μm diametre composed of water soluble substances are highly effective as CCN. This includes both sulphate aerosols and organic aerosols from biomass burning. Direct observations of the impact of CCN on cloud albedos have been reported. Estimates from modelling studies of the indirect radiative effects of aerosols vary widely. Jones *et al.* calculated the indirect global-mean effect since pre-industrial times to be about -1.3Wm^{-2}, whilst Kaufman and Chou found a forcing of -0.45Wm^{-2}. Despite the level of uncertainty, indirect negative radiative forcing is believed to be comparable to the direct forcing.

3

Climate Change

PHYSICAL EVIDENCE FOR CLIMATIC CHANGE

Evidence for climatic change is taken from a variety of sources that can be used to reconstruct past climates. Reasonably complete global records of surface temperature are available beginning from the mid-late 19th century. For earlier periods, most of the evidence is indirect—climatic changes are inferred from changes in proxies, indicators that reflect climate, such as vegetation, ice cores, dendrochronology, sea level change, and glacial geology.

HISTORICAL AND ARCHAEOLOGICAL EVIDENCE

Climate change in the recent past may be detected by corresponding changes in settlement and agricultural patterns. Archaeological evidence, oral history and historical documents can offer insights into past changes in the climate. Climate change effects have been linked to the collapse of various civilizations.

GLACIERS

Glaciers are considered among the most sensitive indicators of climate change, advancing when climate cools and retreating when climate warms. Glaciers grow and shrink, both contributing to natural variability and amplifying externally forced changes. A world glacier inventory has been compiled since the 1970s, initially based mainly on aerial photographs and maps but now relying more on satellites. This compilation tracks more than 100,000 glaciers covering a total area of approximately 240,000 km^2, and preliminary estimates indicate that the remaining ice cover is around 445,000 km^2. The World Glacier Monitoring Service collects data annually on glacier retreat and glacier mass balance From this data, glaciers worldwide have been found to be shrinking significantly, with strong glacier retreats in the 1940s, stable or growing conditions during the 1920s and 1970s, and again retreating from the mid 1980s to present. The most significant climate processes since the middle to late Pliocene (approximately 3 million years ago) are the glacial and interglacial cycles. The present interglacial period (the Holocene) has lasted about 11,700 years. Shaped by orbital variations, responses such as the rise and fall of continental ice sheets and significant sea-level changes helped create the climate. Other changes, including Heinrich events, Dansgaard–Oeschger events and the Younger Dryas, however, show how glacial variations may also influence climate without the orbital forcing.

Glaciers leave behind moraines that contain a wealth of material—including organic matter, quartz, and potassium that may be dated—recording the periods in which a glacier advanced and retreated. Similarly, by tephrochronological techniques, the lack of glacier cover can be identified by the presence of soil or volcanic tephra horizons whose date of deposit may also be ascertained.

VEGETATION

A change in the type, distribution and coverage of vegetation may occur given a change in the climate; this much is obvious. In any given scenario, a mild change in climate may result in increased precipitation and warmth, resulting in improved plant growth and the subsequent sequestration of airborne CO_2. Larger, faster or more radical changes, however, may well result in vegetation stress, rapid plant loss and desertification in certain circumstances.

ICE CORES

Analysis of ice in a core drilled from a ice sheet such as the Antarctic ice sheet, can be used to show a link between temperature and global sea level variations.

The air trapped in bubbles in the ice can also reveal the CO_2 variations of the atmosphere from the distant past, well before modern environmental influences. The study of these ice cores has been a significant indicator of the changes in CO_2 over many millennia, and continues to provide valuable information about the differences between ancient and modern atmospheric conditions.

DENDROCLIMATOLOGY

Dendroclimatology is the analysis of tree ring growth patterns to determine past climate variations. Wide and thick rings indicate a fertile, well-watered growing period, whilst thin, narrow rings indicate a time of lower rainfall and less-than-ideal growing conditions.

POLLEN ANALYSIS

Palynology is the study of contemporary and fossil palynomorphs, including pollen. Palynology is used to infer the geographical distribution of plant species, which vary under different climate conditions. Different groups of plants have pollen with distinctive shapes and surface textures, and since the outer surface of pollen is composed of a very resilient material, they resist decay. Changes in the type of pollen found in different layers of sediment in lakes, bogs, or river deltas indicate changes in plant communities. These changes are often a sign of a changing climate. As an example, palynological studies have been used to track changing vegetation patterns throughout the Quaternary glaciations and especially since the last glacial maximum.

INSECTS

Remains of beetles are common in freshwater and land sediments. Different species of beetles tend to be found under different climatic conditions. Given the extensive lineage of beetles whose genetic makeup has not altered significantly over the millennia, knowledge of the present climatic range of the different species, and the age of the sediments in which remains are found, past climatic conditions may be inferred.

SEA LEVEL CHANGE

Global sea level change for much of the last century has generally been estimated using tide gauge measurements collated over long periods of time to give a long-term average. More recently, altimetre measurements — in combination with accurately determined satellite orbits — have provided an improved measurement of global sea level change. To measure sea levels prior to instrumental measurements, scientists have dated coral reefs that grow near the surface of the ocean, coastal sediments, marine terraces, ooids in limestones, and nearshore archaeological remains. The predominant dating methods used are uranium series and radiocarbon, with cosmogenic radionuclides being sometimes used to date terraces that have experienced relative sea level fall.

VARIABILITY OF CLIMATE CHANGE

VARIABILITY

Variability — that is the outstanding characteristic of temperate-zone weather. Even the brief sea log just quoted shows how weather in the tropics is much the same from place to place and day to day; temperate-zone weather, conversely, is diverse and changeable. But had our voyage extended far enough northward, it would have reached another region of comparative weather sameness — cold sameness instead of warm sameness -the arctic zone.

The temperate zone can have no stable and typical weather of its own. To southward (in the northern hemisphere) is the steady, warm, moist climate of the tropics. To northward is the fairly steady, mostly cold, fairly dry climate of the arctic. (This picture is of course repeated, with reverse directions, in the southern hemisphere.) The temperate zone is a buffer region between tropic and arctic climates, a battleground on which the gigantic atmospheric forces of heat and moisture advance and retreat, an unstable compromise between irreconcilable extremes, swayed this way and that in its evanescent weather as tropic forces (warm air masses) or arctic forces (cold air masses) momentarily gain ascendancy.

TROPIC CLIMATE

Tropic wind-and-climate belts migrate somewhat, north and south, with the apparent north-south seasonal swing of the sun. To get within-the-tropics weather in its most typical form we must go well south of the Tropic of Cancer, nearly down to the equator — say to Panama in about lat. 8° N. In these latitudes there are only two main seasons — a winter dry season under the influence of the steady northeast trades, and a summer wet season brooded over by the fitful calms of the doldrums.

The outstanding features of tropical-marine climate are that the average temperature is practically constant from month to month (nor does it ever vary, by more than a few degrees, from day to day); and that the cloudiness and rainfall, always appreciable, reach high values during the wet-season doldrums invasion of summer and early fall.

Upper winds over Panama, at half-mile altitude, average northeast about 8 m.p.h. in the rainy season and north-northeast about 20 m.p.h. in the dry. At two miles altitude there is less seasonal change — the average wind only varies from east 10 m.p.h. (rainy) to cast-northeast 12 m.p.h. (dry). Upper-air characteristics over the Caribbean Sea, quite typical of tropical-marine climate in general, were studied during the 1939 hurricane season by a United States Weather Bureau expedition to Swan Island.

This four-mile-long islet, named after a nineteenth-century pirate, lies almost exactly in the middle of the Sea at about lat. 16° N., between Jamaica and British Honduras. Within four months, observers Rhamlow and Leech released some two hundred and fifty pilot balloons and one hundred and twenty radio-sounding balloons. They found that the trade wind extended well aloft, though with diminishing force, here as in Panama — prevailing easterlies all the way up to about two and a half miles altitude. But above this level, the upper winds are mostly out of the west and increase with altitude.

Surface temperatures at Swan Island, and all over the Caribbean area, average about +75° to 80° F. the year round. But at the base of the stratosphere, ten miles up in these latitudes, it is cold indeed — about — 105° to — 110° F. Thus in the thick tropical troposphere there is an average lapse rate of something like — 3½° F/1000 ft (about — 6° C/Km). Together with plenty of warmth and moisture in the lower levels, this lapse rate means that conditional instability must extend through considerable air layers, particularly in the vertical-sun-warmed rainy season of summer and fall — and it is this conditional instability, together with inexhaustible supplies of moisture from the warm ocean, that produces the daily torrential showers and thundershowers, the regular afternoon downpours and the moist, misty mornings which can make equatorial regions such a continual sweat-bath for ill-adapted white men.

On or close to the ocean this deep-tropical weather is bearable, or perhaps even comfortable. But the steaming jungles inland, where temperatures can rise unchecked and where all breeze is choked off by the tangled uprearing of dense vegetation, are sometimes aptly described as 'green hells.' As Julian Duguid wrote '... a rich eternal garment of green, dappled with golden sunspots... a dense, fever-stricken thicket shimmering in the heat with a perpetual glassy haze....' However, by no means all the land areas within the tropics are covered with jungle — some of them are sandy and arid in the extreme, and doubly hot though not so muggy. (The word 'jungle' itself, originally, meant only 'waste land' in India.) And even in the fecund jungle areas of the tropics, such as Panama and northern South America, vegetation waxes lush and green with the rainy season, wanes sparse and tawny with the dry season.

There are also irregular year-to-year variations in moisture and prevailing winds that change the face of tropical nature. I got off a ship on the Colombian coast in August, 1939, and walked inland for a while, expecting to wade through wetseason green tangles of vines and trees peopled with tropic creatures — but the soil proved to be dry and sandy, the jungle a dried-up shrinkage of scrub, the creatures hidden away from the glaring vertical sun.

ARCTIC CLIMATE

Before Admiral Peary of the United States Navy marched five hundred icy miles on snowshoes from Ellesmere Land to the North Pole in the early spring of 1909, and since his time, many expeditions have entered the northern arctic regions. No latter-day explorer, incidentally, has ever surpassed Peary's record of intelligent planning and courageous, competent performance. But the airplane has rung startling changes in arctic travel.

In May, 1937, four large air transports set down four Russian scientists and a dog, together with a little portable cabin and food and fuel for a year or more, on one of the uncertain drifting ice floes that cover the Arctic Ocean around the pole. Leader Papanin, astronomer-aerologist Federov, hydrologist Shirsov, and radioman Krenkel had hoped to remain at or near the earth's axis for a whole year of observations, but their ten-foot-thick ice floe soon began to drift (as Peary or Fridtjof Nansen could have foretold), at first very slowly, across the two-mile-deep polar basin towards Greenland.

Then it moved, less gradually, southward along the east coast of that high continent, and finally sailed at twenty-five miles a day on southward. The floe-campers were rescued by icebreakers and airplanes twelve hundred miles south of the pole in February, 1938. Papanin's expedition found the summer temperatures of the polar basin hovering mostly around + 30° to 46° F.; clear spells of a day or two alternated with dull, overcast, rainy-to-wetsnowy weather. Winds varied from calm to light or moderate, mostly out of the north or northeast. Much of the ice was melting in the warmer spells, adding to the troubles and perils of the ice-campers. There was always the danger that their floe would disintegrate and dump them unceremoniously into the Arctic Ocean.

When winter came they had drifted several hundred miles south of the pole itself, but winter climate around the polar basin is probably pretty much the same everywhere. Temperatures ranged all the way from 50° F. below zero to 30° or more above, in accordance with the sweep of unlike air masses and cyclones, which occasionally extends well into the Arctic. For several months in midwinter the polar region is, of course, in continual darkness, with the sun visible at first in the spring only as a pale horizon glow around noon. On the thirty-first of December, strangely enough, the Russian wind-powered radio reported a temperature of + 43° F. over the arctic ice! But within a few hours, the mercury had tobogganed to a more usual value of -31° F.

The general north-polar region is, of course, about midway on the most direct (great-circle) air routes between Europe and Asia, and between eastern United States and Asia. Some years hence, perhaps, great 500-mile-an-hour strato-transports will be winging daily across these icy-watery wastes. Compared with other climatic regions, the polar area offers good flying conditions — during at least two or three days out of a week, that is, and particularly at higher levels.

Spring, when the polar high-pressure area lies symmetrically over the geographic pole, and not over the continental 'cold poles' in Greenland and northern Siberia, is the best flying season. A sealed-cabin, supercharged stratoliner, in any case, would be well up in the low arctic stratosphere (above thirty thousand feet) most of the time. Aside from Ellesmere Land at the north tip of America, one of the best fixed bases for observing the climate of the north-polar basin is the small island of Franz Joseph Land in lat. 81° N. long. 60° E., about six hundred land miles from the pole. In 1932-33 the Russian scientist Moltchanov (incidentally one of the pioneers in the development of radio-sounding balloons) made both surface and upper-air observations on Franz Joseph Land for a year. Like the floe-campers, he found average summer surface temperatures around +30° to 40° F. Winter temperatures averaged around -10 F.

Above the ground, for the first mile, there was usually a temperature inversion amounting to 5° F. or so; above that height the lapse rate or vertical temperature gradient was always surprisingly uniform at -3° or -4° F. per thousand feet up to the low polar 'tropopause,' marking the beginning of the stratosphere at about five miles altitude in winter, six miles in summer. Temperatures at the stratosphere base averaged about -75° F. in winter, about -40° F. in summer. July and August were the warmest months, as usual in the northern hemisphere; December was the coldest month at the surface, February coldest aloft. The southern hemisphere in general, being mostly water, shows more typical wind-and-climate belts than the northern hemisphere, which is all cluttered up with land masses that get too warm in summer and too cold in winter. By the same token, or a reverse one, the antarctic climate, founded upon a high, icecovered continent symmetric around the geographic pole, is more typically 'polar' than the arctic climate.

During the 1933-35 Byrd occupancy of 'Little America' in about lat. 79° S., long. 164° W., aerologists Haines and Grimminger recorded a great deal of antarctic weather. Mostly the average temperature was around +20° to 25° F. in summer (January), and around -35° to -40° in winter (July) — considerably colder, that is, than the north-polar climate. The coldest summer temperature was +12° on January 1; the warmest, +37° on December 22. In winter the mildest reading was +23° on July 27; the coldest was -71° on July 21. During a roaring blizzard in July the temperature fell to -65° with the wind sixty miles an hour out of the southeast — and wind, rather than temperature, is what really hurts in cold climates. These southeasterly blizzards in Antarctica correspond to the northeasterly gales around the north-polar basin.

High winds in Antarctica sometimes come from the north, just as north-polar gales can arise out of the south, but the colder blizzards from somewhere around the South Pole are more usual. Laurence M. Gould, Byrd's geologist and deputy in 1928-30, writes of one terrific southerly gale in the Rockefeller Mountains that first blew away the heavily packed snow wall in front of his firmly anchored airplane, and then finally blew the plane itself away and wrecked it. Some time before the maximum wind came, the airplane's air-speed metre indicated a steady velocity of ninety miles per hour, with gusts well over a hundred.

CONQUEST OF CLIMATE

The decade from 1900 to 1910 signalized an. amazing conquest of the tropics as a livable environment for civilized man. The years after 1940, likewise, may see a parallel conquest of the arctics. Within the tropics, except in certain areas plagued by extreme and wearisome heat, general living conditions were easy enough — too easy for man's own good, in fact; and aside from air transport, radio communication, mechanical refrigeration, and other new conveniences, the conquest was nearly all medical. Such dread warm-country diseases as typhoid, cholera, and yellow fever were finally vanquished — at least in the case of selected, well policed areas.

The work of Walter Reed, his assistant surgeons, and the heroic American soldiers who voluntarily entered the valley of 'yellow jack's' macabre shadow - all this really conquered the tropics and equatorial regions for white men, and contributed more than engineering and steam shovels to the triumphant completion, in 1913, of the world's greatest inter-ocean canal. Within the arctics and around the poles, on the contrary, disease is unknown unless brought to the scene by human carriers, but living conditions (without many bothersome precautions) are unduly hard and exacting. Aside from the extremes of dark and cold and gale and blinding snow, the chief hazards are: difficulty in finding your way around (particularly in the long, cold, dark arctic night); difficulty in surviving ordinary mischances such as falling into water, badly tearing clothes, and so on, particularly if alone; difficulty in getting adequately balanced diet and sufficient ultra-violet radiation. Yet even now, with our immense progress in transportation and communication, some of these difficulties can be largely overcome at a properly organized arctic station. Clothing must, of course, be of arctic type — layers of wool or fur inside for warmth, layers of canvas or leather outside for wind-breaking. Adequate shelter, plenty of fuel, and fresh food can now be transported easily by airplane to almost any arctic site. On location, howling arctic gales will provide plenty of electric power for lighting and radio communication, and perhaps for heating as well. Many of the dark terrors of the outer arctic night, which have heretofore held men prisoned for months in their base camps (except in the full of the moon) can perhaps be conquered by such devices as more powerful and efficient portable searchlights (using super-hot filaments, say, or possibly fluorescent light-sources); portable radio-telephone sets for constant two-way communication with and direction-finding on one or more base stations; more compact and accurate instruments, and simpler methods, for celestial navigation.

Against the white, too-uniform snow-glare of the polar day, improved goggles (perhaps of polarizing type) may be employed. Lighter, stronger and warmer tents and clothing can perhaps be devised. Much of the heavy local leg-work of arctic exploration can now be done with snow-mobiles, either tread-driven or propeller-driven. By means of vitamin diets, ultra-violet sunlamps and radio-diathermy, human beings near the poles will perhaps be enabled to maintain a high state of health during the long arctic night, and to overcome quickly any respiratory diseases that do develop. Eventually we may even see arctic health resorts, rest camps and play camps. Ellsmere Land in northern Canada (Peary's jumping-off place), less than three hours' flying time from the North Pole, is only twelve hours by present slow air liner from New York City. 'Polar Dude Camp on Ellsmere Land — spend next weekend here in the dark restfulness of the polar night under the flickering curtain of the glorious aurora borealis; pack ice exploring parties for the more active' — and so on. Such an advertisement may not startle our children twenty years hence.

TEMPERATE CLIMATE

We have said that the temperate zone is, in a climatic sense, nothing better than a battleground between tropic and arctic weather forces — between warm and cold air masses. Nevertheless the temperate zone (and in particular, the north-temperate zone) is that part of the earth in which civilized man evolved, and is probably the region most favorable to his continued progress and development. On the *average*, of course, temperate temperatures are pleasantest and best for all human activities — neither too hot nor too cold — though this average may rarely be matched by the vagaries of daily weather. And even the extreme moodiness and changeability of temperate weather has its advantages. Variety in weather, at least as to the four- or five-day spells and sequences so common in middle latitudes, perhaps satisfies some deep, innate rhythmic urges in man which cannot be allowed to atrophy without intellectual and cultural and physical loss.

However such obscure causes may or may not operate, it is certain that most human progress up from the dumb-ape level — most human art and ingenuity and imagination, most human science and engineering and social organization — has been best nurtured so far, and probably will be best nurtured in the future, in temperate latitudes. Among all the various meteorological factors that go to make up a climate, temperature is the one that most directly influences human health, happiness, and efficiency. How do various temperatures really feel to average human beings? Different degrees of heat of course feel quite different to different people, or to the same people under different conditions — in particular, under varying conditions of exertion. But it is possible to indicate average human reactions to temperature, and they are somewhat as follows: High humidity makes warm temperatures feel warmer, by retarding evaporation from the body. High humidity also makes cold temperatures feel colder, by making the clothing conduct heat better away from the body.

Wind makes cold temperatures feel colder, by removing the cushion of warm air that tends to form around the body in the absence of wind. This wind-cooling effect is particularly powerful at below-zero temperatures. On the other hand, wind makes warm temperatures up to 100° F. feel less warm, by increasing evaporation from the body. But if the temperature is over 105°, wind increases the discomfort.

The optimum temperature range, 68°-70° F., feels more comfortable if the air is normally moist. For health and comfort, indoor relative humidity should be around sixty per cent. But in the average home or office, in winter, it is seldom more than twenty per cent. Evaporating pans on radiators help some, but not very much. Constant-duty electric evaporators, preferably combined with electric lights, might well be used in temperate-zone houses during the winter.

Today's temperature anywhere is indicated by the thermometer; tomorrow's (which in temperate or arctic latitudes may be shockingly different) is foretold by a good forecast. Beyond these limits, our best present guide to future temperatures is the 'normal.' To get the normal temperature for any particular day of the year at any particular place, the average temperatures for that day, observed over as long a span of years as possible, are all averaged together. The normal temperature is, of course, merely a fictitious quantity seldom or never duplicated by any actual daily average; for climatically speaking, we dwellers in the temperate zone are always traveling now south, now north with the ebb and flow of warm and cold air masses.

Is our North American temperate-zone climate changing? This question keeps recurring in weather talk, and most of the oft-quoted answers foist off something about 'colder winters in Grandfather's day'— a notion perhaps promoted by oldsters in support of their claims to general superiority. The rising generation, contrariwise, have been inclined to charge the whole idea off to dotage and general boastfulness.

A definitive answer has been given by J. B. Kincer of the United States Weather Bureau: For a good many years up through 1936 there has been a decided tendency to warmer and drier weather, the trends being especially marked during the past quarter of a century, notwithstanding an occasional bad flood or a severely cold winter. Take the winter season, for example: With the exception of those for 1917-18 and 1935-36, the winters for the past twenty-five years or more, considering the country as a whole, have been rather uniformly warmer than normal, and, on an annual basis, every year since 1929 has had above normal temperature.

Also, in the matter of rainfall, there has been an equally marked tendency to droughts in recent years. However, an examination of the longer weather records of the country, going back one hundred years or more, indicates that this does not represent a permanent change of climate, but rather a warm, dry phase of our normal climate, to be followed, doubtless, by a cooler, wetter phase, when there will be more rain in summer and lower temperatures in winter....

Since the turn of the century, Mr. Kincer goes on to say, there has been a distinct yearly trend towards warmer winters and drier summers all over the world — and this in spite of a few exceptionally cold winters here and there. In other words, there are long (say fifty-year) cycles in climate, just as there are short (say five-day) cycles in weather.

If all these cycles were only considerate enough to maintain constant length, amplitude, and form, both weather forecasting and climate forecasting would be child's play. But the form, intensity, and length of weather and climate changes are all complexly variable. In any case the semipermanent (say 500-year) picture of our American climate shows, apparently, no appreciable change. But the geologic-time picture, measured in thousands or millions of years, would again show marked climatic cycles — the advance and retreat of the various ice ages.

MITIGATION OF CLIMATE CHANGE

Climate change mitigation is action to decrease the intensity of radiative forcing in order to reduce the potential effects of global warming. Mitigation is distinguished from adaptation to global warming, which involves acting to tolerate the effects of global warming. Most often, climate change mitigation scenarios involve reductions in the concentrations of greenhouse gases, either by reducing their sources or by increasing their sinks.

Scientific consensus on global warming, together with the precautionary principle and the fear of abrupt climate change is leading to increased effort to develop new technologies and sciences and carefully manage others in an attempt to mitigate global warming. Most means of mitigation appear effective only for preventing further warming, not at reversing existing warming. The Stern Review identifies several ways of mitigating climate change. These include reducing demand for emissions-intensive goods and services, increasing efficiency gains, increasing use and development of low-carbon technologies, and reducing fossil fuel emissions.

The energy policy of the European Union has set a target of limiting the global temperature rise to 2 °C compared to preindustrial levels, of which 0.8 °C has already taken place and another 0.5–0.7 °C is already committed. The 2 °C rise is typically associated in climate models with a carbon dioxide-equivalent concentration of 400–500 ppm by volume; the current level of carbon dioxide alone is 383 ppm by volume, and rising at 2 ppm annually.

Hence, to avoid a very likely breach of the 2 °C target, CO_2 levels would have to be stabilised very soon; this is generally regarded as unlikely, based on current programmes in place to date. The importance of change is showed by the fact that world economic energy efficiency is presently improving at only half the rate of world economic growth.

THE EFFECTS OF CLIMATE CHANGE ON ECOSYSTEMS

One way that scientists gain understanding of how global warming will affect ecosystems is to analyse the effects of past climate on paleoecosystems. A paleoecosystem is an ecosystem that existed in a former geologic time period. By relating vegetative cover to past climates, models can be developed. Once a reliable model is created, input variables such as CO_2 can be varied and the results analysed. An example of what the effect would be like on populations of Douglas fir in the northwestern United States if the CO_2 content in the atmosphere were double what it was before the Industrial Revolution was modeled by the U.S. Geological Survey and is shown in the illustration that follows. In a study conducted by the U.S. Global Change Research Programme in Washington, D.C., first in 2001, then updated in 2004, in trying to predict what the effects of future climate change would have on ecosystems, they concluded that"climate change has the potential to affect the structure, function, and regional distribution of ecosystems, and thereby affect the goods and services they provide." They based their conclusions on a modeling and analysis project they conducted called the Vegetation/Ecosystem Modeling and Analysis Project.

This project was used to generate future ecosystem scenarios for the conterminous United States based on model-simulated responses to both the Canadian and Hadley scenarios of climate change. The VEMAP was subsequently used in a validation exercise for a Dynamic Global Vegetation Model by Oregon State University and the U.S. Forest Service in 2008. Their MC1-DGVM was used as the input data in both the VEMAP and VINCERA. Their MC1 was run on both the VEMAP and VINCERA climate and soil input data to document how a change in the inputs can affect model outcome. The simulation results under the two sets of future climate scenarios were compared to see how different inputs can affect vegetation distribution and carbon budget projections.

The results indicated that"under all future scenarios, the interior west of the United States becomes woodier as warmer temperatures and available moisture allow trees to get established in grasslands areas. Concurrently, warmer and drier weather causes the eastern deciduous and mixed forests to shift to a more open canopy woodland or savannatype while boreal forests disappear almost entirely from the Great Lakes area by the end of the 21st century. While under VEMAP scenarios the model simulated large increases in carbon storage in a future woodier west, the drier VINCERA scenarios accounted for large carbon losses in the east and only moderate gains in the west. But under all future climate scenarios, the total area burned by wildfires increased." The similarities of the two models served to validate the VEMAP project.

The Hadley model was developed by the Hadley Centre for Climate Prediction and Research in England. Also referred to as the Met Office Hadley Centre for Climate Change, it is based at the headquarters of the Met Office in Exeter. It is the key institution in the United Kingdom for climate research. It is currently involved not only with understanding the physical, chemical, and biological processes within the climate system, but also with developing working models to explain current phenomena and to predict future climate change. It also monitors global and national climate variability and change and strives to determine the causes of the fluctuations. The Canadian climate model was developed by the Canadian Centre for Climate Modelling and Analysis. The CCCma is a division of the Climate Research Branch of Environment Canada based out of the University of Victoria, Victoria, British Columbia.

Its specific focus is on climate change and modeling. In the past nine years, the CCCma has produced three atmospheric and three atmospheric/oceanic general circulation models, making them one of the international leaders in climate change research. What they found was that over the next few decades climate change in the United States will most likely lead to increased plant productivity as a result of increasing levels of CO_2 in the atmosphere. There will also be an increase in terrestrial carbon storage for many parts of the country, especially the areas that become warmer and wetter. The southeast will most likely see reduced productivity and, therefore, a decrease in carbon storage. By the end of the 21st century, many areas of the country will have experienced changes in the distribution of vegetation. Wetter areas will see the growth of more trees; drier areas will have drier soils that will cause forested areas to die off and be replaced by savanna/grassland ecosystems.

Modeling the vegetation evolution and adaptation is more difficult. The study focused on two time periods: 2025-2034 and 2090-2099. In the near term, biogeochemical changes are expected to dominate the ecological responses. Biogeochemical responses include changes based on the natural cycles of carbon, nutrients and water. The responses are affected by changing environmental conditions such as temperature, precipitation, solar radiation, soil texture, and atmospheric CO_2. It is these natural cycles that affect carbon capture by plants with photosynthesis, soil nitrogen processes, and water transfer. These biogeochemical factors are what influence the production of vegetation. In the results from the near-term biogeochemical model, the scientists concluded that there would be an increase in CO_2. They estimate that currently the average carbon storage rate is 66/Tg/yr.

The Hadley model predicted that carbon storage rates by 2025-2034 would increase to 117/Tg/yr. The Canadian model estimated CO_2 to increase 96 Tg/yr. The Canadian model projected that the southeastern ecosystems will lose carbon in the near term because they predict the climate there will become hot and dry. The biogeography models look at the changing landscape based on changes in CO2, evapotranspiration, vegetation establishment, and competition between species, growth rates, and life cycle/mortality rates. In this model, scientists at both the Hadley and Canadian Centre for Climate Change agree that vegetation will be able to freely move from one location to another.

Changes in vegetation distribution will vary from region to region as follows:

- *Northeast*: Forests remain the dominant natural vegetation, but forest mixes will change. There will also be some increase in savannas and wetlands.
- *Southeast*: Forests remain the dominant ecosystem, but mixes change. Savannas and grasslands encroach on forests, especially towards the end of the 21st century. Drought and wildfires contribute significantly to forest destruction.
- *Midwest*: Forests remain the dominant land cover, but changes in species type occur. There will be a modest expansion of savannas and grasslands.

- *Great Plains*: Slight increase in woody vegetation.
- *West*: The areas of desert ecosystems shrink, and forest ecosystems grow.
- *Northwest*: Forested areas grow slightly.

A separate study conducted by the Canadian government predicts the following ecosystem changes as a result of changing climate over the next 100 years:

- *Coastlines*:
  - Flooding and erosion in coastal regions
  - Sea-level rise
- *Forests*:
  - Increase in pests
  - Increased levels of drought and wildfire
- *Plants and animals*:
  - Warmer temperatures could make water supplies more scarce, having a negative impact on plants and animals, not giving them time to adjust.
- *Crops*:
  - In some areas, warmer climate may allow a three- to five-week extension of the frost-free season, which could benefit commercial agriculture.
  - In other areas, drier soils and lack of water will have a negative impact on agricultural productivity.
- *Wells*:
  - The quality and quantity of drinking water may be threatened by increasing drought.
- Harsh weather:
  - Winter storms, floods, drought, heat waves, and tornadoes could become more frequent and severe.
- *Fisheries*:
  - Populations and ranges of species sensitive to changes in water temperature will be negatively affected.
  - Salmon harvests will be lower in the Pacific.
  - Changes in ocean currents may have a negative impact on the fisheries in the Atlantic.
- *Lakes and rivers*:
  - Water levels will decline under the influence of drought, negatively affecting drinking water quality.
  - Use of lakes for transportation, recreation, and fishing, and the ability to generate electricity may be curtailed under droughtlike conditions.
  - Other areas that may have an increase in precipitation may experience flooding, rising sea levels, and severe storms.

In February 2005, the Met Office in Exeter, England, issued a report titled"Avoiding Dangerous Climate Change." The objective of the study was to determine what levels of CO_2 were considered the tipping point for dangerous climate change with harmful effects on ecosystems and what actions could be taken now to avoid such an outcome. In the report, then prime minister Tony Blair stated:"It is now plain that the emission of greenhouse gases... is causing global warming at a rate that is unsustainable." Environment Secretary Margaret Beckett stated,"The report's conclusions would be a shock to many people.

The thing that is perhaps not so familiar to members of the public... is this notion that we could come to a tipping point where change could be irreversible. We're not talking about it happening over five minutes, of course, maybe over a thousand years, but it's the irreversibility that I think brings it home to people." The report, published by the British government, says there is only a small chance of greenhouse gas emissions being kept below"dangerous" levels. It warns that the Greenland ice sheet could melt, causing sea levels to rise by 23 feet over the next 1,000 years. It also warns that developing countries will be the hardest hit. The report also states, based on the vulnerability of many of the world's ecosystems, that the European Union (EU) has adopted a target of preventing an increase in global temperature of more than 3.3°F (2°C). Some believe even that may be too high. The report states:"Above two degrees, the risks increase very substantially," with"potentially large numbers of extinctions" and"major increases in hunger and water shortage risks... particularly in developing countries." In order to meet their goals, British scientists have advised that CO_2 levels should be stabilized at 450 parts per million (ppm) or below. Currently the atmosphere contains 380 ppm. In response to this, the British government's chief scientific adviser, Sir David King, said that was unlikely to happen. He stated,"We're going to be at 400 ppm in 10 years' time. I predict that without any delight in saying it." Myles Allen, an expert on atmospheric physics at Oxford University, said that:"Assessing a 'safe level' of carbon dioxide in the atmosphere was 'a bit like asking a doctor what's a safe number of cigarettes to smoke per day.'"

The report does conclude, however, that there are technological options available to reduce CO_2 emissions that will need to be used. The study also concluded that the biggest obstacles involved with using these new technologies, along with renewable resources of energy and"clean coal," are the current economic investments and traditionally strong bond to the oil industry, cultural attitudes that oppose change, and simple lack of awareness by many people. Various conservation organizations currently involved in the battle against global warming, such as the Union of Concerned Scientists (UCS), the Defenders of Wildlife, and the World Wildlife Fund (WWF), also support these ideas.

4

Climate Change, Farming and Forestry

TIME SCALE OF CLIMATIC CHANGE

The importance of considering different time scales when investigating climate change has already been identified. Climate varies on all time scales, in response to random and periodic forcing factors. Across all time periods from a few years to hundreds of millions of years there is a white (background) noise of random variations of the climate, caused by internal processes and associated feedback mechanisms, often referred to as stochastic or random mechanisms. Such randomness accounts for much of the climate variation, and owes its existence to the complex and chaotic behaviour of the climate system in responding to forcing. An essential corollary of the existence of random processes is that a large proportion of climate variation cannot be predicted.

Of far more relevance are the periodic forcing factors, for by understanding their mechanisms and the impacts they have on the global climate, it is possible to predict future climate change. How the climate system responds to periodic forcing factors, however, is often not clear. If it is assumed that the climate system responds in a linear fashion to periodic forcing, variations in climate will exhibit similar periodicity. If, however, the response of the system to forcing is strongly non-linear, the periodicities in the response will not necessarily be identical to the periodicities in the forcing factor(s). Frequently, the climate responds in a fashion intermediate between the two.

There are many climate forcing factors spanning an enormous range of periodicities. The longest, 200 to 500 million years, involves the passage of our Solar System through the galaxy, and the variations in galactic dust. These may be considered to be external forcing mechanisms. Other long time scale variations (10^6 to 10^8 years) include the non-radiative forcing mechanisms, such as continental drift, orogeny (mountain building) and isostasy (vertical movements in the Earth's crust affecting sea level). The response of the climate system to this combination of forcing factors itself depends upon the different response times of the various components of the system. The overall climatic response will then be determined by the interactions between the components.

The atmosphere, surface snow and ice, and surface vegetation typically respond to climatic forcing over a period of hours to days. The surface ocean has a response time measured in years, whilst the deep ocean and mountain glaciers vary only over a period spanning hundreds of years. Large ice sheets advance and withdraw over thousands of years whilst parts of the geosphere (e.g. continental weathering of rocks) respond only to forcing periods lasting hundreds of thousands to millions of years.

The response of the climate system to episodes of forcing can be viewed as a form of resonance. When the time period of forcing matches most closely the response time of a particular system component, the climatic response will be greatest within that component.

Milankovitch forcing, for example, with periods of tens of thousands of years will be manifest in the response of the ice sheets, and the overall response of the climate system will be dominated by changes within the cryosphere. In addition, longer response times of certain components of the climate system modulate, through feedback processes, the short term responses. The response of the deep ocean to short term forcing (e.g. enhanced greenhouse effect, solar variations, for example, will tend to attenuate or smooth the response of the atmosphere.

Throughout the remainder, it should be recognised that a range of time scales applies to climate forcing mechanisms, radiative and non-radiative, external and internal, and to the response of the different components of the climate system.

CLIMATE SENSITIVITY

The concept of feedback is related to the climate sensitivity or climate stability. It is useful to have a measure of the strength of various feedback processes which determine the ultimate response of the climate system to any change in radiative forcing. In general terms, an initial change in temperature due to a change in radiative forcing, $\Delta T_{forcing}$, is modified by the complex combination of feedback processes such that:

$$\Delta T_{final} = \Delta T_{forcing} + \Delta T_{feedback}$$

where $\Delta T_{feedback}$ is the temperature change resulting from feedback and ΔT_{final} is the overall change in temperature between the initial and final equilibrium states. The degree to which feedback processes influence the final climatic response is a measure of the sensitivity of the climate system.

Equation can be rewritten as follows:

$$\Delta T_{final} = f\Delta T_{forcing}$$

where f is called the feedback factor. When only one feedback mechanism is operative the solution to equation 6 is simple, assuming both f and $\Delta T_{forcing}$ are known. When more than one feedback is operative, matters become more complicated. For two feedbacks, the net effect is given by:

$$f = f_1f2/ (f_1 + f2 - f_1f2)$$

where f_1 and f2 are the feedback factors of the two feedback processes. Clearly, the feedback factors are neither additive nor multiplicative. A feedback operating alone with a factor of 2 would double the initial climatic response to forcing. If a second feedback with factor 1.5 acted with it, the overall feedback would be enhanced by a factor of 6. It can be seen, then, that a combination of feedback processes could dramatically affect the climate as it responds to only a small change in radiative forcing.

The climate's sensitivity can be mathematically determined in another way. From satellite observations, it has been shown that changes in global temperature are approximately proportional to changes in radiative forcing. If we assume an instantaneous† change in climate, from one equilibrium state to another, then:

$$\Delta Q = \lambda \Delta T$$

where ΔQ is change in radiative forcing (expressed in terms of the net downward radiative flux at the top of the troposphere), ΔT is the change in global temperature, and λ is a measure of the climate sensitivity.

Based on equation, the climate sensitivity is usually expressed in terms of the temperature change associated with a specified change in radiative forcing, usually a doubling of the atmospheric carbon dioxide content. Thus, the carbon dioxide doubling temperature, ΔT_{2x}, is given by,

$$\Delta T_{2x} = \Delta Q_{2x}/\lambda$$

where ΔQ_{2x} is 4.2 Wm^{-2}. The magnitude (and sign) of ΔT_{2x} will depend on λ, the climate sensitivity, which is determined by the net affect of the climate feedback processes. Despite extensive climate modelling over the last 2 decades to understand the problem of contemporary global, it is this parametre that is proving hard to define numerically.

As explained earlier, the idea of static equilibrium and instantaneous climatic response represents an unrealistic situation. To take into account the dynamic and transient nature of the climate's response to forcing, a more complicated equation linking ΔT and ΔQ is required if the evolution of the response with time is to be determined.

In the case:

$$\Delta Q = \lambda \Delta T + Cd\Delta T/\delta t$$

Here, the change in radiative forcing, DQ, is balanced by:

- The change in the outgoing radiative flux at the tropopause caused by the response of the climate system including feedback; and
- The energy stored in the system, $C\delta\Delta T/\delta t$, where C is the system's heat capacity and t is time.

The latter term in equation simulates the time-dependent nature of the climate system's response. The main contributor to the system's heat capacity is the world's oceans. The heat capacity of water is large compared to that of air and is therefore able to store much more energy. Additionally, the high heat capacity means that the oceans take time to heat up (or cool down) and hence slow the surface temperature response to any change in radiative forcing: the transient response will always be less than the equilibrium response.

Solution of Equation leads to the definition of the response time of the climate system, τ, such that:

$$\tau = C/\lambda$$

If the heat capacity of the climate system is large, the response time is large. Equally, if the climate sensitivity is small the response time is large. λ is also known as the radiative damping coefficient. Here, analogy is made between the response of the climate system and an oscillating spring. If a spring has a high damping coefficient, it will stop oscillating soon after it has been set in motion. Similarly, if the radiative damping coefficient is large, the climate will respond quickly and τ will be small.

CLIMATE CHANGE AND AGRICULTURE

Climate change and agriculture are interrelated processes, both of which take place on a global scale. Global warming is projected to have significant impacts on conditions affecting agriculture, including temperature, carbon dioxide, glacial run-off, precipitation and the interaction of these elements. These conditions determine the carrying capacity of the biosphere to produce enough food for the human population and domesticated animals. The overall effect of climate change on agriculture will depend on the balance of these effects. Assessment of the effects of global climate changes on agriculture might help to properly anticipate and adapt farming to maximize agricultural production.

At the same time, agriculture has been shown to produce significant effects on climate change, primarily through the production and release of greenhouse gases such as carbon dioxide, methane, and nitrous oxide, but also by altering the Earth's land cover, which can change its ability to absorb or reflect heat and light, thus contributing to radiative forcing. Land use change such as deforestation and desertification, together with use of fossil fuels, are the major anthropogenic sources of carbon dioxide; agriculture itself is the major contributor to increasing methane and nitrous oxide concentrations in earth's atmosphere.

IMPACT OF CLIMATE CHANGE ON AGRICULTURE

Despite technological advances, such as improved varieties, genetically modified organisms, and irrigation systems, weather is still a key factor in agricultural productivity, as well as soil properties and natural communities. The effect of climate on agriculture is related to variabilities in local climates rather than in global climate patterns. The Earth's average surface temperature has increased by 1 degree F in just over the last century. Consequently, agronomists consider any assessment has to be individually consider each local area. On the other hand, agricultural trade has grown in recent years, and now provides significant amounts of food, on a national level to major importing countries, as well as comfortable income to exporting ones. The international aspect of trade and security in terms of food implies the need to also consider the effects of climate change on a global scale.

A study published in Science suggests that, due to climate change, "southern Africa could lose more than 30% of its main crop, maize, by 2030. In South Asia losses of many regional staples, such as rice, millet and maize could top 10%". The 2001 IPCC Third Assessment Report concluded that the poorest countries would be hardest hit, with reductions in crop yields in most tropical and sub-tropical regions due to decreased water availability, and new or changed insect pest incidence. In Africa and Latin America many rainfed crops are near their maximum temperature tolerance, so that yields are likely to fall sharply for even small climate changes; falls in agricultural productivity of up to 30% over the 21st century are projected. Marine life and the fishing industry will also be severely affected in some places.

Climate change induced by increasing greenhouse gases is likely to affect crops differently from region to region. For example, average crop yield is expected to drop down to 50% in Pakistan just as to the UKMO scenario whereas corn production in Europe is expected to grow up to 25% in optimum hydrologic conditions.

More favourable effects on yield tend to depend to a large extent on realization of the potentially beneficial effects of carbon dioxide on crop growth and increase of efficiency in water use. Decrease in potential yields is likely to be caused by shortening of the growing period, decrease in water availability and poor vernalization.

In the long run, the climatic change could affect agriculture in several ways:

- Productivity, in terms of quantity and quality of crops
- Agricultural practices, through changes of water use and agricultural inputs such as herbicides, insecticides and fertilizers
- Environmental effects, in particular in relation of frequency and intensity of soil drainage, soil erosion, reduction of crop diversity
- Rural space, through the loss and gain of cultivated lands, land speculation, land renunciation, and hydraulic amenities.
- Adaptation, organisms may become more or less competitive, as well as humans may develop urgency to develop more competitive organisms, such as flood resistant or salt resistant varieties of rice.

They are large uncertainties to uncover, particularly because there is lack of information on many specific local regions, and include the uncertainties on magnitude of climate change, the effects of technological changes on productivity, global food demands, and the numerous possibilities of adaptation. Most agronomists believe that agricultural production will be mostly affected by the severity and pace of climate change, not so much by gradual trends in climate. If change is gradual, there may be enough time for biota adjustment. Rapid climate change, however, could harm agriculture in many countries, especially those that are already suffering from rather poor soil and climate conditions, because there is less time for optimum natural selection and adaption.

CLIMATE, WATER, AGRICULTURE AND DEVELOPMENT

Exposure to a high degree of climate risk is a characteristic feature of rainfed agriculture in the drylandsi of sub-Saharan Africa and parts of South Asia, which were largely bypassed by the Green Revolution, and where poverty and food insecurity remain most prevalent. Agriculture represents over 90% of water withdrawals in India and sub-Saharan Africa (SSA). In SSA, 93% of agriculture is rainfed, representing 70% of the population's employment and 35% of GDP.

SSA also has the least-developed water storage infrastructure needed to manage the variability of rainfall. As a result, economic development in SSA remains particularly vulnerable to the vagaries of rainfall. While climate change may increase the challenge, already present climate variability is a major impediment to economic growth in SSA as a result of the large fraction of economies that agriculture represents and its vulnerability to climate anomalies.

Of the 183 million hectares of agricultural land in SSA, only about 9 million is under some form of water management. While investment in expanding irrigation seems an auspicious way to improve agricultural productivity, estimates place the area of additional irrigation that would be profitable investments as part of dam-based schemes at only 3 million hectares.

For economic reasons then, small scale water management and storage approaches are likely to be a key component of efforts to increase agricultural productivity. The same preliminary IFPRI report estimates that small scale approaches might be profitable investments on an additional 38.2 million hectares. However, smaller-scale water management systems are best prospect for improving productivity under near-normal or moderately belownormal rainfall conditions. They are much less capable of managing climate extremes, such as floods and droughts.

Farmers will continue to face considerable climate risk, and extreme events can reverse development gains made over many years. A single drought or flood could set back all the agricultural development progress that result from improved local water management. We term this remaining climate risk, "residual risk," and recommend that a strategy for investing in agricultural water management should include a multi-pronged approach to dealing with the full range of climate variability, including not only moderate years but also the extremes.

CLIMATE VARIABILITY IS AN OBSTACLE TO AGRICULTURAL DEVELOPMENT

A growing body of evidence links unmitigated hydroclimatic variability to poor economic growth in developing countries. Cross country analysis show that in most poor countries, climate variability is high, infrastructure in lacking and GDP is correlated with rainfall.

In a study of climate impacts on economic growth of the countries of sub-Saharan Africa, drought was found to be the dominant climate risk, having a significant negative effect on GDP growth in one third of the countries. Droughts are also the world's most expensive disaster, destroying the economic livelihood and food source for those dependent on the agricultural sector or their own food production.

The World Bank estimates that economic growth in Ethiopia is reduced by one third due to hydrologic variability. A single drought over a 12-year period is estimated to reduce economic growth during the whole period by 10%. Drought impacts in Kenya associated with the La Nina of 1998 to 2000 resulted in losses totaling 16% of GDP. Floods destroy infrastructure, disrupt transportation and economic flows of goods and services and can lead to contaminated water supplies and the outbreak of waterborne disease epidemics, such as cholera.

In the Mozambique floods of 2000, over 2 million people were affected and damages were estimated at 20% of GDP. Flood damages in Kenya associated with the El Nino event of 1997-1998 were estimated at 11% of GDP. The effect of these hydrologic extremes can be devastating in any country, but especially in those with enhanced vulnerability due to high dependence on agriculture and deficient infrastructure. This is the case throughout SSA.

Climate exerts a profound influence on the lives of poor rural populations who depend on agriculture for livelihood and sustenance, who are unprotected against climate-related diseases, who lack secure access to water and food, and who are vulnerable to hydrometeorological hazard. Climate variability is arguably the dominant source of consumption risk in smallholder rainfed agriculture in the dryer environ-ments of much of sub-Saharan Africa and India.

Within farming communities, because the relatively poor have less capacity to buffer against climate risk through own assets or financial markets, they tend to experience disproportionate livelihood risk in the face of climate variations. Current understanding of the mechanisms by which climate variability impedes agricultural development, and hence efforts to ensure food security and reduce rural poverty, provides insights into opportunities for intervention. The various mechanisms by which climate risk impacts households combine with other factors to trap rural populations in chronic poverty. A dynamic poverty traps occurs when there is a critical threshold of household assets, below which individuals are unable to accumulate the necessary resources to escape poverty.

The tendency for risk tolerance to decrease with decreasing resource endowment contributes to the higher opportunity cost of climate risk for the relatively poor, and hence the locally-increasing marginal productivity that contributes to the existence of the multiple equilibria associated with a poverty trap. Ex-post coping response to severe or repeated climate shocks can push households to divest productive assets to a point below the poverty trap threshold.

Climate risk also impacts institutions in a manner that further constrains economic opportunities and hence reinforce poverty traps at the household level. Examples include the increasing cost of food crisis relief competing with agricultural development for shrinking donor resources, and the widespread reluctance of lenders to serve smallholder rainfed farmers.

While much is known about how climate risk impacts agriculture, less is known about the magnitude of the impacts or the magnitude of the livelihood benefits of feasible opportunities for managing climate risk. Despite the known impacts of current climate risk and growing concern about future climate change, climate risk management remains conspicuously absent from many analyses and regional development strategies.

We speculate that this is because the agricultural community has long regarded climate as part of the environmental baseline and not a resource with options for management. Development strategies that recognize climate risk generally limit intervention to expanded irrigation or improved water management. The Comprehensive Africa Agriculture Development Programme (CAADP), endorsed by the African Union, cites "vagaries of climate and consequent risk that deters investment" as one of six key challenges to achieving a productive and profitable agricultural sector across Africa.

Its strategy for addressing this constraint emphasizes "extending the area under sustainable land management and reliable water control systems". However, recent assessments and strategies, such as the Comprehensive Assessment of Water Management in Agriculture, increasingly recognize that feasible investments in water management for the drylands of SSA and parts of South Asia are necessary but not sufficient for overcoming the challenge of climate risk to development. The infeasibility of achieving a high level of water control across the vast dryland farming regions of Africa in the near to medium term, and increasing stress on groundwater and surface water resources in much of India point to the need to exploit every opportunity to deal with the residual climate risk that water control systems cannot mitigate.

CLIMATE CHANGE WILL INTENSIFY AGRICULTURAL WATER CHALLENGES

Although the effects of climate change from anthropogenic forcing on the use of water resources in the world remain difficult to project, anticipated climate change combined with other drivers of change is likely to intensify current agricultural water management challenges in Africa and India.

The effects of population growth and increasing water demand, which are often but not always coupled, are likely to be a more significant source of water stress than climate change when considering changes to mean precipitation and run-off. Increasing temperatures in all regions are expected to increase evaporative demand, which would tend to increase the amount of water required to achieve a given level of plant production if crop phenology and management are held constant.

However if cultivars and planting dates were to remain unchanged, accelerated crop development in response to temperature increases would tend to have the opposite effect on water requirements. Increased temperatures are also expected to increase evaporative losses of surface water resources. The magnitude and even direction of projected changes in precipitation are quite uncertain. Based on how many of the 21 climate models used in the 4th IPCC assessment (AR4) predict increases vs. decreases in annual and seasonal rainfall across Africa and Asia), decreased annual rainfall is very likely in Mediterranean North Africa and likely in much of Southern Africa particularly for the southern winter. Increases in rainfall are likely for much of Eastern Africa particularly in the northern winter, and for the summer monsoon over much of peninsular and eastern India.

These climate models are divided between wetting and drying trends in most of West Africa and western India. The two models within this set that best captured the extended dry period (1970s-1990s) in the West African Sahel predict opposite response to projected greenhouse gas forcing.

There is, however, a consensus that climate change will tend to increase the variability of rainfall and decrease the natural storage provided by snowpack and glaciers, such as in the Himalaya that feed the ricewheat belt of the Indo-Gangetic Plains. The concern pertaining to climate change impacts should not overshadow the present challenges that climate variability poses to agricultural development.

While the future impacts of climate change remain uncertain, climate variability persists as a challenge to development and as an impediment to meeting the Millennium Development Goals. Over the next decades, while climate change trends may begin to have some effect, droughts and floods associated with climate variability will continue to ravage vulnerable communities in developing countries. Fortunately, there is much that can be done now to reduce that vulnerability.

FARMER PARTICIPATORY EXPERIMENTS AND PERCEPTIONS OF CROP

Farmers use their perceptions of crop losses caused by pests to decide when to spray insecticides. This can lead to overestimating the seriousness of highly visible pests or damage symptoms. In making these decisions, farmers often rely on heuristics, or rules of thumb. Developed through experience and guesswork as to possible outcomes, heuristics may have inherent faults and biases. Farmers' decisions about leaf folder infestations in rice provide a case in point. Many farmers spray to control this pest, even though it does not cause yield losses, especially when it attacks in the early crop stages. Farmers' reactions to visible damage or insect presence may be caused by faults in their heuristics.

An approach for solving this problem is to analyse how farmers make these decisions, develop a corrective heuristic, frame it as a hypothesis and motivate farmers to participate in an experiment to test it.

APPROACH

Researchers at IRRI and ViSCA initiated participatory experiments in collaboration with technicians from the local Department of Agriculture and village leaders in Leyte, the Philippines. Half-day group meetings were held in each village for 10–25 invited farmers and a facilitator. The meetings began with general discussions about rice growing and related problems. The damage and losses that they caused, methods of control and their costs and effectiveness. Researchers facilitating the meetings eventually led the discussions as to whether control was needed at all and to the benefits of not spraying.

Next, volunteers were invited to test the heuristic 'We do not need to spray insecticides in the first 30 days after transplanting'. The volunteer farmers marked out an area of about 100 square metres in their fields that would not receive any insecticides during the first days of the crop cycle. They followed their usual practices in the rest of their fields. At the end of the season, participants reported their results in a workshop and received a certificate of participation.

Farmers from both the participating village and neighbouring ones were invited to the workshop. Pre- and post-experiment surveys were conducted to monitor changes in farmers' beliefs and intentions. These variables included farmers' beliefs, intentions, spray frequencies, timing and targets, yields, inputs and other management practices.

Rice yields in about 80 per cent of the experimental plots were equal to or greater than in the main plots. The number of insecticide applications fell from three to two per season. The percentage of farmers applying insecticides in the first 30 days of crop growth fell from 70 per cent to 20 per cent.

Reflections

The benefits of such experiments are that they are usually inexpensive and easy to conduct and they facilitate farmer learning by actively 'testing' a new idea, making participants more likely to adopt successful innovations. The approach provides a mechanism for scientists to learn about farmers' decision constraints, determine research needs, use research information and 'distil' them into testable hypotheses for farmers. It provides a means of exploring changes in farmers' beliefs, behaviour and practices. It is particularly useful for introducing a 'new' idea to a community.

Participatory experiments of this kind may have some disadvantages. They can prove expensive if the process is to be conducted over a large population. The use of media to motivate farmers' participation has been successful in Vietnam and may be an alternative to face-to-face training. Farmers participating in the experiments may risk losses and require compensation. The approach aims to introduce a testable hypothesis to farmers and thus may be viewed by some as 'top-down'.

The presence of scientists may influence farmers differently. Thus scientists applying this approach will need to acquire and use facilitating skills. Peers may view this type of participatory experiment as agronomically 'weak' because some of the controls may not be easily implemented. Because the main objective of the approach is to evaluate farmers' responses to new ideas rather than agronomy, data collection of variables on belief, behaviour and practice changes needs to be emphasized.

The adoption of innovations such as seeds and machines is often discussed in the literature.

Less is known about the adoption or adaptation of information into farmer decisions. Because much of resource management is in the form of information and how to adapt and integrate it into decisions and practices, investing in decision research, an emerging field of applied social psychology, will enhance delivery and communication information.

FARMERS' ABILITY TO MANAGE A DEVASTATING PLANT DISEASE – POTATO LATE BLIGHT

Resource-poor farmers have substantial difficulty in managing the diseases that affect their crops. Potato late blight (LB) is particularly devastating for small-scale producers. Because of recent worldwide migrations of more virulent and fungicide-resistant strains of the pathogen, potato farmers face a problem that behaves differently than before. Poor farmers have little knowledge of the disease, in part because the organism that causes it is essentially invisible. Late blight is usually managed through the use of fungicides, some of which are suspected carcinogens. In developing countries, effective disease management strategies are best devised locally, because of the tremendous variation in human, environmental, host and pathogen factors among potato agroecosystems.

As the result of decades of resistance breeding, potato varieties and breeding lines with promising levels of resistance are available. Although efforts are made to breed for durable resistance, varietal diversification is desirable to reduce the erosion and breakdown of resistance. Getting improved varieties to the farmers is, however, a significant challenge because of the limitations inherent in a vegetatively propagated crop. Deployment of promising breeding lines in marginal and heterogeneous environments without formal seed systems is particularly difficult. Participatory approaches are essential because they help integrate varietal selection with other elements of disease and crop management strategies, and they contribute to the improvement of informal seed systems.

The Food and Agriculture Organization (FAO) developed the farmer field school approach (FFS) for training in IPM. Since 1997, CIP has been working with several research and extension institutions to develop and implement FFSs with farmer groups in the Andes and elsewhere. The original FFS approach was found to require substantial adaptation for potato. In potato production, farmers make many of their key decisions before the start of the growing season. The variety that they choose and their source of seed are key issues in managing LB and other potato pests. We incorporated a substantial element of farmer participatory research (FPR) into the FFS format, and therefore designated our approach 'FPR-FFS' to distinguish if from the FFSs that focus principally on training.

APPROACH

To initiate the development of an FPR-FFS focused initially on potato late blight, CIP convened a series of local and national meetings and an international workshop to develop a strategy and to define available materials. An FPR-FFS curriculum, embodied as a field guide for facilitators, was drafted. A baseline study on LB was conducted in Ecuador, Peru, Bolivia and Uganda. This study confirmed that LB is the most important production problem for potato farmers and provided insight into farmers' knowledge and practices.

In an FPR-FFS, a group of about 10–35 farmers from a given locality meets regularly over the course of one to three cropping seasons (or longer), usually twice a month for a half-day session. Helped by a trained facilitator (usually from a local NGO), the farmer group conducts field experiments and hands-on learning activities. They use direct experimentation and observation to improve their knowledge, and use this expertise to improve their crop and pest management. The field experiments have included testing of promising varieties and/or breeding lines, testing different fungicide strategies for varieties with different levels of resistance, working with varieties derived from true potato seed and comparing IPM and conventional practice. By sharing data among communities through field days and workshops, the groups amass substantial data and can proceed with decisions such as varietal selection with relative confidence.

The programme has expanded in Peru and elsewhere. In 1997, CIP and CARE-Peru initiated FFSs on a pilot scale in four communities of San Miguel, Cajamarca, in northern Peru. Eight FPR-FFSs were conducted during 1998–1999 in San Miguel, and 13 were conducted in 1999–2000. In parallel, pilot-scale FPR-FFSs were established elsewhere in Peru, as well as in Ecuador, Bolivia, China, Bangladesh, Uganda and Ethiopia through the collaboration of researchers, extension organizations (mainly NGOs) and farmer groups. Parallel efforts by other organizations in coordination with CIP led to further efforts in potato. Facilitators from Bolivia, Ecuador and Peru were formally trained in FFS methods through the FAO, and national IPM projects emphasizing the FFS methodology were established in Peru and Ecuador.

Reflections

Farmers have much to learn about the microbial world. Because they cannot see the organism that causes plant disease, they do not understand disease processes well. They are poor at diagnosis and inefficient at managing the diseases that affect their potato crops. However, given the opportunity, they are quick to learn and improve their management decisions. They are keen to try new varieties and are appropriately conservative about making decisions regarding varietal change. Participatory evaluation gives farmers a meaningful basis on which to make decisions about varietal choice.

The FPR-FFS approach is demanding on both farmers and researchers. Farmers must be strongly motivated to improve their potato production if they are to participate successfully. Because potato is a high-value crop and the losses caused by LB are often devastating, LB is a suitable entry point. However, farmers face numerous problems with their potato crops and other agricultural enterprises, and prefer integrated approaches that allow them to cope with multiple problems at a given time. The FPR-FFS method forces researchers, who often have narrow technical interests, to expand their horizons. Researchers and farmers have complementary roles in the evaluation of potential new varieties. Linkages between research and extension organizations increase the potential impact of knowledge-intensive technology.

In the Andes, the FPR-FFS has attracted many young men while older men and women with young children often found participation more challenging. This could result, in part, from the limited role that women play in potato production in the northern Peruvian Andes. However, women have a stake in successful crop and disease management for potato, and in selection of appropriate potato genotypes. In participatory evaluation of varieties and breeding lines conducted through the FPR-FFS, the opinions of men and women participants were sometimes significantly different. This reinforced the importance of involving both men and women in the activity. Efforts are being made to improve the training curriculum to enhance its utility for female participants.

The FPR-FFS approach is still evolving, and from the outset has had much in common with the CIAL methodology. With support from the International Fund for Agricultural Development (IFAD) and the Organization of Petroleum Exporting Countries (OPEC) Fund for International Development, pilot-scale FPR-FFSs are now being established in seven countries, through collaboration among researchers, NGOs and farmer groups. More emphasis is being placed on gender analysis (assisted by the PRGA programme), and the impact of the FFS is being assessed (PRGA and the World Bank). Preliminary observations indicate that the FFS is highly effective in stimulating farmer learning and varietal diffusion.

FARMERS' RISK MANAGEMENT STRATEGIES

Small farms of less than 5 hectares account for about 70 per cent of southern Africa's maize production. Although new technologies are available for improving production in smallholder maize systems, widespread adoption faces major constraints – particularly the constant threat of drought and declining soil fertility. In Zimbabwe and Malawi, the soils in smallholder areas tend to be sandy, with limited organic matter, low nutrient content and low water-holding capacity. Farmers have only limited access to organic manure and cannot usually afford inorganic fertilizers.

The threat of drought, combined with fluctuating market prices, means that farmers are gambling on an uncertain yield and economic return. Therefore, to be attractive to farmers, new technologies for improving soil fertility must be able to reduce production risk. They must also be compatible with farmers' diverse and complex livelihood strategies. To support the development of appropriate soil fertility technologies, CIMMYT's Risk Management Project (RMP) evaluates their biophysical and socioeconomic performance through a combination of computer crop modelling and farmer participatory research in Malawi and Zimbabwe.

APPROACH

The RMP employs both hard (quantitative) and soft (qualitative) approaches to explore the links between agroecosystems and the socioeconomic environment. A participatory research subproject conducts systems diagnostics, identifies stakeholders, determines farmers' soil and climate taxonomies, describes farm families' livelihood strategies and fosters farmer experimentation with soil fertility management practices. A modelling subproject collects data to validate the computer model and fosters its use to examine the long-term biophysical performance of soil fertility management practices under specific soil and climate conditions.

By integrating the two sets of activities, RMP can use farmers' soil and climate taxonomies to develop soil and climate profiles for running the model. Further, the model can be used to evaluate farmer-developed technologies, and farmers can evaluate the model's outputs and participatory technology field trials within the context of their livelihoods and risk management strategies.

This enables farmers with diverse resource endowments and diverse livelihood strategies to verify options within their own environments.

The RMP collaborates with focus groups from the Universities of Zimbabwe and Malawi, national agricultural research programmes and the Africa Centre for Fertiliser Development. Researchers and farmers evaluate soil fertility technologies being developed by the focus groups. Through this integrated approach, researchers can draw on one another's experience and that of farmers. The RMP also has links with ICRISAT and CARE. The CIMMYT and ICRISAT centres jointly fund researchers and field activities and share information. Farmer groups established by CARE link the RMP to communities and provide a social framework for broader dissemination of successful technologies.

One of the activities is to design a framework for running simulation models based on farmers' soil management practices. The goal is to develop an interface that permits discussion of outputs and key management variables, involves farmers in assessing scenarios developed by the model and enables them to ask questions of the model.

Focused planning meetings are conducted to enhance the team's organization. Participants develop common work plans and research frameworks for all project stakeholders. The RMP began with a macro-systems diagnostics approach, which enabled it to identify key stakeholders, secondary data and partners for implementing project activities and to identify appropriate techniques for fieldwork. The RMP has thus created a strong network of research partners throughout the region.

Fieldwork activities concentrated on forming or strengthening farmer groups at two sites in Zimbabwe and one in Malawi. Farmers, extension staff and researchers formed new groups for the 1999–2000 crop season. Activities included participatory wealth ranking, the development of farmer taxonomies of soils and climate, inventories of management options and practices for different resource endowments and varying soil and climatic conditions.

A key collective learning and decision support tool is the use of participatory modelling maps of agroecosystem resource flow or resource allocation maps (RAMs).

These demonstrate key household and community resources and inter-relationships within a systems context. Farmers and researchers together develop the maps and use them to record, monitor and analyse data and decision-making, which then enables them to understand the systems' context of soil fertility. The methodology development for the modelling and FPR interface is considered an iterative and dynamic process, with a diversity of tools and techniques being used, refined, adapted or discarded as the process and project stakeholders require. This is an immensely creative and ambitious research agenda. This kind of interaction is highly unusual but holds great promise for a more effective evaluation of soil fertility management technologies, under highly variable and risky climate conditions.

Reflections

Biophysical crop modelling enables evaluation over time, including different variables such as climate, soils and management. However, the models are only as good as the data and the agenda that are used in their development and are highly data-intensive to establish. Models are often narrowly focused compared to farmers' multi-faceted and complex livelihoods and cannot handle socioeconomic variables. Some factors affecting crop and soil performance are also beyond the model's capabilities. The model-to-farmer interface integrates disciplines and can bring modellers and farmers together with quick feedback.

This can be used to promote collaborative learning and as a decision support tool because it makes it easier to identify and target key research priorities. It has some knowledge gaps with outputs highly subject to interpretation as to by whom and how the research agenda is driven. Linking modelling with GIS and farmer land types for scaling up is being explored but entails expensive start-up costs and questions of data ownership.

On-farm participatory testing of technologies allows farmers more freedom to experiment and helps bring stakeholders together. It serves to develop a better understanding of farmers' priorities and natural resource and socioeconomic factors affecting technology performance, also capitalizing on farmers' indigenous knowledge. On-station testing is easier to manage and provides rigorous controls and designs that reduce variables affecting crop performance. However, socioeconomic variables associated with management are excluded and a long time frame is needed. This integrated research method and process enables a learning forum to be developed whereby the very different mindsets and approaches of all the actors involved in the research can begin to envision a common environment. They can then develop a platform to work together on common problems and solutions within a participatory process

FARMER PARTICIPATORY RESEARCH METHODS

In Malawi and Zimbabwe, ICRISAT is developing new research methods for improving soil fertility. The Institute is also building partnerships among national scientists, extension advisors in NGOs and the public sector and farmers. The aim is to improve the ability of national research programmes to develop 'best bet' NRM technologies for poor farmers by obtaining their input at earlier stages. New technologies are needed that improve human nutrition while enhancing soil management and enabling communities to rehabilitate degraded environments. The main innovations introduced so far are intensification with long-duration and indeterminate legumes and integrated use of organics and inorganics. In Malawi, farmers are testing and adapting options such as maize grown in rotation with a doubled up legume intercrop of pigeonpea and another grain legume. Combining small amounts of fertilizer with manure and pigeonpea residues is also being tested.

APPROACH

A novel aspect of the programme is its evaluation of several participatory approaches applied in parallel in different villages. The results are compared with baseline data from villages having no known relationship with researchers or farm advisors from NGOs or extension services. This approach enables researchers and farm advisors to address their concern that farmer adoption of fertilizer and integrated nutrient management has been practically nil, despite a decade of on-farm research and the recent focus on participatory research and extension as well as training-for-transformation empowerment approaches.

Costs, as well as benefits, will be assessed for different methods of partnering. Some researchers and senior extension staff are also concerned that extension rarely reaches female-headed households and women farmers, nor do their concerns enter into agronomic research.

The methods being compared include farmer empowerment approaches led by NGOs, extension-led demonstration trials and farmer participatory research, each conducted in a different village in the case study areas. The study is determining which of the approaches are best at building institutional linkages and improving the peer relationship among stakeholders, for different locations. All of the partners involved evaluate the effectiveness and costs of each approach. Researchers are also determining how well each approach addresses the needs of female-headed households. Project partners conducted comprehensive surveys to provide a baseline for, and developed methods of, comparison. They agree that the comparison should indicate which methods are working best, as reflected in perceived cost-effectiveness and the satisfaction of researchers, extension advisors and farmers. Farmer adoption and adaptation of technologies, farmer empowerment and improved soil management will be assessed for each method.

The 'mother–baby' trial design is one method for improving communication between farmers and researchers that has so far proved successful.

The approach was originally created to facilitate farmer collaboration in testing soil fertility technologies. Further, it was conceived as a practical means for researchers to rigorously incorporate farmer evaluations of technologies at every step in the development process. The approach links together two trial types: a replicated one that fits researchers' needs and a simpler trial that meets farmer needs, rather than attempting to compromise and meet all objectives in a single trial.

Researchers first design 'best-bet' technologies, attempting to take into account farmers' priorities and resources. Then mother–baby trials are planted in each participating village. The 'mother' is a replicated experiment designed by a researcher. Farmers plant and manage the baby trials, which are single replicates of the mother trial. For this purpose, each farmer selects a best-bet technology from the mother trial and adjusts the level of inputs and equipment according to his or her preferences. Farmer evaluation is documented through surveys, community discussion and ranking exercises, which facilitate researcher incorporation of farmer input. The trials have led to improved spontaneous experimentation among farmers and give researchers and extension advisors the chance to observe and learn from farmers.

Reflections

The larger context of the different case study areas where the approaches are being tested is difficult to assess. For example, the market opportunities and historical extension–farmer relationships may vary markedly among the areas and determine the relative success of different methods. The trial and demonstration approach probably has the least emphasis on farmer knowledge, while facilitation of farmer learning and experimentation, the farmer-led approach, has the most. The changes in the researchers and extension staff to encompass a broader, more participatory approach is proving difficult to document. Further steps, such as how to facilitate farmer training and communication with other farmers on guidelines, are not explicitly part of any approach although this may develop. Farmer-to-farmer communication is probably the most important means of technology dissemination and is not treated explicitly in any of the approaches, except perhaps the farmer empowerment approach (which some professionals view as too costly for large-scale work). This point is difficult to evaluate but should become clearer with time. A survey is underway to evaluate researcher and extension attitudes and beliefs regarding effective ways to communicate with farmers and work together to develop soil management options and to improve farmer experimentation. Surveys documenting farmer assessment of the process and farmer adoption are also on-going.

All of the partners involved in the methods' comparison systematically evaluate methodology approaches and efforts to facilitate farmer–researcher-extension linkages and technology best-bet options. Also, the concerns of women farmers and female-headed households are specifically addressed. This took time to build, in part because almost all of the researchers and extension staff involved were men and no gender sensitivity training or discussion was attempted at initial stakeholder meetings.

Attention to including women farmers in the technology development process was almost nil at first but has increased over time, particularly at the Malawi sites. This adds value to on-going efforts in the area. The mother–baby trial and farmer empowerment approaches attempt to facilitate farmers' learning basic research principles, to expose farmers to a range of new options and to empower them to value their own knowledge. This appears to have improved communication among farmers, researchers and extension staff.

A conundrum is that extension staff and researchers mostly see the trial demonstration approach as the only cost-effective way to scale up dissemination.

Yet this approach does not facilitate farmer experimentation or joint learning. Staff at NGOs also state that their work on farmer and community empowerment is only cost-effective in isolated areas. They believe that they need to go to trials and demonstrations to reach more people, although both groups express frustration with the lack of effectiveness of demonstrations led by extension or farm advisors.

The goal of this case study is to measure long-term impact and changes in how research and extension staff conduct their work. Stakeholders designed this village-based comparison of methods and are involved in evaluating the pros and cons of each participatory method over time. Meetings are held annually for assessment. Strong links are built because researchers, extension staff and NGOs were already carrying out most of this work and through the case study we are attempting to facilitate self-reflection on the value of different approaches to the stakeholders. In Malawi, 400 farmers are assessing best-bet technologies at seven sites around the country through baby trials and their own experimentation. In the process, they are satisfying researchers' need for sound quantitative experiment results.

CLIMATE CHANGE AND ECOSYSTEMS

Unchecked global warming could affect most terrestrial ecoregions. Increasing global temperature means that ecosystems will change; some species are being forced out of their habitats because of changing conditions, while others are flourishing. Secondary effects of global warming, such as lessened snow cover, rising sea levels, and weather changes, may influence not only human activities but also the ecosystem.

For the IPCC Fourth Assessment Report, experts assessed the literature on the impacts of climate change on ecosystems. Rosenzweig et al. concluded that over the last three decades, human-induced warming had likely had a discernable influence on many physical and biological systems. Schneider et al. concluded, with very high confidence, that regional temperature trends had already affected species and ecosystems around the world. With high confidence, they concluded that climate change would result in the extinction of many species and a reduction in the diversity of ecosystems.

- *Terrestrial ecosystems and biodiversity*: With a warming of 3°C, relative to 1990 levels, it is likely that global terrestrial vegetation would become a net source of carbon. With high confidence, Schneider et al. concluded that a global mean temperature increase of around 4°C by 2100 would lead to major extinctions around the globe.
- *Marine ecosystems and biodiversity*: With very high confidence, Schneider et al. concluded that a warming of 2°C above 1990 levels would result in mass mortality of coral reefs globally.
- *Freshwater ecosystems*: Above about a 4°C increase in global mean temperature by 2100, Schneider et al. concluded, with high confidence, that many freshwater species would become extinct.

Studying the association between Earth climate and extinctions over the past 520 million years, scientists from the University of York write, "The global temperatures predicted for the coming centuries may trigger a new 'mass extinction event', where over 50 per cent of animal and plant species would be wiped out." Many of the species at risk are Arctic and Antarctic fauna such as polar bears and Emperor Penguins. In the Arctic, the waters of Hudson Bay are ice-free for three weeks longer than they were thirty years ago, affecting polar bears, which prefer to hunt on sea ice.

Species that rely on cold weather conditions such as gyrfalcons, and Snowy Owls that prey on lemmings that use the cold winter to their advantage may be hit hard. Marine invertebrates enjoy peak growth at the temperatures they have adapted to, regardless of how cold these may be, and cold-blooded animals found at greater latitudes and altitudes generally grow faster to compensate for the short growing season.

Warmer-than-ideal conditions result in higher metabolism and consequent reductions in body size despite increased foraging, which in turn elevates the risk of predation. Indeed, even a slight increase in temperature during development impairs growth efficiency and survival rate in rainbow trout. Rising temperatures are beginning to have a noticeable impact on birds, and butterflies have shifted their ranges northward by 200 km in Europe and North America. Plants lag behind, and larger animals' migration is slowed down by cities and roads. In Britain, spring butterflies are appearing an average of 6 days earlier than two decades ago.

A 2002 article in Nature surveyed the scientific literature to find recent changes in range or seasonal behaviour by plant and animal species. Of species showing recent change, 4 out of 5 shifted their ranges towards the poles or higher altitudes, creating "refugee species". Frogs were breeding, flowers blossoming and birds migrating an average 2.3 days earlier each decade; butterflies, birds and plants moving towards the poles by 6.1 km per decade. A

2005 study concludes human activity is the cause of the temperature rise and resultant changing species behaviour, and links these effects with the predictions of climate models to provide validation for them. Scientists have observed that Antarctic hair grass is colonizing areas of Antarctica where previously their survival range was limited.

Mechanistic studies have documented extinctions due to recent climate change: McLaughlin et al. documented two populations of Bay checkerspot butterfly being threatened by precipitation change. Parmesan states, “Few studies have been conducted at a scale that encompasses an entire species” and McLaughlin et al. agreed “few mechanistic studies have linked extinctions to recent climate change.” Daniel Botkin and other authors in one study believe that projected rates of extinction are overestimated.

Many species of freshwater and saltwater plants and animals are dependent on glacier-fed waters to ensure a cold water habitat that they have adapted to. Some species of freshwater fish need cold water to survive and to reproduce, and this is especially true with Salmon and Cutthroat trout. Reduced glacier run-off can lead to insufficient stream flow to allow these species to thrive. Ocean krill, a cornerstone species, prefer cold water and are the primary food source for aquatic mammals such as the Blue Whale. Alterations to the ocean currents, due to increased freshwater inputs from glacier melt, and the potential alterations to thermohaline circulation of the worlds oceans, may affect existing fisheries upon which humans depend as well.

The white lemuroid possum, only found in the mountain forests of northern Queensland, has been named as the first mammal species to be driven extinct by global warming. The White Possum has not been seen in over three years. These possums cannot survive extended temperatures over 30 °C which occurred in 2005. A final expedition to uncover any surviving White Possums is scheduled for 2009.

FORESTS

Pine forests in British Columbia have been devastated by a pine beetle infestation, which has expanded unhindered since 1998 at least in part due to the lack of severe winters since that time; a few days of extreme cold kill most mountain pine beetles and have kept outbreaks in the past naturally contained. The infestation, which has killed about half of the province’s lodgepole pines is an order of magnitude larger than any previously recorded outbreak and passed via unusually strong winds in 2007 over the continental divide to Alberta. An epidemic also started, be it at a lower rate, in 1999 in Colourado, Wyoming, and Montana. The United States forest service predicts that between 2011 and 2013 virtually all 5 million acres of Colourado’s lodgepole pine trees over five inches in diametre will be lost. As the northern forests are a carbon sink, while dead forests are a major carbon source, the loss of such large areas of forest has a positive feedback on global warming. In the worst years, the carbon emission due to beetle infestation of forests in British Columbia alone approaches that of an average year of forest fires in all of Canada or five years worth of emissions from that country's transportation sources.

Besides the immediate ecological and economic impact, the huge dead forests provide a fire risk. Even many healthy forests appear to face an increased risk of forest fires because of warming climates. The 10-year average of boreal forest burned in North America, after several decades of around 10,000 km^2 has increased steadily since 1970 to more than 28,000 km^2 annually. Though this change may be due in part to changes in forest management practices, in the western U.S., since 1986, longer, warmer summers have resulted in a fourfold increase of major wildfires and a sixfold increase in the area of forest burned, compared to the period from 1970 to 1986. A similar increase in wildfire activity has been reported in Canada from 1920 to 1999.

Forest fires in Indonesia have dramatically increased since 1997 as well. These fires are often actively started to clear forest for agriculture. They can set fire to the large peat bogs in the region and the COreleased by these peat bog fires has been estimated, in an average year, to be 15% of the quantity of COproduced by fossil fuel combustion.

MOUNTAINS

Mountains cover approximately 25 per cent of earth's surface and provide a home to more than one-tenth of global human population. Changes in global climate pose a number of potential risks to mountain habitats. Researchers expect that over time, climate change will affect mountain and lowland ecosystems, the frequency and intensity of forest fires, the diversity of wildlife, and the distribution of water.

Studies suggest that a warmer climate in the United States would cause lower-elevation habitats to expand into the higher alpine zone. Such a shift would encroach on the rare alpine meadows and other high-altitude habitats. High-elevation plants and animals have limited space available for new habitat as they move higher on the mountains in order to adapt to long-term changes in regional climate. Changes in climate will also affect the depth of the mountains snowpacks and glaciers. Any changes in their seasonal melting can have powerful impacts on areas that rely on freshwater run-off from mountains. Rising temperature may cause snow to melt earlier and faster in the spring and shift the timing and distribution of run-off. These changes could affect the availability of freshwater for natural systems and human uses.

ECOLOGICAL PRODUCTIVITY

- A document by Smith and Hitz, it is reasonable to assume that the relationship between increased global mean temperature and ecosystem productivity is parabolic. Higher carbon dioxide concentrations will favourably affect plant growth and demand for water. Higher temperatures could initially be favourable for plant growth. Eventually, increased growth would peak then decline.

- The IPCC, a global average temperature increase exceeding 1.5–2.5°C would likely have a predominantly negative impact on ecosystem goods and services, *e.g.,* water and food supply.
- Research done by the Swiss Canopy Crane Project suggests that slow-growing trees only are stimulated in growth for a short period under higher CO_2 levels, while faster growing plants like liana benefit in the long term. In general, but especially in rainforests, this means that liana become the prevalent species; and because they decompose much faster than trees their carbon content is more quickly returned to the atmosphere. Slow growing trees incorporate atmospheric carbon for decades.

FOREST MANAGEMENT AND USE

UTILIZATION

The original uses of the forest were generally unorganized and unplanned. The natives knew their trees and chose carefully the species and individuals best suited for house posts, canoes, paddles, fishnet floats and other articles to meet their daily needs. The breadfruit was recognized as one of the most valuable of timber trees (propagated vegetatively and existing in endless clonal strains), but was seldom cut, because its fruits were used for food. Coconut also served numerous purposes. Other trees of the strand were selected and many species in the coastalplain forests of the high islands were discriminately chosen, although reports concerning the same species from different island regions vary markedly. Before World War II organized forestry and forest exploitation can hardly be said to have existed in Oceania. There were a few sawmills on the larger islands, but they served immediate local needs only. During the war exploitation was stepped up to a high pace to serve the armed forces, especially in the larger Melanesian islands. As in so many tropical regions, the number of species here is great and bewilderingly complex and many of the species have native names only and are unknown to scientific classification. The wartime operations helped to advance our knowledge of the dendrology and physical characteristics of these forests farther than throughout all previous history, but unfortunately most of such information exists only in unpublished reports and in the minds of the foresters who were directly involved.

RESEARCH

The forestry and agricultural organizations are carrying forward only limited progrmmes of basic forestry research. Usually these organizations are under pressure to serve immediate needs and hence their investigations are oriented toward the planting of exotic tree species of proven worth elsewhere and toward the control of actively destructive factors, such as feral mammals.

On the other hand, scientific research in Oceania is beginning to advance rapidly along general lines and this should benefit forestry in the long run.

Forestry in Oceania involves the management of lands originally or potentially covered with trees and such products of forest land as water and wildlife may locally be more important than commercial timber. For these reasons a sound forestry programme must go beyond the range of conventional silvicultural activities.

It must include interpretations of the vegetation of pre-European and pre-native times and also serious studies of the remaining samples of natural and semi-natural vegetation, which in some instances have very desirable characteristics.

The artificially induced forest types, such as those from plantings, should also be investigated with regard both to their continuing or self-perpetuating characteristics and to their values in other respects than for timber production. Because of the extraordinary complexity of the native and introduced floras and because of the wide range of distinctive environmental conditions within relatively small areas, forestry in Oceania offers unsurpassed challenges to the ingenuity of the practical researcher.

VALUE OF THE FORESTS

Timber holds first place among the forest products of New Guinea and Fiji and may do so in the future in the other continental islands, as is generally true on the continents. On the oceanic volcanic islands, however, the native forests bear very few species or individual trees of such size and quality as are suitable for commercial timber and experiments are constantly progressing with introduced plants. The aborigines use wood for house posts, paddles, canoes and sundry minor articles, for which individual trees are selected when needed.

In Hawaii the chief value of the Forest Reserves is in relation to watersheds, because of the demands of the sugar and pineapple plantations and of urban communities and the same probably also holds true for the other high islands under more intensive use.

Where there is mangrove vegetation, it is extensively exploited for fuel and in Fiji is under a 40-year management cycle, as we have seen.

On most of the islands the main use of secondary growth is for firewood and charcoal. For fence posts, highly important in all grazing regions, the naturalized Prosopis chilensis is one of the chief sources of supply. There is no pulp industry of importance, although recent developments in the use of bagasse are significant. Rubber has been tried in the southwestern islands, but not successfully. Minor forest products are of no more than local importance. For the New Guinea native the principal value of the forest lands is as agricultural soil and once the grasses take over, he moves on, or vanishes.

OWNERSHIP AND ADMINISTRATION

Organized forestry, private or governmental, exists in only a few of the island groups. During the Japanese administration of Micronesia, forestry activities were accelerated, with local sawmill operations, minor exports to Japan and a certain amount of reforestation, especially with Casuarina equisetifolia. The present forestry situation in American Samoa and Micronesia has recently been surveyed.

New Caledonia, Fiji (where a Forest Department was established in 1938) and Hawaii have, or have had, some government administration, with Hawaii far in the lead.

In Hawaii a Board of Commissioners of Agriculture and Forestry has existed since 1903. Its activities have centered largely around the establishment of reserves almost entirely for watersheds, the elimination of the hordes of feral mammals, fencing, a very elaborate planting programme that has involved a vast number of alien forest species and the controversial practice of introducing game animals. In the early days, forest plantings were sometimes made in total disregard of basic silvical principles and failures were excessive. For a long time, success was gauged on survival and rapid growth and undoubtedly this is a chief factor in the case of a timber species that is to be harvested. For self-perpetuating forests, however, this policy may be very deceptive. An intensive report is much to be desired on all these forest plantings of the past, not only with regard to the survival of planted individuals, but also as to whether different species are capable of spreading naturally. Some species may be found to act as "nurse crops" that is, to prepare the way for the introduction of another semi-natural plant community closer to the original self-perpetuating forest.

DESTRUCTIVE FACTORS AND SHIFTING CULTIVATION

The activities of the shifting cultivator have contributed much to the establishment of certain of the present plant communities. Owing to its good accumulation of humus adapted to crops, high forest is invariably cleared and burned and thousands of cubic feet of valuable timber go up in smoke every year. This method of cultivation, which ignores sound agricultural principles, is responsible for covering large areas with secondary forest (jungle) of no commercial value, or else with grasses. Accelerated soil erosion in such areas is enormous, especially on steep slopes.

Tradition and lack of suitable land to satisfy all chena (shifting) cultivators have prevented the government from taking drastic steps. To localize shifting cultivation, however and thereby to prevent the wholesale devastation of valuable forests, the government has divided the chena land into blocks, each of which is cultivated once in ten or twelve years and has also made available irrigable land for permanent cultivation. Areas subject to heavy chenaing are now being reforested on a co-operative basis according to the "taungya forest system," which was first developed in Burma in 1856 and has been adopted in Java and other countries of Southeast Asia. Under this system government land is leased out rent-free for the raising of short-term food crops and in return the lessee raises approved forest species along with his crops.

COMMERCIAL PLANTATIONS

Clearing for commercial plantations has been the second most important destructive human factor. Since 1873 more than half a million acres have been opened up for tea plantations and since 1876 nearly 700,000 for rubber (Hevea brasiliensis). Most of these areas were previously covered by forest. Soil erosion on the plantations is heavy and has adversely affected rivers, tanks, irrigation channels and the plantations themselves.

ILLICIT FELLING

A third factor altering the composition of plant communities in most forest areas is illicit felling and the extraction of poles for agricultural and building purposes. Illicit felling is locally looked upon as an honest trade, one that has come down from father to son. Extraction of poles, seldom properly supervised, tends to destroy nearly all valuable timber species in the round-pole stage. These sizes are extensively used for fencing cultivated land and the construction of semipermanent buildings. Despite the fact that "the round pole of today is a log of tomorrow," removal of poles in other than silvicultural operations is still permitted.

Forest History and Administration

When Ceylon was ruled by the Sinhalese kings the entire forest estates of the island belonged to the Crown and large areas were cleared for agriculture based on irrigation. After the eighth century, however, the forest came back, owing to the destruction and neglect accompanying continuous wars with invaders from the continent and to the ensuing ravages of malaria. The forest at first reclaimed the Anuradhapura area, including the large city of the same name and, a few centuries later, the city of Polonnaruwa.

But in the later forest history of Ceylon "man the destroyer" features prominently. Export of timber began with the Portuguese occupation in the sixteenth century.

The Dutch destroyed much forest for the cultivation of cinnamon, in which they carried on a lucrative trade and after 1830 the British undertook timber exploitation on an enormous scale. After several million cubic feet of the best timbers had been exported, a forest policy was initiated under British rule in

1885 and since then conservation and propagation of the forests have been promoted by the government.

The total forest cover is approximately 70 per cent of the island's area. Forest administration is at present the responsibility of four different bodies. The major portion of the forests is administered by the Forest Department, with a Conservator as head. The Department of Wild Life administers forests in the national reserves and sanctuaries for the flora and fauna. The Gal Oya Development Board, an autonomous body created for the multipurpose development of the Gal Oya valley, administers the forests in that area and the Revenue Department administers village forests set apart for domestic needs. Private forests form a negligible part of the total forested area.

Care of the Forests

Timber extraction for domestic and commercial purposes has been heavy in the past, but the inroads made during and since World War II have contributed most toward the present depletion of timber stocks. A considerable period of rest and cultural treatment is necessary for the majority of the forests, except where inaccessibility, lack of drinking water in the dry seasons and labour difficulties have prevented exploitation and thus helped to protect the stock. In the tropical evergreen zone the forests of the Sinharaja region are still intact. The stocks here, which consist mostly of Dipterocarpus zeylanicus (hora), could, on a rough estimate, produce half a million cubic feet yearly. In the dry-forest zone, the main source of valuable timbers, the only unworked forest is one of about half a million acres in the Northern Division (northwest of Trincomalee).

SILVICULTURAL OPERATIONS

Creepers and unwanted trees are removed by felling, girdling, or poisoning and thinning is confined to plantations. Plantations in the dry and intermediate zones were started in 890, as part of a plan for the co-operative reforestation of areas opened for shifting cultivation. About 15,000 acres of good forest have been planted. Species introduced with varying success are Artocarpus integrifolia (Oak), Swietenia macrophylla (mahogany), Tectona grandis (teak), Azadirachta indica (margosa), Vitex pinnata (milla), Felicium decipiens (pihimbiya), Alstonia macrophylla (havari-nuga) and Berrya cordifolia (halmilla).

In some areas the forest stock has been improved by planting Berrya cordifolia in small clearings scattered within the original forest. These operations have been fairly successful. The afforestation of the grasslands (patanas) of the central hills comprising the montane zone was begun in 1929 and approximately 5,000 acres have been successfully planted with exotics (Australian gums and some conifers). Although no previous tests were made of the suitability of the species to local conditions, the choice has mostly proved good. The species that have shown themselves most adaptable are Eucalyptus robusta, E. microcorys F. Muell. and E. saligna Smith.

After trials the planting of E. globulus,E. acmenioides Schauer and E. maculata was abandoned. Some areas in the montane zone have been planted with conifers, mostly Cupressus macrocarpa and trials have been carried out with Juniperus procera Hochst ex. Endl., Araucaria and Pinus spp. Acacia melanoxylon R. Br. has been successfully raised, but only on a small scale. Conifer planting was abandoned for some years owing to the extreme susceptibility of the plantations to damage by fire. A fresh experimental attempt, however, is being made with Pinus caribaea Morelet. Except where strong winds prevail, all plantations on up-country grasslands show good promise. With the cooperation of the FAO, the Forest Department has embarked on an extensive scheme of mechanized afforestation of the patanas and low forests and at present the target is a minimum of 2,000 acres per year.

Chief Timber Uses

In Ceylon timber is used not only for building construction, which makes heavy demands, but also in the manufacture of plywood, furniture, chests, safety matches, toys and railway sleepers. The demand for nondurable light hardwoods is on the increase. A plywood factory run by the government utilizes such species as Doona congestiflora (kokun), Calophyllum bracteatum (walu-kina), Palaquium petiolare (kirihembiliya) and Myristica dactyloides (malaboda). More than 150,000 cubic feet were used for this purpose in 1952. Species consumed in the manufacture of chests, toys and matches are all nondurable varieties such as Alstonia scholaris (ruk-attana), Canarium zeylanicum (kekuna) and Antiaris toxicaria (riti).

The furniture industry, on the other hand, uses the most valuable durable timbers and there is a yearly demand for approximately 200,000 sleepers (600,000 cubic feet), which can be made only from primary hardwoods. Wood is also extensively used as fuel in the processing of tea and rubber and for domestic purposes, but this requirement should eventually be met by the development of hydroelectric schemes, one of which (Laxapana) is already under way.

The timber position of the island as a whole is unsatisfactory, extraction at present exceeding the yearly increment. The population of Ceylon, now well over 7,000,000, is growing rapidly and at the present rate of increase will double itself in another 25 years. Concurrently the demand per capita for timber, now approximately 10 cubic feet, is also rising.

The problem of future supplies cannot be solved solely by afforestation and better forest management, which can never keep pace with the demand. During the last dozen years there has been an alarming increase in the import of timber, the value of which has jumped from Rs. 193,000 in 1942 to Rs. 21,000,000 in 1952. The only solution is the utilization of species that have hitherto been treated merely as forest weeds. The Gal Oya Development Board has taken the first step in this direction by installing a plant for the pressure treatment of timber and even secondary species (light hardwoods) are now used successfully for building purposes. The Forest Department has also started experiments in timber treatment in a small experimental pressure plant and in open tanks and steps are being taken to install a plant similar to that of the Gal Oya Development Board.

With a view to the utilization of waste timber material from sawmills and carpentry workshops and of branch wood and inferior species, efforts are now being made to develop building bricks from a mixture containing timber and cement and production on a small scale has already been begun.

Governmental and Public Interest

Of recent years both government and the public have shown an increasing awareness of the value of forests as a national asset and of the important role they play in the country's economy. To raise the standards of forestry a forestry field-training school was established in 1951 with two training centers, one in the wet and the other in the dry zone and considerable research work is carried on by the Forest Department. The Soil Conservation Act, passed in 1951, empowers the Government to enforce anti-soil-erosion measures even on private lands. All these steps augur well and it is expected that in the not too distant future the present unsatisfactory timber position will be remedied.

There are three national parks, two of which-Ruhunu and Wilpattuhave been in existence since 1937. These abound in the fauna of the island and provide excellent relaxation and recreation for young and old. Ruhunu in particular, which is only 176 miles from the port of Colombo, has earned popularity in many parts of the world and even casual visitors to the island find time to spend a few hours here. The third and most picturesque park, recently created under the Gal Oya scheme, shows promise of great popularity. The island has some of the most beautiful scenery in the world and forest work is made especially interesting when the forester finds deep in the forest, as he often does, ancient ruined temples and other monuments to a splendid past.

EXTRACTION AND REGENERATION

In Ceylon today modern machinery and methods applied in biblical times are found side by side. Owing to inadequate staff and consequent lack of supervision, timber contractors used to apply the "cut what you like" policy and the present generation is paying heavily for the sins of its forbears. Now, however, the extraction of timber is usually done on a selective basis and precautions are taken to leave sufficient trees of valuable species as seed bearers.

The heavy intermixture of species renders the removal of valuable timber, which seldom represents more than 5 to 10 per cent of the stock, both difficult and costly. The best forest tractor, one that can not only move easily between bushes and unwanted trees but also bring logs to the roadside and load them into trucks, is still the elephant and elephants are also employed in almost every sawmill yard.

Regeneration of valuable species after extraction is quite satisfactory, although species of low commercial value, such as Xylopia parvifolia (netaw), Ailanthus triphysa (wal-bilin) and Canarium zeylanicum (kekuna), in the wet zone and Ficus spp. and Hemicyclia sepiaria, in the dry zone, tend to overdevelop in areas that have been heavily exploited. Clear-cutting or excessive opening usually results in areas being overrun by forest weeds such as Macaranga peltata (kenda) and Trema orientalis (gedumba).

OCEANIA

Ten thousand islands -- thirty million square miles of sea -- less than 2 per cent of the surface land. One island alone, New Guinea, has three times the land area of all the others. Those others are as if an area equal to that of Great Britain had been fragmented and scattered over one third the world. Such is Oceania. Yet its islands and their forests, specked on some of the earth's most critical airways and seaways, of strategic importance in peace and war, command attention. The origins and relationships of the floras of Oceania have challenged the ingenuity, imagination and fact-finding capacities of botanists.

A drop of only 60 feet from present sea level would join New Guinea and Australia. Such a united continent of Papualand is believed to have existed 30,000 years or less ago. As a result of this or earlier unions, the forest flora of New Guinea bears a marked Australian character, especially as to casuarina, eucalyptus and oak. A lowering of only 60 or 70 feet more would expose the Sunda Shelf, uniting Southeast Asia with Sumatra, Java and Borneo. Such a union in the past must have brought the continental flora far east before its overseas migrations were launched.

THE EFFECTS OF MAN

Black Men and Brown Men. The first peoples to be pushed out of Asia into the islands, either by deterioration of climate during the last glacial period or by more aggressive tribes, or both, were Negroids, Ainoids and Veddoids. As they entered Melanesia ("black islands"), the black fuzzy-haired Negroid element was predominant. Hunters and food-gatherers of an Old Stone Age culture, they probably had no marked effect on the forests of their islands.

Four to five thousand years ago, the New Stone Age culture spread to the Negroid islands of Oceania, a culture that turned the people into farmers practicing a destructive shifting cultivation that has persisted to this day and is of great significance from the standpoint of world forestry. To these folk the forest was and still is a hindrance. They cut and burn the trees and plant crops for a few years, until soil fertility presumably lessens. Then the site is abandoned and a new forest is cut. In the meantime the old site grows up to secondary forest. In a decade this is recleared and reused, but for a shorter period. Fire is omnipresent. Rhizomatous grasses invade, which successive fires only encourage by burning the tree seedlings.

The grassland, encroaching on the surrounding forest even without agriculture, is unmanageable and the farmer moves away permanently to new forest sites. New Guinea today is a patchwork of forest and grassland wherever (which is almost everywhere) a "burn-dry season" occurs, from sea level to the frost regions above the mossy forest. Forest remains only in swamps, along stream sides and where the natives have not yet entered, as in a no-man's land between warring tribes.

When such tribes annihilate each other, the forest replaces the grassland, to be destroyed by some future tribe. Since the roving agriculture is a one-way process, especially with increasing population, this destruction of much timber of high commercial value is a problem of first importance to forestry. In addition, it poses a more general problem: the Melanesian in effect destroys his own habitat, for the grassland is useless to him.

After the New Stone Age had spread to Melanesia, but still before the Christian Era, paler, more Mongoloid people invaded the islands to the north, the area now known as Micronesia ("tiny islands"). And then the far-flung islands from Hawaii to Samoa and New Zealand were populated by the handsome race known as Polynesians ("of many islands"), who conquered vast distances in frail canoes. Ethnologists now consider these folk to have migrated through Melanesia toward the east and to have emerged with their present bronze skin.

The theory that they came from Peru, by the Kon-Tiki route is not fashionable. All these invasions were barely completed by A.D. 1400, the date of the first entering of Hawaii.

The Micronesians and Polynesians radically altered the forests of the high islands with regard to both flora and vegetation. They brought with them a limited number of plants; these included a few weeds, but for the most part were food plants, of which the coconut has done the most to change the landscape. It is now dominant on practically every low island and at the lowest elevations of the high islands too. What the forest was originally, or would be now if left to itself or planted to other trees, is a matter of conjecture and future experimentation.

A shifting upland cultivation, such as that of Melanesia, was not part of the culture on the high volcanic islands and consequently the mountain forests were but seldom entered. On the other hand, fire -- as always with primitive peoples -- was uncontrolled and often started for pleasure and hunting. It is the author's opinion that on many islands fires burned as frequently and as far up the hills as the vegetation and the climate would permit, destroying not only large areas of original forest but also the evidence of present forest potentialities.

White Men and Yellow Men. In 1520 Magellan passed through the straits that bear his name and, the following year, landed at the Mariannas. Thus was opened a new chapter in the exploitation of Oceania, a chapter of conflict that continues today between two fundamentally opposed philosophies of social living, a conflict that every forester faces squarely when he tries to administer tribal forest lands.

The original natives lived in small social units. Theirs was an agricultural and subsistence-barter economy, based on few material wants. With the crudest of tools, they fashioned a satisfying existence on infertile specks of land. Living under a form of "communalism," the individual had no incentive to amass personal wealth, for he would only have to share it with his tribe. To this day and for these reasons it is difficult to lure natives to forest work, unless they make a total break with their home villages.

In striking contrast, the new invaders were members of large social units. Theirs was a complex money economy, based on many and evergrowing material "necessities" Under the guise of "individual liberty," each man had every incentive for personal gain and thus wave after wave of Occidentals has passed over Oceania, each intent on its own brand of personal gain: souls, morals and modesty; loot and trade. The invaders streamed in from Europe, the United States and Japan and, when native muscle was not co-operative, Hindus, Chinese, Malayans, Filipinos, Javanese, Tonkinese and Puerto Ricans were herded in. World War II, however, left Oceania with a new political geography and under a "trusteeship" philosophy that has shown itself more altruistic and more humanitarian than history has usually known. Such elements make the Oceania of today the unusual stage that the forester must know if he is to act effectively.

Sandalwood, 1790-1850, is a finished case history of forest "mining" Undoubtedly, the value of island sandalwood was discovered independently on several occasions and the period of its exploitation is filled with double-crossings. Many a chief succumbed to the money lure, assuming island monopolies and forcing his people to work at gathering the wood in the unhealthy rain forests. Either by nature or with the connivance of the gatherers, who are said to have pulled up all the seedlings they saw, commercial sandalwood disappeared. Apparently the plant has some silvicultural peculiarities, for it has not reappeared in any quantity, either naturally or by artificial means.

The introduction of large naturalized mammals is another aspect of the Occidental invasion that is having a continuous and most important effect on the forests. Only the continental islands possessed a native fauna of large animals; the oceanic islands apparently developed their biotic communities in the complete absence of hoofed animals. It appears that the native forests have extremely shallow root systems, critically located in a moist surface layer. The slightest disturbance to the ground vegetation, as by grazing or browsing, kills the trees, in turn opening the adjacent forest to drying and destroying it, with an effect almost of a chain reaction. Cases are on record in Hawaii where grazing stock was put into a forest for a matter of a few weeks and then removed, but the forest died thereafter. If such destruction were on land planned for grazing, no unexpected problem would result.

Unfortunately the early white settlers introduced cattle, sheep, goats and horses. The islanders placed a taboo upon them, but some escaped to the hills and showed remarkable fertility. In addition, missionaries and traders often released animals on small uninhabited islands, for the sustenance of shipwrecked individuals. The destruction of the native vegetation by these feral animals and by pigs, earlier naturalized, has been tremendous. Wholesale slaughter of thousands of animals has since taken place in some islands; and to this day one of the most useful activities of a local forest service is the building and upkeep of border fences, not only to keep out adjacent high-quality stock, but to prevent reinvasion by feral animals.

Although rabbits have in general been less of a problem, Christophersen has made an interesting report on their history in Laysan, in the Hawaiian Islands. Until 1903 this little island had a flora of 28 species and a solid vegetation of grass and shrubs. In 1911 rabbits were swarming "over the island by thousands" By 1923 only four species of plants could be found and Laysan was described as a "desert of sand," with no mention of rabbits, alive or dead-another example of an introduced mammal annihilating itself by destroying its own habitat.

Much more than half of the flora of some of the islands is composed of species that have been introduced and naturalized since the white invasion. Moist uplands have been less affected, but arid lowlands -- with their original vegetation already destroyed-were fair prizes for hordes of weeds. In many islands the bulk of the lowland vegetation is now composed of alien species and one must search carefully for the indigenes.

Furthermore, on destruction of the native forest by grazing, the land is taken over almost at once by alien grasses and shrubs which, as on the grasslands of New Guinea, form a strong deterrent to the return of the native trees. This situation has given rise to egregious comparisons between the native flora and the native cultures that disintegrate before the onslaught of aggressive aliens. The contrary, however, would seem to prevail in the case of the flora. The writer (1942), in accord with Allen's interpretation for New Zealand (1936), has presented field evidence in support of the thesis that most of the alien plants are pioneer species expanding into "open" lands and forests that have already been disturbed, whereas the native trees are largely of end-stage (climax) species that will hold their own against most of the invaders. The critical need in reforestation is for a "nurse tree," native or alien, which can naturally invade the grassland and scrub, forming a growth that can in turn be invaded by the native forest trees.

Cimates

Oceania comprises three main kinds of islands:

- "Oceanic coral low islands," composed entirely of coral limestone and presumably capping submerged volcanic deposits;
- "Oceanic volcanic high islands," composed of active or extinct volcanoes, arising from the floor of the ocean basin and usually complicated by fringing reefs;
- "Continental high islands," composed of sedimentary strata overlying a granite base, usually complicated by volcanoes and coral reefs.

The low islands partake of what might be called regional climates, controlled by the warm moist trade winds blowing diagonally from the horse latitudes (at about 30° N and S) westward toward the equator. These winds, as they pass over warm, low, sandy islands, may be sufficiently heated so that no rain falls, even when there is precipitation on the surrounding ocean. In general, within five degrees of the equator there is a dry zone, which includes the Phoenix and Gilbert islands and at higher latitudes a moister zone.

At still higher latitudes, the seasonal shifting of the trade-wind belt with the prevailing westerlies modifies the climate, as in Hawaii. In the western Pacific, especially in New Guinea, the influence is predominantly that of monsoons, related to the greater seasonal warming and cooling of large land masses than of the adjacent ocean.

The climates of high islands may vary markedly even within distances of less than a kilometre. The moist trade winds, striking a volcanic island, are deflected upward, where they cool to condensation. Passing over and beyond the summit, they drop, are warmed and dry up. The result is a series of concentrically arranged precipitation belts, placed askew on the contour pattern.

The centre of highest precipitation is slightly to windward of the peak and each rainfall belt is tipped down to windward so that at any one elevation there is less rainfall and a drier vegetation on the lee than on the windward side of the island. Each valley and ridge adds its minor complications. The majority of these climates exhibit a great variety of physiognomic vegetation types, depending largely on human activities, especially management practices. Consequently, any climatic classification which invokes a single vegetation type as the "climax" for its zone runs immediately into inconsistencies.

Soils

Few generalizations can yet be drawn on the soils of Oceania. Those of the continental islands are probably as varied and as complex as in any continental area, though they are today little known.

The low limestone islands have invariably a thin and poor soil, occasionally favored by a layer of organic matter from the vegetation or from bird colonies. More has been published on the soils of the high volcanic islands, especially as regards agricultural Hawaii. It is probable that many of the forest soils are ferruginous laterites. Aside from a surface crust, sometimes present, aluminum and silicon increase downward, iron concentrates at an intermediate position and titanium remains abundant at the surface.

Soil erosion can be critical, often spectacular, but as yet has failed to arouse the governmental authorities, as in other parts of the world. Perhaps this is because most of the erosion has been in unpopulated areas from the start and has not resulted in the displacement of contemporary populations and agricultures.

Much of it apparently was consummated before the advent of the white men, or else immediately after-owing to the naturalization of grazing mammals -- but before European land managers began to look more critically at the land. Within a few miles of Honolulu there was a valley forest a few years ago half buried with debris from the hills. Elsewhere entire coastal plains are said to have been formed in recent years and relict soil patches hanging on rocky mountains attest to past conditions. Unlike that in temperate regions, the decomposition of the volcanic rocks in the tropics may be rapid and some lands may be amenable to management practices that would restore seminatural forests to areas now completely barren. The research on this remains to be done.

Island Types

However, that these are not only incomplete, but are not indicative of potential forest land, since there is much grassland and savanna that could be converted to forest under rational management practices. Such grasslands include not only those of New Guinea and the Solomons, but also those of Hawaii, Guam and even Yapy and Palau.

THE COMING OF THE PLANTS

Although the last word has by no means been said, it would appear that the lesser Pacific islands obtained their original plants from across the seas, with all the hazards and uncertainties and failures of a man seeking to cross in a rowboat and trying to perpetuate himself if he could land. Hurricanes, rafts of floating logs, stray birds and other "accidents," continuing to this day, have contributed to the origins of the present floras.

There is now a consensus that the floristic invasion of Oceania has come from three main regions: the Americas, most obviously in Hawaii; a now vanished Antarctic continent, with the remnants of its flora found in southern South America, South Africa, Australia and New Zealand; and Indo-Malaysia, which in turn is basically Asiatic. Despite the prevailing easterly winds, however, the tropical cyclones from the west have been most influential. As one proceeds from the western borders of the Pacific, the flora becomes more and more impoverished in genera and appears obviously to be a fraying-out, an attenuation of the Asiatic plant world, as the chances of successful immigration become less and less over the tremendous distances of water.

Fosberg has analyzed the Hawaiian flora with regard to its origins, basing his discussion on the number of assumed original immigrants. For the seed plants, he finds that 40.1 per cent are Indo-Malaysian, 18.3 per cent American, 16.5 per cent Antarctic and the remainder boreal, cosmopolitan, or obscure.Offsetting in part this paucity in genera and larger categories, there has been an extraordinary speciation in certain groups.

Whatever be the causes, the original immigrants, including trees, have sometimes indulged in endless morphologic variation, often with each valley developing its recognizable form.

Certainly many of the named forms are ecologic modifications and many more would interbreed and lose their identity if placed in a related population; but many others are undoubtedly established genetically and the recognition of them as desirable strains in forest trees is an open field for silvicultural research.

Immigration and speciation are probably continuing now as in the past, but on too slow a time scale to be of practical importance. The striking effects of introductions by man will be considered later. The above discussion holds for what one might term the "inland flora," for species that cannot tolerate salt, die when submerged in salt water and shun the strand. It is these species which pose the problem of overseas migration to the botanists.

The "strand flora," entirely to the contrary, seems to have evolved capable of overseas migration. Buoyant seeds that withstand many days of floating -- even as germinated seedlings -- are found in totally unrelated plant families. This flora is remarkably constant in all the islands -- evidence of the success of these adaptations; and this flora is adapted not only to migration but also to growth under saline conditions, either in soil or air. Indeed, the plants are frequently incapable of terrestrial migration (though many grow well where planted). For all these reasons, the strand flora forms the entire vegetation on countless low islands. On the high islands it tends to predominate only in the plant communities near the seacoast.

Among all these native plants, as well as among naturalized species, there are innumerable kinds of trees of distinct value to the forester, especially in the islands close to Southeast Asia. No manual of dendrology exists for this territory, although Kraemer has published a volume for part of Malaysia, including some of Melanesia. The book describes more than 100 important species, said to be about one per cent of the total tree flora.

OCEANIC LOW ISLANDS

Where volcanic or other materials have formed the sea floor at depths of less than 150 feet in tropical waters, they have been subject to colonization by coral and related organisms. These have built up calcareous deposits as living reefs at, or slightly below, sea level. Surmounting such a reef, generally at its margins, an atoll is usually a more or less circular group of small islands, produced by wave and wind action, enclosing a "lagoon" To the geomorphologist atoll islands seem ephemeral, subject to rapid destruction, especially during storms. Forestry practice must reckon with the hazards to which such islands are subject.

Table. Oceania: Land Area and Forest Area for Certain Islands, 1950-1953 (In Thousand Hectares)

	Land Area	Forest Area Accessible	Inaccessible	Total
British Solomon Islands, 1953	2,890	20	2640	2,660
Fiji, 1953	1,828	140	815	955
French Oceania, 1950	360	50	65	115
Hawaii, 1950	1,665	450	325	775
New Caledonia, 1952	1,582	10	70	80
New Guinea (Australian) and Papua, 1951	47,400	3,00	30,000	33,000

As a habitat for plants, atoll islands, despite their almost infinite number, their human populations and their emotional appeal, do not hold much for the forester interested in quantity. Their vegetation normally comprises a series of belts from sea to lagoon. Back of the sea beach is a scrub 2 to 5 meters high. Next is a narrow forest zone of Messerschmidia argentea and Pandanus spp. In the Marshalls and Carolines Terminalia litoralis is found in this Messerschmidia zone.

The greater part of the typical atoll is a forest which today is almost entirely composed of Cocos nucifera, topped sometimes by huge breadfruit trees (Artocarpus altilis) where the rainfall is heavier. The timber of breadfruit is excellent in almost every respect, but the tree is seldom cut since it serves primarily for food. It would appear that the original forest vegetation was composed of such trees as Pisonia grandis, Pandanus spp., Ochrosia parviflora, Pipturus argenteus, Hibiscus tiliaceus, Messerschmidia argentea, CaloPhyllum inophyllum, Barringtonia asiatica, Eugenia spp. and others. Pisonia may be the sole dominant on dry islands, such as Vostok. Some of the isolated eastern atolls may have floras composed of as few as five species, or even less, seemingly the result of chance migration.

Along the lagoon beach of most islands is a narrow strip of scattered trees such as Cordia subcordata, Hernandia ovigera, Terminalia catappa, Barringtonia asiatica, Thespesia populnea and, in rocky places, Pemphis acidula. Mangrove species, where found at all, occur in small depressions in the centers of the islands.

CONTINENTAL ISLANDS

The continental islands of Oceania include Fiji, New Caledonia, the New Hebrides, the Solomons, the Bismarck Archipelago and New Guinea. In them are found not only all the varieties of forest characteristic of coral low islands and oceanic volcanic islands, but also their own complexities, due to larger floras, more varied rock strata and topographic features and climatic regimes ranging from slight to complete dominance by the monsoons. Nevertheless, the same basic pattern of arid, humid, cloud and super-cloud zones occurs, if not in the orderly arrangement found on a single high volcano in the tradewind belt.

OCEANIC HIGH ISLANDS

The oceanic high islands display endless variations, but basically they consist of volcanoes. They not only occur in chains and groups but also in such total isolation as that of Easter Island. The original formations were pahoehoe and aa lavas, with bombs, ash, cinders, dust and other ejecta. Erosion may have advanced to such a degree as to leave knife-edge ridges and deep valleys (sometimes amphitheater-headed, because of greater precipitation upstream), footed by coastal plains. Ancient coral reefs, developed at previous higher sea levels, may appear along an island's flanks. Soil maturation has progressed variably from the different parent materials, with the development of distinctive profiles. Despite this profusion of types, a pattern emerges of interest to the forester.

This zones, arranged consecutively from sea level to summit. Each zone, except in the driest, coldest and cloudiest sites, has a characteristic assemblage of forest, savanna and grassland, with certain apparent correlations with contributing factors, although implications of causality are still premature. Sunlight attains a maximum at sea level on the lee side of the typical island and decreases upward, with a marked diminution at the beginning of the socalled "cloud zone," a minimum in the mossy forest and then a rapid increase to a second, high-altitude maximum.

Precipitation varies inversely with sunlight, being low at sea level, maximal in the cloud zone and again low at very high elevations.

Cloudiness, although it may occur at any elevation, is of significance only within the cloud zone, where its intensity increases rapidly to a maximum at its upper level. Altitude, perhaps because of the ease of obtaining data regarding it, is mentioned ad extremum by botanists and foresters, but does not seem to exert any direct effects, at least upon the details of vegetation. The distribution of floras in relation to vegetation zones has been considered only for Oahu,

Hawaii. Despite its significance in forest management, both in control of natural stands and in planting progrmmes, foresters have almost totally neglected this subject. Its importance is illustrated by the fact that in any one zone the predominant trees may be either (a) those that are there capable of maximum individual development and would not thrive in an adjacent zone, or (b) those that have been crowded out of an adjacent zone by competition with others and would actually develop more luxuriantly if planted in the adjacent zone with the competition lessened.

As a working hypothesis for Pacific high islands, areas characterized by six different floras:

- The strand flora;
- A desert flora, which may exist on some islands where there is less than 35 inches of precipitation a year (some widely naturalized Opuntias may occur in this area);
- An arid-region flora on islands having roughly less than 50 inches of rain, with a maximum floristic complexity at the higher precipitation levels and attenuating out below to mingle with the desert plants (tremendous numbers of alien plants have been naturalized in this floristic area);
- A humid-region flora occupying a region of rain forests, with high percentages of endemic species and few naturalized aliens (many of the plants pass upward into the cloud zone, where present);
- A cool-tropical cloud-region flora;
- A flora or floras above the frost line. On some islands, above the cloud zone but below the frost line, the humid-region and arid-region floras are repeated, but in reverse order, as on the island of Hawaii.

The significance of an effective frost line on the islands of Oceania remains to be investigated. One might expect, judging from continental areas, that it would form a fairly distinct separation of tropical from nontropical conditions. The situation in New Caledonia and in the New Hebrides, Society, Bismarck and Solomon islands is unknown. In oceanic Hawaii the flora at high altitudes appears to be nothing more than an extremely impoverished attenuation of plants primarily tropical, with almost no temperate American origins-mute testimony of the failure of aerial migration from that direction. Continental New Guinea, on the other hand, has an extensive super-frost-line flora, but its geographical affinities are poorly known. All the super-frost-line regions hold great potentialities for the introduction and naturalization of temperate forest species.

Each vegetation zone possesses a complex of topographic variations, ridges, valleys and plains that make difficult the recognition of zonal differences. In those cases where a so-called rain-forest zone is predominant on coastal plains and a "montane-forest zone" on foothills and slopes, one has reason to wonder whether a climatically significant zonal difference actually exists. The precise mapping of the boundary between two distinct zones has been attempted only on Oahu. Here the line was found to pass through a wide physiognomic variety of plant communities and to vary in altitudinal range by almost 1,000 feet, bulging seaward on the ridges, where the cooler temperatures may have offset the lower rainfall to permit an equivalent moisture effectiveness.

TEMPERATE FORESTS

Temperate forests are those that lie within the temperate, or midlatitude areas on Earth, such as the United States. In terms of global warming and carbon dioxide forests play a crucial role. They store an enormous amount of CO_2: The trees, other vegetation, and soils within forests help to drive the global carbon cycle by sequestering CO_2 through photosynthesis. As ecosystems, many forests are endangered today—mainly by human practices. Forests are being cut down so that the land can be used for other purposes, such as grazing land or agriculture. Other forests are being logged—hundreds of acres of trees are being harvested for wood products. Mining exploration and operations are encroaching into other forests. Urbanization and the construction of transportation networks to connect urban and industrialized areas have also claimed forested lands. Some forests have been deliberately set on fire. With all this activity occurring in the world's forests, this valuable resource is being lost at a rapid rate. Not only is carbon storage no longer available, but when trees are burned, all the stored carbon is released back into the atmosphere.

CARBON SEQUESTRATION

If forests are managed properly, they are one ecosystem that could be used to combat the effects of global warming. The experts at the Union of Concerned Scientists, forests could reduce net carbon emissions by the equivalent of 10–20 per cent of projected fossil fuel emissions through 2050. In the United States alone, measures to increase carbon storage in forests could sequester an additional 44–88 million tons of carbon annually. The greatest potential carbon sequestration in forests is in the tropical and subtropical regions on Earth. Carbon sequestration is viewed as a feasible way to offset global warming.

The UCS, forest sequestration, also called forest based mitigation, could occur via the following two strategies:

1. conservation of existing forests
2. increasing forests' carbon-absorbing capacity by
   a. Planting more trees,
   b. Promoting the natural regeneration of forests,
   c. Improving forest management practices in order to increase biomass.

Once again, the UCS has stated that if forest sequestration is done correctly, it can slow climate change and the effects of global warming. For instance, based on their data, U.S. forests in the Pacific Northwest and Southeast could double the carbon they store if forest managers lengthened the time between harvests and allowed older trees to remain standing. Working with Canada and Mexico to attain the same goals would also help. Another suggestion proposed by the UCS is to create a system that allows private companies to get credit for reducing carbon when they purchase and set aside natural forests for conservation. There are abundant environmental benefits that come along with conservation, such as protecting biodiversity and maintaining clean watersheds.

FOREST VALUES

One of the problems the UCS has identified is that the public must be better educated as to the valuable services that forests provide, such as biodiversity, carbon storage, and water purification. One of the major issues that must be faced today is the economics-when a quick sale can be made on the lumber or other goods taken from a forest, it is hard to convince people to turn away and choose to protect an environment that helps everyone but is intangible. The market-based versus environmental- based issues create some of the biggest challenges in dealing with global warming today.

EFFECTS OF GLOBAL WARMING

Not all forests will have the same outcomes under the influence of global warming and changes in precipitation. Some forests will die back, while others may extend their ranges. Amounts of CO_2 will vary, adding its influence to the mix as well. Whatever the outcome of a particular forest, if its population cannot adapt or migrate with the changes, it will face extinction. The IPCC, at least one-third of the world's remaining forests may be negatively affected by climate change during this century. Global warming may force plant and animal species to migrate or adapt faster than is physically possible, disrupting entire ecosystems.

The IPCC also predicts that forests will have changes in fire intensity and frequency and increased susceptibility to insect damage and diseases. The World Wildlife Fund, climate scientists use a variety of methods to predict the impacts of climate change on forests. On a global or large regional scale, they can predict shifts in ecosystems by combining biogeography models with atmospheric general circulation models (GCMs) that project changes in an environment where the CO_2 content has doubled. They also use biogeochemistry models to simulate the carbon cycle, flow of nutrients, and changes in precipitation, soil moisture, and temperature to study ecosystem productivity. They even have global models that simulate worldwide changes in vegetation composition and distribution. The IPCC states that global warming both directly and indirectly affects forests during climate change. The direct impacts of warmer temperatures, changing rainfall patterns, and severe weather events can already be seen in certain tree and animal species. Even small changes can affect forest growth and survival-especially of the portions that lie along the outer edges of the ecosystem, where the conditions are marginal. When it gets hotter, more water is lost through evapotranspiration, which causes drier conditions and decreases plants' use of water. Water temperatures can also throw off the timing of flowering and fruiting for plants and adversely affect their growth rate. Forests will also be threatened when the seasonal precipitation patterns they have been used to change and water is not supplied when it is needed, causing drought conditions and stress, or supplies too much where it cannot be assimilated, causing flooding and mudslides.

The Natural Resources Canada, the age and structure of a forest will also play an important role in determining how quickly it will respond to changes in moisture conditions. Mature (older) forests have well-established root systems, which means that they can tolerate drought better than younger forests or forests that have been disturbed in some other way, such as through disease infestation. Species type also plays a part-some species are more resistant than others. Models have been run simulating the impact of elevated levels of CO_2 on forests. The Natural Resources Canada, the results of these simulations are not as straightforward or conclusive as desired. Results are much more problematic. They found that higher levels of CO_2 improve the efficiency of water use by some plants, and some plants have been shown to adjust to higher CO_2 levels, but their absorption rates decrease over time.

It is also hard to model CO_2 effects when other greenhouse gases may be involved, such as ozone (which works against CO_2) or nitrogen oxide (which can enhance tree growth). In the short term (50-100 years), changes due to global warming will be focused on ecosystem function. In the long term, shift of forest types will be more significant. The IPCC, boreal (northern) forests will feel the most impact because their ecosystem will be greatly reduced since warming is expected to be most significant in the polar regions. Some of the most vulnerable temperate forests will be the island, or isolated, forest communities, such as the fragmented forests encroached upon by urban and agricultural areas. There is nowhere for them to migrate. Forests located in high elevations on mountain systems face a similar threat-as they migrate upward, eventually there will be nowhere for them to go. Individual species that are indigenous to small geographic areas or have limited seed dispersal will also be threatened and endangered.

The WWF, the following are the five major categories of changes expected in temperate forest ecosystems as a result of global warming:

1. *Disturbance*: Forest ecosystems will become more disturbed by extreme weather incidence and change in rainfall and temperature.

Forests will become more fragmented and isolated, permanently altering the ecosystem.

2. *Simplification*: If global warming is severe enough, it will cause slower growing species to be replaced by faster-growing, short-lived weeds and other invasive species. This is a problem because it degrades the landscape, creating a speciespoor forest-only a few invasive species dominate instead of the original rich, diverse mixture of many species.
3. *Movement*: Migration of species is expected both in latitude (towards the poles) and in elevation (up mountains). How fast individual species will be able to migrate is still unknown.
4. *Age reduction*: With all the stresses put on forest trees with global warming, the old growth stands are expected to die off, leaving younger stands in their place. This will have a negative impact on biodiversity.
4. *Extinction*: The most vulnerable forest habitats could disappear forever.

Under global warming conditions, surviving forests of the future will look very different from those of today. Although changes will vary in degree from area to area, all forests will be affected.

CLIMATE CHANGE IMPACTS

The forests in the United States and Canada are already showing the effects of global warming. Over the past century, a 1.67°–3.3°F (1°–2°C) increase in air temperature and changes in precipitation have already been documented. Experts believe that higher levels of CO_2 have already caused dieback of forested areas along the Pacific and Atlantic coasts. It is expected that temperate forests in the United States will migrate northward from 62–329 miles (100–530 km) over the course of the next century. If the air temperature warms 3.3°F (2°C) over the same period, models have predicted that tree species will have to migrate 1–3 miles (1.6–4.8 km) per year, which is too fast for most temperate species, except for those whose seeds are carried by birds over greater distances. It is expected that grasslands will dominate many of these areas. The wildlife inhabiting the temperate forests will also be affected by loss of habitat, drought, and wildfires. Diseases will also weaken them. The National Assessment Synthesis Team (NAST) of the U.S. Global Change Research Programme, some forest models predict an overall increase in forest productivity with increased temperatures.

They also clarify that other environmental factors, such as severe weather, could offset any productivity. NAST also states that a great deal of biodiversity change is expected in the United States. The maple/beech/birch forests in the Appalachian Range from New England to West Virginia could completely disappear. In the Upper Great Lakes region, aspen, birch, and red spruce could be obliterated. If global warming occurs gradually, models suggest that oak/pine and oak/hickory forests will replace these threatened areas. If, however, conditions are bad enough, entire vegetation species could be lost. NAST also found that warming in cooler areas—such as the northern United States and western mountains near the Canadian border— will increase tree richness and provide a better habitat for reptiles and amphibians. These same models also predict a decrease in bird and mammal richness in the eastern United States. NAST models have predicted that temperate forests in Scandinavia will move northward and replace other forest communities.

RECENT DEVELOPMENTS

Four examples of modern-day forested areas suffering the effects of global warming follow. In July 7, 2006, report in USA Today, the frequency and size of large forest fires have increased dramatically in the last 20 years, and global warming is being blamed as part of the cause. The western region of the United States has seen an increase in wildfire occurrences since 1987; the wildfire season is now two and a half months longer than it was back then. Part of the problem is that spring warm-up is happening earlier and more quickly now than it has in the past. This sudden temperature transition melts the winter snow accumulation in the mountains, turning it into heavy spring run-off. Early green-up of grass, shrubs, and other vegetation promotes an abundance of biomass. When the summer months heat up significantly—there can be several days with temperatures in excess of 100°F (37.8°C)—and the winter melt has already run its course through the rivers and into reservoirs, the landscape becomes extremely arid in the hot, dry air, and the abundant biomass that grew during the ample spring run-off becomes extremely dry, posing a high fire hazard. Anthony Westerling, a researcher at the Scripps Institution of Oceanography of the University of California, San Diego, "We didn't set out to make a climate change argument, but it's easy to see how a further rise in temperature under climate change would result in more frequent wildfires in these severe years.

You get early snowmelt, the soil and vegetation dry out sooner, and you get a lot more fires, burning longer and getting bigger." A group of scientists from the University of California, San Diego, and the University of Arizona studied 1,166 large forest fires that were each a minimum of 1,000 acres (405 ha) in size from 1970 to 2004 on national forest and parklands. They determined that in the latter part, from 1987 to 2004, there were four times as many forest fires and more than six times as much total land area burned. The fires in the latter part were also more difficult to control and lasted longer. The duration of fires began averaging five weeks.

The most famous fire that occurred during this time was the Yellowstone fire in 1988. One-third of Yellowstone National Park burned, and the fire was not extinguished until it finally burned itself out in the fall. Tom Swetnam at Arizona's Laboratory of Tree-Ring Research says that climate is the principal factor causing large wildfires, especially in the northern Rockies. He backs up his conclusion by pointing out that the number of large fires in Wyoming, Idaho, and Montana has risen 60 per cent since 1987. The combination of rising temperatures and drying is making forests highly susceptible to wildfires.

TOURISM AND RECREATION

One recreational activity that may be hard-hit by the effects of global warming is the ski industry. Already, many of the glaciers in the Alps in Europe are melting, threatening the future of ski resorts. Because of already warmer conditions and changed weather patterns, many winter recreational resorts have to use snow machines to make artificial snow. In the New York Times, Shardur Agrawala, a climate specialist, noted that alpine resorts, especially those at low altitudes, are developing other revenue opportunities that do not require snow, such as spas, restaurants, shopping centres, swimming pools, convention centres, and tram rides geared for the entertainment of hikers and sightseers. Some resorts in Europe claim global warming has already begun affecting tourism. Roman Codina, the managing director of a mountain lodge in Zermatt, Switzerland, testified about a surge of visitors during the summer months-tourists from the hotter European cities seeking relief in the cooler mountains. Ski resorts in the United States stand to face a similar dilemma. Already, resorts create artificial snow to establish a snow base in years when there is not an adequate accumulation to open the ski resorts by Thanksgiving in November-a traditional start of the ski season. Artificial snow is also used as a backup for ski competitions during droughtlike years.

SPREAD OF INSECTS

A report in National Geographic News on November 5, 2001, a specific species of mosquito, Wyeomyia smithii, has gone through an evolution and been able to adapt to the climate changes caused by global warming. William Bradshaw, an evolutionary biologist at the University of Oregon, has found that this tolerance can occur in as little as five years' time. Because winters are gradually getting warmer, the mosquitoes are now breeding later in the year. Compared with their hibernation habits 30 years ago, northern populations of the mosquito have adapted to milder winters and become dormant later in their annual cycle. In fact, hibernation today is about nine days later than it was in 1972. The problem with this altered breeding and hibernation schedule is that it has an impact on the natural food chain and ecological system. If the timing of the component in a system gets out of synch, then food sources may not be available when predators need them, upsetting the entire balance of the ecosystem.

WILDLIFE

One noteworthy contribution to atmospheric methane-a serious global warming gas in the atmosphere-is, surprisingly, from livestock. In a report May 13, 2002, on National Geographic News, scientists in New Zealand determined that livestock in their country were adding to the global warming problem. Katherine Hayhoe, an atmospheric scientist at the University of Illinois, remarks,"New Zealand is unique in that more than 50 per cent of its greenhouse gas emissions arise from methane released by enteric fermentation." Enteric fermentation is methane that is produced as part of the digestive process of cows, sheep, and other animals. It is released when the animals burp, which is, surprisingly, a lot. There are roughly 10 million cattle and 45 million sheep in New Zealand, and their collective burps add up to 90 per cent of New Zealand's methane emissions. The U.S. Environmental Protection Agency (EPA), livestock in the United States only account for approximately 2 per cent of methane emissions. To combat this problem, scientists in New Zealand have discovered that by feeding livestock plants that are high in condensed tannins they produce up to 16 per cent less methane. Tannins are naturally occurring plant polyphenols that bind and precipitate proteins. They are found in legume forages and some grasses. Scientists view this as a significant step to reduce the effect livestock are having on the global warming problem. Michael Tavendale, the researcher who made the discovery,"It's early, but this is very encouraging news that will give our research new impetus and offer positive opportunities to New Zealand farmers for controlling the problem." This diet has other benefits as well: It increases livestock weight gain, improves milk yields, and decreases parasites. As a comparison, a 200-cow dairy herd has the"petrol equivalent" of 6,400 gallons (24,000 l) of gasoline, the gas necessary to drive an average car 124,274 miles (200,000 km).

CONTROVERSIAL FINDINGS

A study conducted by the Carnegie Institution in Washington, D.C., and the Lawrence Livermore National Laboratory in California led their scientists to conclude that forests in temperate regions actually could worsen global warming. Using complex modeling software to simulate changes in forest cover and their effects on global climate, the results they obtained from their model were surprising. Ken Caldeira from the Carnegie Institution remarked,"We were hoping to find that growing forests in the United States would help slow global warming. But if we are not careful, growing forests could make global warming even worse."

The researchers concluded from their study that it was the tropical forests that cooled the environment because they released great quantities of water during the evapotranspiration process. Forests in the temperate regions, however, not only did not release as much water, but also absorbed a lot of sunlight, heating up the Earth. To show this point, during one of the computer simulations, they covered the Earth with forests north of the 20-degree latitude line. They found that the surface air temperature responded by rising more than 6°F (3.6°C). In another iteration, covering the entire planet's land area with trees increased temperatures only about 2°F (1.2°C). The warming was not evident at first. During the initial phase, there was cooling because the trees took up carbon dioxide, which offset the absorption of sunlight. But this effect did not last. The absorption of energy lasts forever, but as trees mature, they sequester less carbon.

Based on this simulation, the involved scientists concluded that planting a forest in the United States could cool the Earth for a few decades but would lead to planetary warming in the long term. The scientists determined that a global replacement of current vegetation by trees would lead to a global warming of 2.4°F (1.4°C). If areas were replaced with grasslands, it could lead to a cooling of 0.7°F (0.4°C). This situation does not hold true for tropical forests, however. Rain forests keep the Earth cool not only by absorbing CO_2 but also by evaporating water.

BOREAL FORESTS

The effects of global warming are felt most at the poles, where it is predicted that temperatures could climb 8.3°–17°F (5°–10°C) or higher over the next century. It is estimated that warming will have a negative effect on the species that live in the ecosystems and that approximately 24–40 per cent of the species living in the boreal forests right now will be lost. Simply put, species that live in the north will be crowded out by species from the temperate regions that will be migrating northward in search of cooler climates. The species that will invade the present boreal forests will be today's temperate forest species and grasslands.

The IPCC, as the boreal forest vegetation is forced out, it will migrate poleward 186–311 miles (300–500 km) in the next century. Proof of this can already be seen in western Canada; plant zones there have already begun shifting poleward. There are major challenges this migrating vegetation will encounter. To begin with, the soils in the tundra region are not fertile and conducive to high-density vegetation or tree growth. The IPCC, they lack the biota necessary for colonization. Specific seed dispersal rate and migration tolerance range are also important factors that might keep trees from being able to survive the poleward migration rate set up by global warming. As an example, white spruce can colonize 62–124 miles (100–200 km) over 100 years, and Scots pine can migrate 2.5–5 miles (4–8 km) every 100 years. A. Solomon and K. Jardine of Greenpeace, the rate of global warming will be about 10 times faster than what is needed for successful species migration at an average migration rate of 16 miles (25 km) each century through natural seed dispersal. There are other factors that will hurt species migration as well, such as habitat fragmentation (small isolated clusters instead of one large cohesive unit) and competition from more hardy species. As temperatures change, it may also affect the timing and rate of seed production, which will affect the growth and strength of trees.

Trees that have limited seed dispersal mechanisms will also suffer, *i.e.,* trees whose seeds are carried long distances by the wind will have a better chance of survival than those whose seeds fall closer to the tree. The ability to adjust to a larger temperature range will also play an important role. Vegetation with narrow temperature tolerances will be vulnerable to extinction. The IPCC has stated that a drastic change in species composition and loss of habitat with even a 3.3°F (2°C) warming near the poles will damage the ability of an ecosystem to function as species richness begins to be killed off. They warn that a decrease in habitat if this were to occur would result in the loss of 10–50 per cent of all the animals living in the boreal Great Basin mountain ranges. The Natural Resources Canada, an average rise in temperature of 1.7°F (1°C) over Canada in the last century has had a negative impact on vegetation.

At mid to high latitudes (45° N–70° N) plant growth and the length of the growing season has increased. In portions of western Canada, there has been a decrease in rainfall as temperatures have risen, and this has hurt the growth of some tree species, such as aspen poplar. In Alberta, aspen are now blooming 26 days earlier than they were 100 years ago. Another major concern for boreal forests in a warmer climate is insect infestations. Insects commonly found in temperate forests, such as mountain pine beetle, will migrate north along with the forests and continue to infect as they move northward. As temperatures climb, droughtlike conditions may develop. If this happens, there will also be greater incidences of wildfire. In the past 40 years, the trend has already been established that as climate warms, wildfires have become more frequent and are burning larger areas.

This overall trend has been seen in places such as southern California and also in boreal locations in Canada and Russia. As global warming continues, longer fire seasons, drier conditions, and more frequent severe electrical storms are projected to increase, causing fire seasons to become more problematic as the climate continues to change. While it is true that some forest species' seeds are actually dispersed by fire, which will aid their migration, and burned litter will add nutrients to the soil, over time reoccurrences of wildfire will fragment established vegetation colonies and make it more difficult for them to migrate. In addition, as older trees burn, they will add carbon to the atmosphere, and as younger trees replace these burned areas, there will be less initial carbon storage capability.

5

Forests' Effect on the Climates

Forests play an important role in regulating the earth's temperature and weather patterns by storing large quantities of carbon and water. This regulatory function has a profound effect on both the local and the global climate.

Locally, trees provide shade, which in turn lowers summer temperatures and prevents the soil from drying out, they reduce heat loss from the ground in winter and reduce storm damage by providing shelter from wind. Globally, forests regulate the global carbon cycle, having a profound effect on the climate. As well as this, deforestation is also contributing to climate change. Indeed the CO_2 released each year from forest loss is higher than that released by our yearly transport emissions.

The continued existence of forests is particularly necessary if we are to halt what is known as runaway climate change. Runaway climate change is the point whereby increases in temperature lead to more GHG emissions which in turn leads to increased temperatures. Examples of this include that increased temperatures will melt ice caps which will lead to huge increases in the release of GHGs; and increases in temperature are projected to negatively affect up to two-thirds of existing forests, thereby exacerbating deforestation and increasing the release of carbon.

THE CLIMATE'S EFFECT ON FORESTS

Global warming, which on a geological timescale is occurring in the equivalent of a split second, is significantly disrupting the intricate and poorly understood web of interactions that governs the very structure and composition of forest ecosystems. This means that around a third of today's forests are likely to change their species composition. A temperature increase of 3°C by 2100 would result in forest ecosystems having to move 500 km towards the poles or 500m in elevation in order to find the same climatic conditions. Such distances are far beyond the average rate of dispersal for individual tree species, let alone entire forest ecosystems.

Early warnings about the consequences of the impacts of climate change on forests have been documented in, among others, The Carbon Bomb: Climate change and the fate of the northern Boreal forests - a 1994 Greenpeace report which states on page 2 that:

- "Studies on the global carbon cycle suggest that boreal forests are not absorbing as much carbon as they did before 1976. As a result, the atmosphere already appears to contain 10-15 billion tonnes of carbon more than it would have if forests had continued to absorb carbon at the pre-1976 rate. If boreal forests continue to decline, estimates suggest that burning and rotting of boreal forests could contribute to the release of up to 225 billion tones of extra carbon into the atmosphere, increasing current levels by a third. This would accelerate the rate of climate change."

While it is possible that the boreal forest could expand into the frozen tundra as temperatures increase, such an expansion would likely be delayed by slow tree migration rates. Even in the long-term, the boreal forest is unlikely to move northward fast enough to compensate for the breakdown of boreal forests at the southern part, turning dense forest into open woodlands and grassland, which in turn will result in a lowered biological diversity and a reduced ability of these ecosystems to store carbon and water.

Other forest ecosystems are faced with a similar fate; according to the Intergovernmental Panel on Climate Change(IPCC), a UN panel of climate scientists, it is likely that many tree species will not be able to change their geographic distribution fast enough to keep up with projected shifts in suitable climate and increased rates of extinctions are expected to occur.

What can be Done?

So trees (or the lack of them) are one of the key problems that must be tackled if we are to minimise the effects of climate change. It is not surprising then that many have also suggested that planting trees (carbon sinks) or reducing emissions from deforestation and degradation (REDD) are among the solutions to our climate crisis. SinksWatch believes however that while urgent action is needed to halt forest loss and restored degraded forests, both of these concepts - REDD and carbon sinks - are a dangerous distraction from tackling climate change. Instead of addressing the root causes of the forest as well as the climate crisis, REDD and carbon sinks are concepts that are linked to a global carbon market where planting trees or reducing deforestation in one place (usually the global South) justifies even more fossil fuel emissions somewhere else (usually in an industrialised country). And many of these carbon sink projects have financed the expansion of large-scale industrial tree plantations.

SinksWatch believes that efforts to tackle forest loss must not justify more fossil fuel emissions because such a trade-off would neither help avert runaway climate change not help save forests in the long run because runaway climate change will also have a major negative impact on the world's forests.

THE CLIMATES OF THE PAST AND PRESENT

The planet on which we live is an oblate spheroid, its equatorial region being warmer than the two polar areas. This temperature gradient, in conjunction with the earth's rotation, sets up a reasonably standard pattern of atmospheric circulation. Since the surface of the earth is neither all land nor all water, but divided into land masses surrounded by even larger areas of ocean, the physical properties of water tend to decrease seasonal contrasts near continental margins relative to those in interior regions. Any enlargement of continental areas would tend to increase the seasonal contrasts, induce steeper barometric gradients and thus heighten the general storminess of their interiors.

Conversely, any decrease in the continental areas, as by a lowering of their general masses with an encroachment of oceanic waters, would tend to bring about an amelioration of climate and a general stabilization of seasonal variations. As the geological record clearly indicates, there have been times when the parts of the continental masses exposed above oceanic waters increased for a while and then shrank, with long periods of relative quiescence between, often accompanied by extensive invasions of the sea.

The periods of increase of the exposed portions of the continents were, on the whole, correlated with periods of mountain building and the development of plateaus. The geological record also indicates that these periods of continental uplift were usually accompanied by intermittent glaciation, at least in the polar regions. Volcanic activity of greater or lesser intensity, especially near the continental margins, also was characteristic of such times.

The fact that we happen to be living in one of these cycles of disturbance focuses our attention on the climatic (and consequent vegetational) features of such periods rather than on the more quiescent and far longer periods that have intervened between them. As Russell has pointed out, man during his entire history has not known what it is to live in a "normal" climate. One may also add that, during, his existence, man has not witnessed what might be termed a "normal" phytogeographic pattern. If the surface of the earth were uniform there would be three parallel zones of higher-than-average precipitation, one in the equatorial region and the other two in the general regions of the 60° parallels of latitude. There also would be four zones of lower-than-average precipitation, two in the polar regions and two centering about 30° N and S latitude. Because of the rotation of the earth and the interaction of factors resulting from the positions of continental masses and oceans — and their unequal distribution — this theoretical pattern is greatly altered. Furthermore, as Rossby has noted, in the northern hemisphere the much greater percentage of land tends to disrupt the expected pattern more than it is disrupted in the southern hemisphere with its much larger areas of ocean.

There is a close correlation between effective precipitation and vegetation. Hence one may draw certain inferences regarding precipitation from a generalized map of the world's vegetation types. In spite of the heat, the precipitation near the equator is such that few desert areas are found there. Apart from polar wastelands, subhumid and essentially treeless regions tend to develop at the western continental margins at about latitudes 20°-30° and to trend inland in a northeastward direction in the northern hemisphere and southeastward in the southern hemisphere. Although local conditions such as mountain ranges modify these relationships, this, on the whole, is the general pattern.

The belts of heavy precipitation to be expected on a uniform-surfaced world at about 60° would also be turned northeastward on the continents of the northern hemisphere and southeastward in the southern hemisphere. In the skewing of these belts they would be depressed about 100 at the western continental margins. Their poleward trends would soon carry them into the polar subhumid areas; thus they would disappear as belts and become more or less regional coastal areas centering at about latitude 50°. Evidences of these are to be found in the northern hemisphere in the areas of heavy precipitation in North America in the Pacific Northwest and also in northwestern Europe. In the southern hemisphere only South America extends into sufficiently southerly latitudes for the southern superhumid area to be fully evident, as in southern Chile; but here no extended area is found, because of the southern Andes and the small total amount of land area available.

Reference to the vegetation map will indicate that the tropics, which we sometimes think of as characterized by heavy precipitation and an extensive development of rain forest, are by no means uniformly so. The less humid areas just north and south of the equatorial regions restrict the band of equatorial rain forest and other factors shift it and break it into segments.

Those areas, loosely referred to as "the tropics," actually are not characterized by a heavy effective precipitation and dense rain forest; the latter is a forest type of relatively limited extent. The montane rain forests sometimes developing on the windward slopes of mountains in frost-free regions are superficially similar to lowland rain forests, but the two would be distinguished in any detailed analysis of forest types. On the basis of theoretical climatology, those longer periods in the history of the world when its surface was relatively quiescent, with the exposed portions of the continents smaller than at present and also with somewhat subdued relief, were times of "normal" climate. Vegetational zonation poleward from the equator was evident, but without the sharp gradients of today. The air may have been slightly more humid than it now is over large areas, but the total rainfall probably was somewhat less.

The latter fact would tend to decrease the extent of the expected "standard" areas of superhumid climate, which were probably little better developed than they now are. Contrarily, with a more uniform air circulation, the subhumid zones entering the western margins of the continents, while in evidence, probably did not give rise to desert conditions except in the interiors of the largest of the land masses. Again on the basis of climatological considerations, it can be postulated that, during periods of world-wide continental depression and quiescence, the frost-free area of the world extended much farther poleward than at present. Theoretically the frost-free area would have reached to the neighborhood of latitudes 60° N and S; actually, judging from plant remains, it may have extended somewhat farther poleward on occasion. However, where one places the frost-free limits in relation to the 60° parallels depends to a certain extent on whether one does or does not believe in the theory of continental displacement.

Lastly, on purely theoretical grounds, the polar regions of such periods would be characterized by a cool but not cold climate. Snow probably fell in winter, but along the margin of any polar sea which existed the climate is not likely to have been much more severe than it now is in southwestern British Columbia, northeastern Nova Scotia, northern Scotland, or southern New Zealand. The abundant plant fossils of both the Arctic and Antarctic regions offer ample testimony that these theoretical speculations are not far wrong; if theory errs, the plants indicate that it may be on the side of conservatism.

The general configuration of the climatic pattern of the lands of a hemisphere during periods of general continental lowering. It may be taken as roughly diagrammatic of the combined land masses of North and South America (together with the present land of Antarctica), or else of the combined land masses of Europe, Asia and Africa; in both cases there is more land in the northern than in the southern hemisphere. During the periods represented, the climate of the world was more generally equable than it is today. With no polar ice, the land masses nearest the poles, as just noted, enjoyed reasonably temperate conditions, with the warm-temperate zone of transition to frost-free climate at about 60°.

The somewhat lower temperatures of what then were the temperate zones produced a condition sufficiently humid to limit the poleward extension of the midlatitude subhumid areas. The superhumid areas, centering at about the 50° parallels, soon dwindled in the interiors upon meeting the zone of low polar precipitation. In spite of the wide latitudinal extent of the frost-free zone, the equatorial region, on the whole, was probably but little warmer than at present. While there was a superhumid equatorial area, the general lowering of precipitation on a world-wide basis did not permit it to be very extensive. Low mountain ranges may have altered the precipitation in particular areas and brought about local or regional departures from the pattern as here outlined.

The corresponding general configuration during periods of continental uplift and disturbance. At such times the frost-free zone was greatly reduced in extent, the temperate regions were moved equatorward and great areas of what we now think of as having a typical "polar climate" developed in the Arctic and

Antarctic regions, accompanied by icecaps and extensive intermittent glaciation. With these changes came a marked enlargement of the subhumid areas, originating at about 30° and a considerable increase in their aridity, especially in the continental interiors, where great wind-scoured deserts were formed. During periods of extensive continental glaciation, the lowering of the temperature along the ice fronts locally — or even regionally — reduced the extent and severity of these subhumid areas, as it did in the Great Basin area of North America during the ice maxima of the Pleistocene. The maxima of periods of disturbance, when the "polar climates" and continental glaciers descended to beyond the 40° parallel, as they did in certain regions several times during the Pleistocene, but a somewhat more equable condition such as we have today. The Permian appears to have been characterized by periods of much more severe climate than was the Pleistocene.

The Forests of the Past

The development of the vegetation types — and especially the forest types — of the past. It should be held in mind throughout that the continents, in those periods when they were of more subdued relief, nevertheless were by no means topographically featureless. The incompletely eroded bases of former high mountain chains might have lingered on, even when the next cycle of mountain building was already beginning. The face of the land was varied and its elevations, if not on the magnificent scale of the present Andes or Himalayas, still were sufficient to bring about modifications in the precipitation and local climatic pattern. And these elevations — the uplands of the past — have played an important role in the development of the world's vegetation, for it becomes evident, when we consider the details, that in great measure it was on such uplands that the new groups of plants underwent their primary evolution.

CLIMATE CHANGE IMPACTS ON FOREST HEALTH

The world's climate is changing. Increased temperatures and levels of atmospheric carbon dioxide as well as changes in precipitation and in the frequency and severity of extreme climatic events are just some of the changes occurring. These changes are having notable impacts on the world's forests and the forest sector through longer growing seasons, expansion of insect species ranges, and increased frequency of forest fires. Climate change can affect forest pests and the damage they cause by: directly impacting their development, survival, reproduction and spread; altering host defences and susceptibility; and indirectly impacting the relationships between pests, their environment and other species such as natural enemies, competitors and mutualists. A deeper understanding of the complex relationships between a changing climate, forests and forest pests is vital to enable those in forest health protection and management to expect and prepare for changes in pest behaviours, outbreaks and invasions. This chapter investigates these relationships and the implications for forest health protection and management.

Specific examples where insect and pathogen lifecycles or habits have been altered by local, national or regional climatic changes are discussed. For example, the mountain pine beetle (*Dendroctonus ponderosae*) has shown decreases in generation time and winter mortality resulting in exponential population growth and major range extension in western North America. Warmer temperatures have resulted in range expansions of pests such as the pine processionary caterpillar (*Thaumetopoea pityocampa*) and the oak processionary caterpillar (*T. processionea*) in Europe and the southern pine beetle (*D. frontalis*) and red band needle blight (*Mycosphaerella pini*) in the US. Similar patterns are expected for other pathogens such as *Armillaria mellea*, and *Phytophthora cinnamomi* in Europe.

In the Democratic Peoples Republic of Korea, decreased winter temperature and lighter snow cover have contributed to reduced mortality and early emergence of overwintering *Dendrolimus spectabilis* (pine moth) with resulting loss of native *Pinus densiflora*.

The purpose of this report is to compile and summarize the present knowledge on the impacts of climate change on forests and forest pests. Since limited research has been dedicated to determining such impacts on forest pests in particular, information on non-forest pests is provided throughout the report to enable a better understanding of the potential impacts of climate change on forest health.

CURRENT KNOWLEDGE AND FUTURE EXPECTATIONS

The world's climate is changing. While there are natural climate variations, the climate change we are most concerned about is in regards to the modifications to the greenhouse effect that human activities have caused. The Intergovernmental Panel on Climate Change (IPCC), in its Fourth Assessment Report, concluded with more certainty that global climate change is unequivocal and it is widely believed to result primarily from the effects of emissions of carbon dioxide (CO_2) and other greenhouse gases (GHGs) such as methane (CH_4) and nitrous oxide (N_2O), from human activities.

This conclusion was based on a number of observations about the Earth's climate including the following.

- Global surface temperature has increased by an estimated 0.74 degrees Celsius (°C) over the past century. Over a 50 year period from 1956 to 2005 the warming trend was nearly twice that for the 100 years from 1906 to 2005. Eleven of the 12 years from 1995 to 2006 rank among the 12 hottest years on record (since 1850, when sufficient worldwide temperature measurements began). Increases are widespread globally and are greater at higher northern latitudes. Over the last 50 years, cold days and nights as well as frosts have become less frequent over most land areas, while hot days and nights and heat waves have become more frequent.

- Consistent with warming, the extent of snow and ice has decreased. Mountain glaciers and snow cover have also declined on average worldwide. The maximum area of seasonally frozen ground in the Northern Hemisphere has decreased by about 7 percent since 1900, with decreases in the spring of up to 15 percent. Satellite data since 1978 show that the extent of Arctic sea ice during the summer has shrunk by more than 20 percent.
- Since 1961, the world's oceans have been absorbing more than 80 percent of the heat added to the climate, causing ocean water to expand thereby contributing to rising sea levels. This expansion was the largest contributor to sea level rise between 1993 and 2003. Melting glaciers and losses from the Greenland and Antarctic ice sheets have also contributed to recent sea level rise. The incidence of extreme high sea level has increased at a number of sites globally since 1975.
- From 1900 to 2005, significant increases in precipitation have been observed in eastern parts of North and South America, northern Europe and northern and central Asia. The frequency of heavy precipitation events has also increased over most areas. In contrast, precipitation has declined in the Sahel, the Mediterranean, southern Africa and parts of southern Asia. Droughts have become longer and more intense worldwide and have affected larger areas since the 1970s, particularly in the tropics and subtropics.
- Evidence of an increase in intense tropical cyclone activity in the North Atlantic has been observed since about 1970 and there are suggestions of similar increased activity in some other regions.

Predictions for the Future

The IPCC has made a number of predictions for future climate change. A warming of about 0.2°C per decade is projected for the next two decades; temperature projections beyond this period depend on specific emissions scenarios. Even if the concentrations of all GHGs and aerosols were kept constant at year 2000 levels, a further warming of about 0.1°C per decade would be expected.

The full range of projected temperature increase, based on six emission scenarios, is 1.1-6.4 °C by the end of the century. The best estimate range of projected temperature increase, which extends from the midpoint of the lowest emission scenario to the midpoint of the highest, is 1.8-4.0 °C by the end of the century. It is very likely that continued GHG emissions at or above current rates would cause further warming and induce many changes in the global climate system during the 21st century that would be larger than those observed during the 20th century.

Geographically the predicted warming trends for the 21st century are believed to be similar to those observed over the past several decades whereby temperature increases are expected to be greatest over land and at most high northern latitudes, and least over the Southern Ocean (near Antarctica) and the northern North Atlantic.

It is very likely that extreme heat, heat waves and heavy precipitation events will become more frequent. Increases in high latitude precipitation are very likely, while decreases are likely in most subtropical regions such as Egypt. It is also likely that future tropical cyclones (typhoons and hurricanes) will become more intense, with higher peak wind speeds and heavier precipitation associated with warmer tropical seas. Snow cover area is projected to contract, widespread increases in thaw depth are projected over most permafrost regions and sea ice is projected to shrink in both the Arctic and Antarctic. In some projections, Arctic late-summer sea ice disappears almost entirely by the latter part of the 21st century.

Climate change is impacting the world's ecosystems and it is expected that the magnitude of these impacts will increase along with temperatures over this century. Many species and ecosystems may not be able to adapt as the effects of global warming and associated disturbances, such as floods, drought, wildfire and insect outbreaks, are compounded by other stresses such as land use change, overexploitation of resources, pollution and fragmentation of natural systems. If the global average temperature increases more than 1.5-2.5 °C, it is believed likely that approximately 20-30 percent of plant and animal species assessed so far will be at an increased risk of extinction. Major changes in ecosystem structure and function, species' ecological interactions and species' geographical ranges, with predominantly negative consequences for biodiversity and ecosystem goods and services are also projected.

IMPACTS ON FORESTS AND THE FOREST SECTOR

Climate change, in particular increased temperatures and levels of atmospheric carbon dioxide as well as changes in precipitation and in the frequency and severity of extreme climatic events, is having notable impacts on the world's forests and the forest sector.

Productivity and Health

Forest productivity and species diversity typically increase with increasing temperature, precipitation and nutrient availability, although species may differ in terms of their tolerance. As a key factor that regulates many terrestrial biogeochemical processes, such as soil respiration, litter decomposition, nitrogen mineralization and nitrification, denitrification, methane emission, fine root dynamics, plant productivity and nutrient uptake, temperature changes are likely to drastically alter forests and ecosystem dynamics in many ways. The impacts of elevated temperatures on trees and plants will vary throughout the year since warming may relieve plant stress during colder periods but increase it during hotter periods.

Moisture availability in forests will be strongly influenced by changes in both temperature and precipitation. Warmer temperatures lead to increased water losses from evaporation and evapotranspiration and can also result in reduced water use efficiency of plants. Longer, warmer growing seasons can intensify these effects resulting in severe moisture stress and drought. Such conditions can lead to reductions in the growth and health of trees although the severity of the impacts depends on the forest characteristics, age-class structure and soil depth and type. Young small plants such as seedlings and saplings are particularly susceptible whereas large trees with a more developed rooting system and greater stores of nutrients and carbohydrates tend to be less sensitive to drought, though they are affected by more severe conditions. Shallow-rooted trees and plants as well as species growing in shallow soils are more susceptible to water deficits. Deep-rooted trees can absorb water from greater depths and therefore are not as prone to water stress. Moisture stress and drought can also impact forest health by enhancing susceptibility to disturbances such as insect pests and pathogens and forest fires.

Higher atmospheric CO_2 levels result in increased growth rates and water use efficiency of plants and trees, so long as other factors such as water and nutrients (*e.g.* nitrogen, phosphorus, sulphur, some micronutrients) do not become limiting. It has been suggested however that this positive effect declines with increasing concentrations (Stone, Bhatti and Lal, 2006). However current free air CO_2 enrichment facilities have observed multi-year growth increases of 23 percent with a 175 ppm enrichment above a 375 ppm ambient CO_2 concentration. Elevated carbon dioxide levels can also result in changed plant structure such as increased leaf area and thickness, greater numbers of leaves, higher total leaf area per plant, and larger diameter stems and branches. It is important to note that plant responses to CO_2 enrichment may differ between species and local environmental conditions which are likely to result in substantial changes in the species composition and dynamics of terrestrial ecosystems. Concurrent increases in concentrations of ground-level ozone (O_3) may lower tree productivity and enhance susceptibility to pathogens, while N_2O may enhance growth in nitrogen-limited ecosystems such as boreal forests.

Distribution

Consistent responses of species and communities to climate change or 'fingerprints' are typically associated with changes in their distribution, particularly at their latitudinal or altitudinal extremes. The distribution of forest plants and trees is expected to shift northwards or to higher altitudes in response to climate warming.

In a recent study, Lenoir *et al.* (2008) compared the altitudinal distribution of 171 forest plant species between two periods 1905-1985 and 1986-2005 in Western Europe and concluded that climate warming has resulted in a significant upward shift in species optimum elevation (altitude of maximum probability of presence) averaging 29 m per decade. This study showed that climate change affects not just species' ranges at their distributional margins but at the spatial core of the distributional range of plant species. The quickest species noted to relocate were those with shorter life spans and faster reproduction cycles such as herbs, ferns and mosses; larger long-lived trees and shrubs did not show a significant shift and are thus under greater threat from the impacts of climate change because they can't adapt to local conditions quickly enough and relocate. Such distributional changes will no doubt result in forest ecosystems that are very different from what we see today. A similar study from 26 mountains in Switzerland reported that alpine flora have expanded toward the summits since the 1940s. Upward movements of tree lines have also been observed in Siberia, the Canadian Rocky Mountains, and New Zealand and northward shifts have been noted in Sweden and Eastern Canada.

The timing of such shifts however will not be solely determined by temperature but will depend on a number of factors such as the rate at which seeds can disperse into new regions that are climatically suitable (*i.e.* with proper moisture conditions, soil characteristics and nutrient availability), possible human interventions to promote movement of species, and changes to disturbance regimes.

Forest Sector

All of these impacts on trees and forests will inevitably have widespread impacts on the forest sector. Changes in the structure and functioning of natural ecosystems and planted forests (due to temperature changes and rainfall regimes) and extreme events and disasters (hurricanes, droughts, fires and pests) will have negative impacts on the productive function of forest ecosystems which in turn will affect local economies. Production patterns and trade in forestry commodities will be altered as species are grown more competitively in higher latitudes and altitudes.

Conversely, markets may be saturated due to increased mortality of trees following pest infestations as has been experienced with the mountain pine beetle in Canada. Decreased forest ecosystem services, especially water cycle regulation, soil protection and conservation of biological diversity, as a result of climate change may imply increased social and environmental vulnerability.

While climate change is likely to increase timber production and lower market prices in general, the increases in production will certainly not be evenly distributed throughout the world; some areas will experience better conditions than others. For example, forests with low productivity due to drought will likely face further decreases in productivity, while areas where temperature limits productivity may benefit from rising temperatures.

Disturbance

Forests are subjected to a variety of disturbances that are themselves strongly influenced by climate. Disturbances such as fire, drought, landslides, species invasions, insect and disease outbreaks, and storms such as hurricanes, windstorms and ice storms influence the composition, structure and function of forests. Climate change is expected to impact the susceptibility of forests to disturbances and also affect the frequency, intensity, duration, and timing of such disturbances. For example, increased fuel loads, longer fire seasons and the occurrence of more extreme fire weather conditions as a consequence of a changing climate are expected to result in increased forest fire activity. A changing climate will also alter the disturbance dynamics of native forest insect pests and pathogens, as well as facilitating the establishment and spread of non indigenous species.

Such changes in disturbance dynamics, in addition to the direct impacts of climate change on trees and forest ecosystems, can have devastating impacts particularly because of the complex relationships between climate, disturbance agents and forests. Either of these disturbances can result in forest susceptibility to other disturbances.

For example, pine forests in Central America became infested with bark beetles, primarily *Dendroctonus frontalis* in association with other *Dendroctonus* and *Ips* species, after suffering damage from Hurricane Mitch in 1998. The beetle outbreaks caused further extensive tree mortality thereby increasing fuel loads in the region's forests which severely increased the risk of wildfires. Making predictions on the future impacts of a changing climate on forest disturbances is made more difficult by these interactions.

IMPACTS ON INSECTS AND DISEASES

Changes in the patterns of disturbance by forest pests (insects, pathogens and other pests) are expected under a changing climate as a result of warmer temperatures, changes in precipitation, increased drought frequency and higher carbon dioxide concentrations. These changes will play a major role in shaping the world's forests and forest sector.

Climate change can affect forest pests and the damage they cause by: directly impacting their development, survival, reproduction, distribution and spread; altering host physiology and defences; and impacting the relationships between pests, their environment and other species such as natural enemies, competitors and mutualists.

Direct Impacts

Climate, temperature and precipitation in particular, have a very strong influence on the development, reproduction and survival of insect pests and pathogens and as a result it is highly likely that these organisms will be affected by any changes in climate. With their short generation times, high mobility and high reproductive rates it is also likely that they will respond more quickly to climate change than long-lived organ-isms, such as higher plants and mammals and thereby may be the first predictors of climate change.

Physiology

Climate influence on insects can be direct, as a mortality factor, or indirect, by influencing the rate of growth and development. Some information on the impacts of increased CO_2, and O_3, is becoming available but only for specific environments and only very partial information is available on changing UVB levels and altered precipitation regimes. For these reasons this report will focus on the impacts of temperature. Temperature is considered to be the more important factor of climate change influencing the physiology of insect pests. Precipitation however can be a very important factor in the epidemiology of many pathogens, such as *Mycosphaerella pini*, that depend on moisture for dispersal.

The magnitude of the impacts of temperature on forest pests will differ among species depending on their environment, life history, and ability to adapt. Flexible species that are polyphagous, occupy different habitat types across a range of latitudes and altitudes, and show high phenotypic and genotypic plasticity are less likely to be adversely affected by climate change than specialist species occupying narrow niches in extreme environments.

Increases in summer temperature will generally accelerate the rate of development in insects and increase their reproductive capacity while warmer winter temperatures may increase overwinter survival. Perhaps the best example of such impacts is the mountain pine beetle (*Dendroctonus ponderosae*) which has been at epidemic proportions in western Canada for several years. Successive years of mild winters have decreased the mortality of overwintering stages and generation time allowing for massive destruction of pines in the region, particularly lodgepole pine (*Pinus contorta*). Decreased snow depth associated with warmer winter temperatures may also decrease the winter survival of many forest insects that overwinter in the forest litter where they are protected by snow cover from potentially lethal low temperatures.

The impact of a change in temperature will vary depending on the climatic zone. In temperate regions, increasing temperatures are expected to decrease winter survival while in more northern regions, higher temperatures will extend the summer season thereby increasing growth and reproduction. Given the more severe environmental control, and greater predicted increases in temperature in boreal and polar regions, the impacts of temperature are expected to be greater on species from those regions than on species in temperate or tropical zones. However, Deutsch *et al.* (2008) suggest that, in the absence of ameliorating factors such as migration and adaptation, the greatest extinction risks from global warming may be in the tropics. Warming in the tropics, though proportionately smaller in magnitude, could have the most deleterious impacts because tropical insects have very narrow ranges of climatic suitability compared to higher latitude species, and are already living very close to their optimal temperature.

Some important forest insect pests have critical associations with symbiotic fungi but limited information is available on how temperature changes may affect these symbionts and thus indirectly affect host population dynamics. In some cases insect hosts and their symbionts may be similarly affected by climatic change while in other cases, hosts and symbionts may be affected asymmetrically, effectively decoupling the symbiosis.

Distribution

Climate plays a major role in defining the distribution limits of a species. With changes in climate, these limits are shifting as species expand into higher latitudes and altitudes and disappear from areas that have become climatically unsuitable. Such shifts are occurring in species whose distributions are limited by temperature such as many temperate and northern species. It is now clear that poleward and upward shifts of species ranges have occurred across many taxonomic groups and in a large diversity of geographical locations during the 20th century. Parmesan and Yohe (2003) reported that more than 1 700 Northern Hemisphere species have exhibited significant range shifts averaging 6.1 km per decade towards the poles (or 6.1 m per decade upward).

The range expansions of many lepidopterans have been particularly well documented. Parmesan *et al.* (1999) reported a poleward shift of 35-240 km for 22 out of 35 non-migratory European butterfly species during the last century. Wilson *et al.* (2005) noted that the lower elevational limits of 16 butterfly species in central Spain had risen on average by approximately 212 m in 30 years, a rise attributed to an observed 1.3 °C rise in mean annual temperature. Wilson *et al.* (2007) showed uphill shifts of approximately 293 m in butterfly communities in the Sierra de Guadarrama of central Spain between 1967-1973 and 2004-2005 as a result of climate warming. Climate change may also weaken the association between climatic and habitat suitability. Franco *et al.* (2006) concluded the importance of climate warming and habitat loss in driving local extinctions of northern species of butterflies in northern Great Britain over the past few decades. Forest pests are also occurring outside historic infestation ranges and at intensities not previously observed. Some examples of forest pest species that have responded or are predicted to respond to climate change by altering distribution include the following.

- A major epidemic of the mountain pine beetle (*Dendroctonus ponderosae*) has been spreading northwards and upwards in altitude in western Canada (British Colombia and more recently, Alberta) for several years.
- Warmer temperatures have influenced the southern pine beetle (*D. frontalis*) resulting in range expansions in the United States.
- The pine processionary caterpillar (*Thaumetopoea pityocampa*) has significantly expanded its latitudinal and altitudinal distribution in Europe.
- The oak processionary caterpillar (*T. processionea*) has shifted its distribution north in Europe during the latter half of the 20th century.
- The European rust pathogen *Melampsora allii-populina* is likely to spread northwards with increased summer temperatures.
- The root rot pathogen *Phytophthora cinnamomi* is predicted to spread into colder regions of Europe and have increased severity with climate change scenarios of increased average temperatures.

The ability of a species to respond to global warming and expand its range will depend on a number of life history characteristics, making the possible responses quite variable among species. Bale *et al.* (2002) suggested that fast-growing, non-diapausing insect species or those not dependent on low temperature to induce diapause, will respond to warming by ex-panding their distribution whereas slow-grow-ing species which need low temperatures to induce diapause (*i.e.* boreal and mountain species in the northern hemisphere) will suffer range contractions.

Range-restricted species, particularly polar and montane species, show more severe range contractions than other groups and are considered most at risk of extinction due to recent climate change. Range shifts may be limited by factors such as day length or the presence of competitors, predators or parasitoids. For example, the range expansion of insects which are very host-specific (specialists) may be limited by the slower rate of spread of their host plant species.

Phenology

Phenology is the timing of seasonal activities of plants and animals such as flowering or breeding. Since it is in many cases temperature dependent, phenology can be expected to be influenced by climate change. It is one of the easiest impacts of climate change to monitor and is by far the most documented in this regard for a wide range of organisms from plants to vertebrates. Common activities to monitor include earlier breeding or first singing of birds, earlier arrival of migrant birds, earlier appearance of butterflies, earlier choruses and spawning in amphibians and earlier shooting and flowering of plants.

Evidence of phenological changes in numerous plant and animal species as a consequence of climate change is abundant and growing. In general, spring activities have occurred progressively earlier since the 1960s and has been documented on all but one continent and in all major oceans for all well-studied marine, freshwater, and terrestrial groups.

Where life cycle events are temperature-dependent, they may be expected to occur earlier and increased temperatures are likely to facilitate extended periods of activity at both ends of the season, provided there are no other constraints present. With increased temperatures, it is expected that insects will pass through their larval stages faster and become adults earlier. Therefore expected responses in insects could include an advance in the timing of larval and adult emergence and an increase in the length of the flight period. Members of the Order Lepidoptera again provide the best examples of such phenological changes. Changes in butterfly phenology have been reported from the UK where 26 of 35 species have advanced their first appearance. First appearance for 17 species in Spain has advanced by 1-7 weeks in just 15 years. Seventy percent of 23 butterfly species in California, USA have seen an advancement of first flight date of approximately eight days per decade.

Changes in phenology - early adult emergence and an early arrival of migratory species - have also been noted for aphids in the UK. Gordo and Sanz (2005) investigated climate impacts on four Mediterranean insect species (a butterfly, a bee, a fly and a beetle) and noted that all species exhibited changes in their first appear-ance date over the last 50 years which was correlated with increases in spring tem-perature.

Parmesan and Yohe (2003) estimated that more than half (59 percent) of 1598 species investigated exhibited measurable changes in their phenologies and/or distributions over the past 20-140 years. They also estimated a mean advancement of spring events by 2.3 days/decade based on the quantitative analyses of phenological responses for these species. Root *et al.* (2003), in a similar quantitative study, estimated an advancement of 5.1 days per decade. Parmesan (2007) investigated the discrepancy between these two estimates and noted that once the differences between the studies in selection criteria for incorporating data was accounted for, the two studies supported each other, with an overall spring advancement of 2.3-2.8 days/decade found in the resulting analysis. However, in this last study, latitude explained only 4 percent of overall variation of phenological changes even though it is strongly associated with the importance of warming trends. This last observation may relate to the importance of the change in climate relative to the natural amplitude of the climate variability.

EFFECT OF FOREST CLEARFELLING ON WATER YIELD

A basic summary of short-term results obtained from paired catchment experiments conducted before 1980 was given by Bosch and Hewlett (1982). Hornbeck *et al.* (1993) and Swank *et al.* (1988) discussed longer-term results, including the effects of multiple treatments, for catchments in the north-eastern and south-eastern USA respectively. Generally, increases in water yield during the first year after treatment are roughly proportional to percentage reductions in stand basal area, provided the latter exceed a threshold of 20-25%. Bosch and Hewlett (1982) made a distinction between coniferous and deciduous hardwood forests, suggesting that increases in flow associated with the cutting of conifers were higher (40mm per 10% change in cover) than for hardwoods (25 mm per 10% change in cover).

However, the scatter around the two tentative regression lines was large, coefficients of determination modest (at 42% and 26% respectively) and no confidence limits were provided. In a later analysis of the same dataset plus an additional 50 studies (mostly from the tropics and Australia), Sahin and Hall (1996) used fuzzy linear regression analysis and obtained the following average changes in flow per 10% change in cover: conifers, 23mm; mixed hardwoods-conifers, 22mm; and hardwoods, 17-19 mm (depending whether annual precipitation exceeded 1500mm or not). Based on this fuzzy regression approach, the overall differences between the respective vegetation groups were therefore smaller than suggested earlier by Bosch and Hewlett (1982). Stednick (1996) approached the problem of the heterogeneity in results for 95 paired catchment studies from the USA by deriving separate regression equations for each of eight geographically homogeneous regions.

In doing so, a clearer picture was obtained of the regional variation in yield increases, which ranged from about 10mm per 10% change in cover in the Rocky Mountains to over 60mm in the Central Plains. Also, some of the regionally applicable regression equations had substantially higher coefficients of determination (*e.g.* 0.65 for the Pacific Northwest or Appalachia, which both had the largest number of experiments), although correlations for other regions were extremely low (*e.g.* 0.01 for the Rocky Mountains). As such, there is something to say for the approach of Trimble *et al.* (1987) who simply pooled the coniferous and mixed hardwood data of Bosch and Hewlett (1982) and added 10 data points of their own, representing reductions in annual flow from large non-experimental ('real world') catchment areas in the Piedmont of the south-eastern USA that had undergone 10-28% reforestation.

The overall coefficient of determination for the combined dataset attained the reasonable value of 0.50 by the inclusion of the Piedmont subset, with 8 of 10 data points from the latter falling within half the standard error of the estimate (compared with only 22 of 55 data points in the Bosch and Hewlett set). This finding is all the more remarkable because half the additional data involved a change in cover of less than 20%, a value below which Bosch and Hewlett (1982) had considered effects on streamflow to be more or less non-detectable (at least on small headwater streams).

Apart from contrasts in tree species and age of the vegetation, much of the variability in the change in stream-flow observed for different sites and years after a change in cover relates to differences in climate, catchment exposure and soil depth. The climatic aspect is aptly summarized by the classic statement of Hewlett (1966) that 'it takes water to fetch water'. The importance of catchment exposure is illustrated by the difference in first-year streamflow gain after cutting differently exposed deciduous hardwood forest catchments at Coweeta in the south-eastern USA.

Flows from northerly exposed catchments increased by about 130mmyear^{-1} but increases from southerly exposed catchments could be as high as 410mm year^{-1}. As for the effect of soil depth, Trimble *et al.* (1963), when discussing differences in duration of the changes in streamflow following forest clearing in West Virginia, stated that 'the deeper the soil, the longer it takes roots of new growth or the expanding root systems of residual plants to occupy the soil mantle. When the soil mantle is reoccupied, transpiration reaches maximum levels again, and increases in streamflow disappear'. The role of soil depth can be illustrated further by examining the timing of the maximum and minimum increases in flow after forest clearance during the year. At Coweeta, where precipitation is distributed evenly over the year and the loamy soils are deep (up to 6 m) and capable of storing large amounts of moisture, the soil does not become fully recharged until early spring (April-May).

As a result, the presence or absence of a forest has little effect on the amounts of flow during this time of year (the soils being full of water anyway). However, at the onset of the growing season, increased flow due to reduced water use after clearcutting becomes evident and the contrast in streamflow from forested and cleared areas increases as the growing season advances. The effect continues well into the winter months (November-January), even though the trees have lost their leaves by then and evaporation is strongly reduced. This reflects the development of a much larger soil water deficit below the forest during the summer that takes an additional 2 months (February-March) to be refilled by the winter rains. Conversely, on catchments with shallower or sandier soils (having less water storage and a lower water retention capacity), soil water recharge and depletion occurs much more rapidly and increases in flow after forest clearance follow the pattern of forest water use much more closely.

The effect of the timing of the cutting and the importance of controlling regrowth are demonstrated by a comparison of the results obtained for catchments 2 and 5 at Hubbard Brook (HB 2 and HB 5). On catchment HB 2, the forest was clearcut during the dormant season and regrowth eliminated by herbicide application the following spring. This produced a first-year increase in water yield of 347 mm. In contrast, on catchment HB 5, the removal of the trees took nearly a full year, while regrowth was not suppressed.

The resulting first-year increase in flow thus amounted to only 152 mm, or 44% of that observed for catchment HB 2. In addition, the gain in streamflow had largely disappeared after 3 years. Such contrasting findings once more illustrate the effect of regenerating vegetation on soil water levels and streamflow. However, it should be noted that regeneration was particularly vigorous in the example from Hubbard Brook as it took place mainly via sprouting. As demonstrated by comparative work in West Virginia, elevated streamflow levels lasted much longer (15-20 years) when regeneration had to originate from seeds because of previous herbicide applications than when sprouting was allowed (6-7 years).

Similarly, when a regenerating hardwood forest at Coweeta was cut a second time after 23 years (when streamflow was still about 80mmyear^{-1} above that observed for mature mixed forest), the initial increase in water yield was nearly identical to that of the first cut (375 vs. 362mm). However, the subsequent decline in (extra) water yield was distinctly faster. The difference was attributable to a more rapid recovery of stand biomass during the second regeneration period in association with the greater sprouting potential of the even-aged forest, which contained numerous small stems prior to the second cut. As a result, the (projected) total duration of the second period of increased flows was estimated to be 18 years shorter than that associated with the first cutting (estimated at about 35 years). Such long durations reflect the time required by the roots to reoccupy the very deep soils at Coweeta compared with those of the north-eastern USA where effects are more likely to last 3-9 years.

A rather different pattern of changes in forest water use and streamflow with time after clearing has been documented for native mountain ash *(Eucalyptus regnans)* forest in south-eastern Australia. Following clearing and burning, there is an initial rise in streamflow lasting 4-6 years. This is followed by a rapid decrease in flows until the regrowth is about 27 years old, after which run-off totals slowly return to predisturbance levels. The latter may take as long as 150 years. Vertessy *et al.* (1998) provided a mechanistic explanation for this long-term pattern in terms of concurrent changes in overstorey and understorey leaf biomass and transpiration rates, as well as changes in rainfall interception and evaporation from the forest floor.

Results from paired catchment experiments in the humid tropics also show a proportionality in flow increase with percentage of forest removed, as observed under humid temperate conditions. Reported first-year increases after clearfelling range from 125 mm in Nigeria to 820 mm in Peninsular Malaysia and show little relation to amounts of precipitation.

This led Bruijnzeel (1996) to suggest that perhaps the results were determined more by the degree of surface disturbance than by climatic or soil factors. However, in view of the large variability in total evapotranspiration among tropical forests, it may be rewarding to reanalyse the data in the light of contrasts in rainfall interception between regions as well.

CLIMATE CHANGE AND FOREST PEST SPECIES

Some examples of forest insect pests, diseases and other pests which have been impacted or are predicted to be impacted by climate change are presented below. Information on non-forest pests is also provided to enable a better understanding of the potential impacts of climate change on forest health.

INSECTS

Coleoptera

Agrilus pannonicus (Buprestidae): A number of Buprestid beetles of the genus *Agrilus* have been linked to oak decline. Incidences of these species have increased worldwide (both in their countries of origin and by international movement) and their impacts are being linked to host tree stress potentially caused by climate change. For example, *Agrilus pannonicus* (=*A. biguttatus* (Fabricius)) has recently been associated with a European oak decline throughout its natural range and has increased in incidence in several countries including France, Germany, Hungary, Poland and the Netherlands, and the UK where it is believed to be contributing to oak decline. Infestations can result in extensive tree mortality which, combined with other factors involved in the decline, can drastically alter the species composition of oak forests.

Dendroctonus frontalis Zimmermann (Scolytidae) - Southern pine beetle: *Dendroctonus frontalis* is considered to be one of the most damaging species of bark beetles in Central America and southern areas of North America. It is a major pest of pines and has a wide distribution occurring from Pennsylvania in the United States south to Mexico and Central America. Populations can build rapidly to outbreak proportions and large numbers of trees are killed. Initial attacks are generally on weakened trees however *D. frontalis* is capable of killing otherwise healthy trees. This beetle kills trees by a combination of two factors: girdling during construction of egg galleries; and the introduction of blue stain fungi of the genus *Ophiostoma.* Because of their short generation time, high dispersal abilities and broad distribution of suitable host trees, the southern pine beetle has the potential to respond quickly and dramatically to any changes in climate.

In October 1998, Hurricane Mitch hit Central America, causing floods and mudslides that ravaged local communities, forests and infrastructure. In the years that followed an unprecedented regionwide outbreak of pine bark beetles, mainly *D. frontalis* in association with other *Dendroctonus* and *Ips* species, destroyed over 100 000 ha of pine forest. As most of the standing dead and felled trees were left on site, fuel loads were drastically increased thereby resulting in extensive wildfires. With climate change expected to increase the frequency and severity of extreme events such as hurricanes, the potential for future devastating impacts on forests from both the initial disturbance and its cascading effects (*i.e.* other disturbances such as pest outbreaks and fire) is quite high.

Warmer temperatures attributed to climate change have also influenced the southern pine beetle resulting in range expansions in the United States. Laboratory measurements and published records of mortality in wild populations indicate that a temperature of -16 °C or less result in almost 100 percent mortality of the pest, thereby limiting its distribution in its current northern range. It was predicted that an increase in temperature of 3 °C would enable outbreaks to occur approximately 178 km farther north than in historical times (Ungerer, Ayres and Lombardero, 1999). Recent outbreaks of the southern pine beetle in northern and high-altitude ecosystems, where they were previously rare or absent, have been attributed to a warming trend of 3.3 °C in minimum winter air temperatures in the southeastern US from 1960-2004. This northern expansion is about as predicted by Ungerer, Ayres and Lombardero (1999).

The southern pine beetle has also possibly adapted ways to increase survival in cooler climes. Tran *et al.* (2007) showed, through field and laboratory studies of a northern population, that prepupae were more cold tolerant (by more than 3 °C) than pupae, adults and feeding larvae, and that the winter life stage structure was strongly biased toward this most cold-tolerant life stage. This tendency to overwinter in a cold tolerant life stage could be a coincidence however rather than a true adaptation.

Dendroctonus ponderosae (Hopkins) (Scolytidae) - Mountain pine beetle: The mountain pine beetle (*Dendroctonus ponderosae*) is the most destructive pest of mature pines in North America, particularly lodgepole pine (*Pinus contorta*). In the western United States, outbreaks have been increasing in area after several years of drought (Tkacz, Moody and Villa Castillo, 2007). A major epidemic of this pest has also been ongoing in western Canada (British Columbia (BC), and more recently, Alberta) for several years and even with large-scale efforts to mitigate the impacts of the pest, millions of trees have been killed. A record of over 10 million hectares of pines were recorded as infested during 2007 aerial overview surveys in BC, with 860 973 ha of this located in provincial parks and protected areas. It has been predicted that if the beetle continues to spread at its current rate as much as 80 per cent of mature pine in BC will be dead by 2013. The large numbers of dead and dying trees have also increased the risk of wildfires.

The problem has been exacerbated by successive years of mild winters, resulting in decreases of mortality of overwintering stages and generation time.

Their life cycle is generally completed in one year; warmer temperatures can result in two generations per year while cooler ones may results in one generation every two years (Amman, McGregor and Dolph, 1990). Drought conditions associated with warmer temperatures have also weakened the trees and increased their susceptibility to the beetles. Warmer temperatures have thus opened up previously climatically unsuitable mature pine stands to the pest.

Dendroctonus rufipennis (Kirby) (Curculionidae) - Spruce beetle: *Dendroctonus rufipennis* is a North American pest of spruce, particularly white spruce (*Picea glauca*) and black spruce (*P. mariana*) in the north, Engelmann spruce (*P. engelmannii*) and sitka spruce (*P. sitchensis*) in the west, and red spruce (*P. rubens*) in the east. It tends to attack weakened or windthrown trees and outbreaks are mostly linked to predisposing factors. As a result it can be expected that the impacts of climate change on trees and forests could enhance spruce beetle outbreaks.

In fact, Hebertson and Jenkins (2008) investigated the impact of climate on spruce beetle outbreaks in Utah and Colorado, USA between 1905 and 1996 and found that historic outbreak years in the intermountain region were related to generally warm fall and winter temperatures and drought conditions. Similarly outbreaks in both Canada (Yukon Territory) and the US (Alaska) appear to be related to extremely high summer temperatures which influenced spruce beetle population size through a combination of increased overwinter survival, a halving of the maturation time from two years to one year, and regional drought-induced stress of mature host trees.

HEMIPTERA

Aphids (Aphididae)

With short generation times and low developmental threshold temperatures, aphids are a group of insects that can be expected to be strongly influenced by environmental and climatic changes. In general, it has been predicted that aphids will appear at least eight days earlier in the spring within 50 years, though the rate of advance will vary depending on location and species. This could potentially result in greater damage to host plants depending on the phenology of host plants and natural enemies.

Zhou *et al.* (1995), for example, investigated the timing of migration in Great Britain for five aphid species (*Brachycaudus helichrysi, Elatobium abietinum, Metopolophium dirhodum, Myzus persicae, Sitobion avenae*) over a period of almost 30 years and concluded that temperature, especially winter temperature, is the dominant factor affecting aphid phenology for all species.

They found that a one degree Celsius increase in average winter temperature advanced the migration phenology by 4-19 days depending on species.

Elatobium abietinum (Walker) (Aphididae)

The green spruce aphid (*Elatobium abietinum*) is also believed likely to benefit from the increase in winter survival, leading to more intense and frequent defoliation of host spruce trees (*Picea* spp.).

This aphid is native to Europe but has also been reported in both North and South America. Infestations in Great Britain have resulted in large losses of spruce foliage and height both during the active infestation and in subsequent years.

Westgarth-Smith *et al.* (2007) showed that warm weather associated with a positive North Atlantic Oscillation (NAO) index caused spring migration of *E. abietinum* to start earlier, last longer and contain more aphids.

Positive NAO values correspond to warmer atmospheric conditions over Great Britain. Since global warming is believed to increase NAO variability, shifting the system to more positive values, this will most likely lead to further increases in aphid activity and more damage to spruce trees and forests in the area.

HYMENOPTERA

***Cephalcia arvensis* Panzer**

The spruce webspinning sawfly is monophagous on spruce (*Picea*) and endemic to the spruce range in Eurasia, where outbreaks have been seldom recorded.

From 1985-1992 however there was a sudden outbreak of the sawfly in the Southern Alps during which populations developed an annual life cycle and grew exponentially, causing repeated defoliations resulting in extensive tree death. *Cephalcia* species generally show low fecundity and have an extended diapause of a few years that is stimulated by low temperatures at pupation time.

The outbreak corresponded to a period of high temperatures and low precipitation and severe water stress for the host trees. As a result, the insect was able to adapt to the new climate resulting in lower mortality, faster development and higher feeding rates of the sawfly. In addition, the sudden increase in population density was not quickly followed by that of natural enemies, thereby allowing for unlimited population growth.

***Neodiprion sertifer* (Geoffroy) (Diprionidae)**

The European pine sawfly *Neodiprion sertifer* is an important pest species on pines in Europe, northern Asia, Japan and North America where it was introduced. It is one of the most serious defoliators of Scots pine (*Pinus sylvestris*) forests in northern Europe. Virtanen *et al.* (1996) suggested that outbreaks of the sawfly on Scots pine in eastern and northern Finland are prevented by low winter temperatures which kill eggs, and predicted that outbreaks would become more common with winter warming. A high variation in freezing avoidance of eggs was also noted which would allow *N. sertifer* to adapt to predicted climate change and spread its distribution northwards.

LEPIDOPTERA

Warmer temperatures have been linked to increasing populations of forest Lepidoptera species.

Butterflies

While only a few species of butterflies are considered to be serious forest pests, some of the best, and most researched, examples of the impacts of climate change on insect distributions and phenology have been butterflies. The geographic ranges of many species have shifted northward and upwards in elevation associated with climate warming, leading to increases in species richness at high latitudes and elevations and in some cases possible local extinction at lower altitudes.

Range expansions in butterflies have been well documented and changes in butterfly phenology have also been reported. In the UK, species have been advancing their flight periods by approximately 2-10 days for every 1 °C increase in temperature (Roy and Sparks, 2000; Menéndez, 2007). Similar changes in phenology as a response to warming has been noted in Spain where butterflies have advanced their first appearance by 1-7 weeks in 15 years and in California, USA which has seen an advancement of approximately eight days per decade (Forister and Shapiro 2003).

Some species-specific examples of the influence of climate change include the following.

- The African Monarch butterfly (*Danaus chrysippus*) has spread northward, establishing its first population in southern Spain in 1980 followed by the establishment of multiple populations along the east coast of Spain.
- Warmer temperatures have increased survival and facilitated a latitudinal and altitudinal range expansion of *Atalopedes campestris* in the western USA.
- Edith's checkerspot butterfly (*Euphydryas editha*) has shifted its distribution northwards and also upwards in altitude in North America. Populations at the northern edge of the species range in Canada and also at higher altitudes within the main range have experienced increased survival whereas populations at the southern edge in Mexico have declined.
- In Europe the speckled wood butterfly (*Parage aegeriae*) has increased its range northward beyond its original, primary host (Logan, Régnière and Powell, 2003).

- The black-veined white butterfly, *Aporia crataegi*, has expanded its altitudinal range in the mountains of the Sierra de Guadarrama of central Spain resulting in local population extinctions at lower warmer altitudes. While climate is becoming less of a limiting factor in its distribution at higher altitudes, it is however limited by the absence of host plants.

Choristoneura fumiferana

The spruce budworm, *Choristoneura fumiferana*, is a major defoliator of coniferous forests across North America. Balsam fir (*Abies balsamea*) is the preferred host but they readily attack white, red and black spruce (*Picea glauca*, *P. rubens*, *P. mariana* respectively) and may even be feed on tamarack (*Larix* spp.) and hemlock (*Tsuga* spp.). Outbreaks of this budworm can persist for 5-15 years with periods of 20-60 years in between. In eastern Canada the period of population cycle has averaged 35 years over the last 270 years. During uncontrolled outbreaks they can kill almost all trees in dense, mature stands of fir.

Climatic influences on life history traits are considered a major factor in restricting outbreaks and as a result, a changing climate is expected to impact the severity, frequency, and spatial distribution of spruce budworm outbreaks (Logan, Régnière and Powell, 2003). The success of the insects in establishing feeding sites in the spring depends on initial egg weights and synchrony of their development with that of buds of their hosts which is strongly influenced by climatic factors (Volney and Fleming, 2000, 2007). This synchronization is critical in initiating outbreaks and thus determining the intensity of damage. However the spruce budworm is able to tolerate some asynchrony between spring emergence and vegetative shoot development as the second instars have adapted morphologically and behaviourally allowing them to mine needles.

In parts of its range, particularly at northern extremes, temperature can also influence the duration of outbreaks as collapses are often associated with the loss of suitable foliage often as a result of late spring frosts. Normal collapse of the outbreak in the core range of host trees is associated with mortality caused by natural enemies late in the larval stage. Natural enemies of the spruce budworm, *C. fumiferana*, are less effective at higher temperatures and therefore climatic factors have the potential to enable massive outbreaks of this pest providing there is suitable availability of host trees.

Epirrita autumnata (Geometridae)

Epirrita autumnata is a holarctic species that has been expanding its outbreak range in some areas. In the Nordic countries of Europe, *Epirrita autumnata* outbreak cycles are typically most prevalent in northernmost and continental birch forests but during the past 15-20 years it has expanded into the coldest, most continental areas previously protected by extreme winter temperatures. This pest overwinters in the egg stage and therefore the level of egg survival is dictated by minimum winter temperatures. Virtanen, Neuvonen and Nikula (1998) investigated the relationship between *E. autumnata* egg survival and minimum winter temperatures in northernmost Finland. They predicted that climate warming would result in a two-third reduction of the area of forests with winter temperatures cold enough to keep *E. autumnata* populations low by the middle of the next century. A rise in winter temperatures therefore will likely increase the area of forest susceptible to damage by the autumnal moth.

Lymantria dispar Linnaeus

The gypsy moth, *Lymantria dispar*, is a significant defoliator of a wide range of broadleaf and even conifer trees. While low population levels can exist for many years without causing significant damage, severe outbreaks can occur resulting in severe defoliation, growth loss, dieback and sometimes tree mortality. Two strains of gypsy moth exist – the Asian strain, of which the female is capable of flight; and the European strain, of which the female is flightless.

The Asian strain is native to southern Europe, northern Africa, central and southern Asia, and Japan and has been introduced into Germany and other European countries where it readily hybridizes with the European strain. It has also been introduced but has not established in Canada, the US and the UK (London). The European strain is found in temperate forests throughout Western Europe and has been introduced into Canada and the US. The gypsy moth is considered a significant pest in both its native and introduced ranges.

There has also been a noted increase in outbreaks in areas previously unaffected by this pest such as the Channel Islands (Jersey) and new areas in the UK (Aylesbury, Buckinghamshire). In Canada the spread of the gypsy moth has so far been prevented by climatic barriers and host plant availability as well as by aggressive eradication of incipient populations (Régnière, Nealis and Porter, 2008). However it is predicted that the gypsy moth will be able to extend its range in North America as a result of higher overwinter survival of egg stages because of milder winters and higher accumulation of day degrees for larval development. Similar predictions have been made for other areas. For example, Pitt, Régnière and Worner (2007) noted an increase in the probability of establishment of the gypsy moth in New Zealand, particularly in the South Island.

Increasing atmospheric concentrations of CO_2 may also influence the severity of gypsy moth outbreaks. Larval performance on host plants grown under elevated CO_2 varies depending on the species, being reduced on some hosts such as aspen and increased on others, such as oak.

Lymantria monacha

Lymantria monacha is a major pest of broadleaved and coniferous trees in Europe and Asia. Defoliation by nun moth larvae can kill host trees especially conifers and has caused extensive losses despite intervention with biological and chemical insecticides. In parts of Europe, the occurrence of outbreaks has increased possibly as a result of the establishment of extensive pine plantations in poor quality areas or as a result of a changing climate. It has been predicted that nun moth will spread northwards in Europe because of higher accumulated day degrees and improved overwinter survival. Using modelling software, Vanhanen *et al.* (2007) predicted that climate warming would shift the northern boundary of distribution north by approximately 500-700 km and the southern edge of the range would retract northwards by 100-900 km.

Operophtera brumata Linnaeus

Operophtera brumata is distributed throughout Europe, North Africa, Japan and Siberia and has also been introduced into Canada and the US. It feeds on a variety of deciduous trees and shrubs including apricot, cherry, apple, plum, blueberry, crab-apple, sweet chestnut, red currant and black currant, oaks, maples, basswood and white elm.

Climate change is impacting the spread of the winter moth. In the Nordic countries of Europe, Jepsen *et al.* (2008) noted that *O. brumata* had been climatically restricted to more southern and near-coastal locations in the regions but warmer temperatures has resulted in expansions in its outbreak area further northeast. While increased temperatures appear to assist the winter moth expand its distribution, it appears that they do not have the same impact on its natural enemies which may allow populations of this pest to grow unchecked.

Climate change has affected the phenology of many species in different ways. In the Netherlands over the past 25 years, early spring temperatures have increased while winter temperatures have not. As a result a climate change induced asynchrony has occurred between winter moths and their host, pedunculate oak, *Quercus robur*, with eggs hatching before bud burst (van Asch and Visser, 2007). Such a situation leaves no food for the larvae resulting in starvation and death. This also has implications for other species that depend on the larvae for food such as the great tit (*Parus major*) which feed *O. brumata* caterpillars to their young. While both egg hatch and bud burst have advanced over the last 25 years, egg hatch has advanced much more leading to a decrease in synchrony from a few days to almost 2 weeks. However others have noted that, while warmer temperatures have lead to earlier egg hatch, the autumnal pupal diapause of the winter moth is prolonged at higher temperatures thereby counteracting the impact and resulting in a life cycle that is not shortened overall. Differences between observations of synchronicity between moth egg hatch and host bud burst may result from regional, intra-specific differences.

Thaumetopoea pityocampa **(Thaumetopoeidae)**

The pine processionary caterpillar, *Thaumetopoea pityocampa*, is considered one of the most important pests of pine forests in the Mediterranean region. It is a tent-making oligophagous caterpillar that feeds gregariously and defoliates various species of pine and cedar. The life cycle of the pine processionary caterpillar is typically annual but may extend over two years at high altitudes or in northern latitudes. At northern latitudes and at higher altitudes, adults emerge earlier.

Climate change is having clear impacts on the distribution of this important forest pest. Battisti *et al.* (2005) reported a latitudinal expansion in north-central France of 87 km northwards from 1972-2004 and an altitudinal shift of 110-230 m upwards in the Alps of northern Italy from 1975-2004 and attributed the expansions to reduced frequency of late frosts, which increases survival of overwintering larvae, as a result of a warming trend over the past three decades. In the last ten years the pine processionary caterpillar has spread almost 56 km northward in France.

During the summer of 2003, the warmest summer in Europe in the last 500 years, *T. pityocampa* exhibited an unprecedented expansion to high elevation pine stands in the Italian Alps, increasing its altitudinal range limit by one third of the total altitudinal expansion over the previous three decades. This unusual and fast spread has been attributed to increased nocturnal dispersal of females during the unusually warm night temperatures.

The gradual warming of the region has allowed the pest to maintain its presence at this altitude because of increased larval survival. In the Sierra Nevada mountains of southeastern Spain, *T. pityocampa* has expanded to higher elevations over the last 20 years as a result of increasing mean temperature. Relict populations of Scots pine (*Pinus sylvestris* var. *nevadensis*) occurring within this newly expanded range of the caterpillar are being increasingly attacked, particularly in warmer years. This range expansion caused by climate change has potentially devastating consequences for this endemic mountain species which is likely to suffer from the direct effects of climate change as well.

Given that the present distribution of the *T. pityocampa* is not constrained by the distribution of its hosts, that warmer winters will increase winter larval feeding activity, that the probability of lower lethal temperature will decrease, it can be expected that the improved survival and spread into previously hostile environments will continue.

Thaumetopoea processionea

Native to central and southern Europe, *Thaumetopoea processionea* is a major defoliating pest of oak. Since the late 20th century it has been expanding its range northwards and is now firmly established in Belgium, Denmark, northern France, and the Netherlands and has been reported from southern Sweden and the UK. It is believed that the northward progression of the oak processionary moth is due to improved synchrony of egg hatch and reduction of late frosts as a result of warmer temperatures.

***Zeiraphera diniana* Guenée**

The larch bud moth, *Zeiraphera diniana*, is a European pest that has been defoliating large areas of larch forests in the Alps every 8-10 years for centuries. It has an annual life cycle, overwintering as an egg on the larch branches and feeding on the needles as soon as the bud breaks. As such synchrony between egg hatch and bud burst is critical. Increased temperatures associated with climate change have affected this relationship leading to asynchrony and reduced incidences of the moth in Switzerland. It has been reported that abnormally high temperatures result in unusually high egg mortality.

IMPACTS OF THE FOREST PLANTATION ESTATE

The potential for forest plantations to partially meet demand for wood and fibre for industrial uses is increasing. According to FRA 2000, the global forest plantation area accounts for only 5 per cent of global forest cover and the industrial forest plantation estate for less than 3 per cent. However, as an indication only, forest plantations were estimated in the year 2000 to supply about 35 per cent of global roundwood and an increase to 44 per cent anticipated by 2020. If plantation development is targeted at the most appropriate ecological zones and if sustainable forest management principles are applied, forest plantations can provide a critical substitute for natural forest raw material supply. In several countries industrial wood production from forest plantations has significantly substituted for wood supply from natural forest resources. Forest plantations in New Zealand met 99 per cent of the country's needs for industrial roundwood in 1997; the corresponding figure in Chile was 84 per cent, Brazil 62 per cent and Zambia and Zimbabwe 50 per cent each. This substitution by forest plantations may help reduce logging pressure on natural forests in areas in which unsustainable harvesting of wood is a major cause of forest degradation and where logging roads facilitate access that may lead to deforestation.

Forest plantations also provide additional non-wood forest products, from the trees planted or from other elements of the ecosystem that they help to create. They contribute environmental, social and economic benefits. Forest plantations are used in combating desertification, absorbing carbon to offset carbon emissions, protecting soil and water, rehabilitating lands exhausted from other land uses, providing rural employment and, if planned effectively, diversifying the rural landscape and maintaining biodiversity.

Not all forest plantation development has positive economic, environmental, social or cultural impacts. Without adequate planning and without appropriate management, forest plantations may be grown in the wrong sites, with the wrong species/provenances, by the wrong growers, for the wrong reasons. Examples exist where natural forests have been cleared to establish forest plantation development or where customary owners of traditional lands may have been alienated from their sources of food, medicine and livelihoods. In some instances poor site/species matching and inadequate silviculture have resulted in poor growth, hygiene, volume yields and economic returns. In other instances, changes in soil and water status have caused problems for local communities. Land use conflicts can occur between forest plantation development and other sectors, particularly the agricultural sector. The negative impacts of forest plantations can draw the focus away from the fact that forest plantation resources are totally renewable and can be economically, socially, culturally and environmentally sustainable with prudent planning, management, utilization and marketing.

SELECTED FOREST PLANTATION TOPICS

On average *Eucalyptus* and *Pinus* species, which dominate industrial plantations in developing countries, have similar MAIs of 10 to 20m per hectare per year. However, many of the popular species of both genera frequently achieve much faster growth rates. Thus *Eucalyptus grandis,* which is the most widely planted *Eucalyptus* species, can achieve 40 to 50 m^3 per hectare per year and in very exceptional conditions with advanced tree improvement 100m^3 per hectare per year. Other widely planted tropical hardwoods including *Casuarina equisetifolia, Casuarina junghuhniana*, *Tectona grandis* and *Dalbergia sissoo* have MAIs of less than 15m^3 per hectare per year and frequently under 10 m^3 per hectare per year.

Climate and site have a very large impact on growth rates. The humid tropics and more fertile sites are more conducive to higher growth rates than locations with long dry seasons or infertile or degraded soil. Teak on many sites in India, for example, frequently has an MAI of 4 to 8 m^3 per hectare per year, partly because of drought combined with poor soils. Some species such as *Gmelina arborea* and some of the *Eucalyptus* species are very site sensitive. *Pinus* spp., in contrast, generally tolerate adverse conditions better and are more flexible with respect to site. Both tree breeding and silviculture have improved growth rates. Good examples are *Eucalyptus grandis* and *E. urophylla* in Brazil and *Pinus radiata* in some countries of the Southern Hemisphere.

Advanced silviculture typically includes improved nursery and establishment techniques such as good site preparation, weed control and judicious use of fertilizer. It has been suggested that growth of teak, for example, could be doubled in Kerala, India and Bangladesh and increased sixfold in Indonesia by adopting these practices.

With coppice species productivity varies with rotation, the first and second coppice rotations usually being more productive than the seedling one. The growth patterns vary among species. For example, very fast growing species such as *Gmelina arborea* can reach a peak MAI in less than 10 years, while *Pinus caribaea* var. *hondurensis* grown in Trinidad reaches maximum MAI at about 25 years and *P. radiata* at over 40 years. With *Cupressus lusitanica* in Costa Rica, the MAI maximum is reached at about 30 years.

Rotation lengths can reflect both end-use and economics. Many fast-growing *Eucalyptus, Acacia* and *Casuarina* species and *Gmelina arborea* are grown on short rotations of under 15 years as they are used primarily for pulp or woodfuel. Usual rotations in Kenya for *E. grandis* are 6 years for domestic woodfuel, 7 to 8 years for telephone poles and 10 to 12 years for industrial woodfuel. In Brazil this species is largely grown for pulp or charcoal on 5 to 10 year rotations. Species being grown for high-value sawlogs usually have longer rotations; teak (*Tectona grandis*) is grown on 50 to 70 year rotations and high-value conifers such as *Araucaria angustifolia* on 40 year rotations. Generally pines are grown on medium-length rotations of 20 to 30 years, unless grown solely for pulpwood, when shorter rotations may be adopted.

Modelling of growth, rotations, harvest yields and product mix by species is important for decision-making in forest management. One of the major obstacles to model development for planners and managers is a lack of suitable data. Data can come from a range of sources, including temporary and permanent sample plots and experiments. Experiments and protocols for obtaining data need to be carefully designed so that reliable information is obtained over the complete range of conditions to which the model is to apply. Tree Growth and Permanent Plot Information System (TROPIS), sponsored by the Centre for International Forestry Research (CIFOR), seeks to coordinate and improve access to tree growth information.

A growth and yield model developed for *Pinus elliottii* plantations in coastal Zululand, South Africa, predicts height, basal area, total stem and merchantable volume and stocking, with age on a stand level for harvest planning. In New Zealand several simulation models have been developed for *P. radiata* which predict similar variables but also include wood quality, harvesting and marketing aspects, which make it possible to link the silvicultural options to industrial use.

Valuable Hardwood Plantations

Long-rotation, slow-growing but valuable hardwood species have special technical properties, such as strength, natural durability, hardness and easy machining and appearance that make them suitable for high-value end-uses such as furniture.

These high-grade hardwoods contrast with short-rotation, fast-growing, lesser-quality woods used for woodfuel, pulpwood or reconstituted products and less demanding building timbers. In tropical countries teak (*Tectona grandis*), mahogany (*Swietenia* spp.) and rosewood (*Dalbergia* spp.) are the main hardwood plantation species, while in temperate countries oak (*Quercus* spp.), ash (*Fraxinus* spp.), cherry (*Prunus* spp.), walnut (*Juglans* spp.), tulipwood (*Jacaranda* spp.) and hard maple (*Acer* spp.) predominate.

Because many valuable hardwood species are difficult to establish because of their ecological requirements or disease or insect susceptibility, focus has been on the easier species to grow, including teak (*Tectona grandis*), Indian rosewood (*Dalbergia sissoo*) and mahogany. In 1995 the global areas of these species were 2 254 000, 626 000 and 151 000 ha, respectively.

Table. Characteristics of Valuable Hardwoods used in Tropical Areas

Use categories	*Desirable wood properties*	*Main end-uses*	*Matching valuable hardwood species*	*Comments*
Decorative timbers	Appearance, consistent quality, dimensional stability, durability, good machining, staining and finishing properties	Quality furniture and interior joinery	*Tieghemella* spp.; *Entandophragma cylindricum*, *Chorophora* spp., *Aucoumea klaineana*, *Afrormosia* spp., *Entandophragma utile*, *Mansonia* spp., *Lovoa* spp., *Khaya* spp., *Swietenia* spp., *Dalbergia* spp., *Aningeria* spp.	Highest value, competition from temperate hardwoods and MDF
High to very high-density timbers	Appearance, strength, high natural durability, availability in large sizes	Principally in construction	*Dipterocarpus* spp., *Lophira* spp., *Chlorophora* spp., *Ocotea rodiaei*	Small share of total tropical timber use
Low to medium-density utility timbers	Appearance, clear grain, natural durability, good machining properties	External joinery, shop fittings, medium-priced furniture	*Shorea* spp, *Hevea brasiliensis*, *Terminalia* spp., *Heritiera* spp.	Most commonly used, prone to competition from substitutes

They accounted for about 10 per cent of total hardwood plantations in the tropics. More than 90 per cent of teak plantations were located in Asia, mainly in Indonesia, India, Thailand, Bangladesh, Myanmar and Sri Lanka. About 95 per cent of rosewood plantations are located in India and Pakistan. The largest mahogany plantations are located in Indonesia and Fiji, which together make up about 80 per cent of the established area. The market preference for large piece sizes, slow growth and very long rotation lengths combine to reduce the attractiveness for commercial investment in these species.

This is only partially counteracted by their value. The low return on capital investment, coupled with the long wait period for this return, has made it difficult to interest private investors without supportive, secure and stable government policies. As markets demand a continuity of supply, plantations need to be on a sustainable scale within a region. Some of the less common species are not known in the marketplace. Other potential market problems are that the timber may be wrongly associated with tropical deforestation and changing fashions that often occur with decorative timbers. Niche marketing is important for valuable hardwoods.

Projections for supplies of timber from existing valuable hardwood plantations indicate that because of the age class distribution and long rotations there will not be a significant increase in supply in the next 20 years. Future promotion of quality hardwood plantations needs to emphasize choice of species with versatile end-uses, market research and development to hold on to niche markets and maintained high standards from production to marketing.

Careful site selection, use of high-quality planting materials of superior genetic origin and good silviculture are important. Planting programmes should be economically viable, environmentally appropriate and socially desirable. Incentives may also be necessary to stimulate private investment because of the long rotations.

Even though valuable hardwood plantations have the potential to reduce the pressure on natural forests, they will not prevent deforestation resulting from agricultural encroachment.

The supply of large quantities of high-value timber could perhaps undermine the value of natural forest stands and so lead to more rapid destruction. Hence it is advisable, where possible, to manage plantations and forest resources and forest products in a complementary manner.

Sustaining Productivity

It is possible not only to sustain but also to increase productivity in successive rotations. This requires clear definition of the end-use objective for forest plantation development and a holistic view in their management. There is a need to integrate strategies for genetic improvement programmes, nursery practices, site and species/provenance matching, appropriate silviculture (site preparation, establishment, weeding, fertilizing, pruning, thinning), forest protection and harvesting practices with prudent management. New Zealand and the southern United States have shown that substantial gains can be made by adopting this holistic approach.

In developing countries where resources may be constrained, highly technical solutions may not be essential but it is critical to get the fundamentals correct: careful species and provenance choice, good nursery stock, site preparation, planting techniques, weed control and, less frequently, fertilizer inputs. Once fast-growing, uniform plantations have been established, later silvicultural tending may become increasingly important, depending on the end-use objective.

Current evidence suggests that plantation production can be sustainable if foresters implement prudent genetic and silvicultural tree improvement programmes and sound management practices. There has been, however, limited long-term research on the subject; there are few definitive studies, limited to few species.

In one of the most promising studies, with *Pinus patula* in Swaziland grown intensively on about 15 year rotations, site productivity was maintained or increased over three rotations. The question of declining growth in teak (*Tectona grandis*) plantations in Indonesia and India remains unclear. How forest plantations are managed affects the chemical and physical properties of the soils and site. However, only recently have long-term studies been undertaken to evaluate these critical factors or processes.

The methods adopted for site preparation (ripping, ploughing, scarifying, bedding, windrowing, controlled burning), establishment (manual, mechanical), weeding (manual, chemical, mechanical), fertilizer application, pruning and thinning (manual and mechanical, for commercial to waste), forest protection and harvesting (manual, mechanical, clear-fell or selection) all affect the pool of nutrients in the ecosystem. Interference with the drainage, litter and recycling of organic matter and change in the physical conditions of soils during these operations are critical to long-term sustainability. Because of litter recycling and the rapid development of tree roots, plantations are used for rehabilitation of fragile and degraded lands prone to soil erosion and excessive water runoff.

Tree plantations often have higher evapotranspiration rates than grassland or agricultural crops and thus change the hydrology of the site. This can be either beneficial (for example by reducing salinity problems in some dryland conditions) or detrimental (if it reduces water required for other uses).

The rare studies of changes of productivity between rotations have concluded that negative changes have primarily been due to inappropriate or inadequate management practices or weed invasions rather than a result of the plantations themselves.

Burning and excessive cultivation in site preparation, soil compaction from mechanical operations, inappropriate harvesting techniques and poor forest protection can contribute to loss of nutrients and soil erosion, with a resultant loss in productivity of forest plantation sites. This cannot be addressed solely by addition of fertilizer, but by the adoption of the whole range of tree improvement, silviculture, protection and harvesting techniques in an integrated forest management strategy.

Plantations and Wood Energy

Woodfuels from plantations or natural or semi-natural forests are particularly important in developing countries, providing about 15 per cent of their total energy demand. Woodfuel provides about 7 per cent of energy demand for the world as a whole and in industrialized countries only 2 per cent. Woodfuel provides more than 70 per cent of energy needs in 34 developing countries and more than 90 per cent in 13 countries. Woodfuel makes up about 80 per cent of total wood use in developing countries and about 89 per cent in Africa

The prediction of a woodfuel crisis in developing countries in the 1980s was based largely on looking at supply and demand from forest plantations and natural forests. The reaction to the expected woodfuel crisis was to plant trees for this purpose, often in the form of traditional plantations. Many programme failures resulted from lack of appreciation for the complexities of bioenergy supply and demand, failure to take into account social aspects and people's needs and poor programme structures. The importance of planted trees on farmland, in villages and homesteads and along roads and waterways as a source of woodfuel supply was underestimated.

Rural communities harvest stems, branches, stumps, twigs, leaves and litter for woodfuels in chronic woodfuel supply areas. In these instances the nutrient recycling process is broken, resulting in degradation of forest plantation sites. In many rural communities in developing countries, woodfuel is considered a public free good, to be foraged from public natural and plantation forests. Often women and children collect the woodfuel at little or no cost. As a result, the growing of private forest plantations specifically for woodfuel, in which development costs and rotation cycles are involved, can be a foreign concept.

Asian studies show that forest-based supply can range from 13 per cent in the Philippines to as high as 73 per cent in Nepal. In many countries less than 50 per cent of fuelwood is from forests.

Globally, non-industrial forest plantations in 1995 were estimated to cover about 20 million hectares. This was almost 17 per cent of the world's total plantation area in 1995. A significant proportion of these plantations were planted for woodfuel and 98 per cent were in developing countries. These plantation figures do not account for trees planted outside the forest on farms or in villages, etc., nor do they consider plantations that were considered agricultural plantations, such as *Hevea* or palm plantations.

In developing countries about one-third of the total plantation estate was grown primarily for woodfuel in 1995. Three-quarters of these plantations were in Asia, where they accounted for 60 per cent of total plantation production. In

Latin America more than half of plantation production went to woodfuel; in Africa and Oceania a larger proportion of plantation production was as industrial wood. However, plantations, in general, provided only a small proportion of total woodfuel used. Uruguay is an interesting exception.

Production of woodfuel from plantations currently makes only a small contribution to energy requirements, although it is very important in some localities and countries. Plantations currently supply 5 per cent of woodfuel. Production from these non-industrial plantations is likely to double over the next 20 years, even with little expansion in area, because the age class distribution is heavily concentrated in young plantations. In an optimistic scenario where planting continues at the same rate as in the past ten years and then gradually declines, a 350 per cent increase in woodfuel production would be anticipated by 2020. By-products from wood-using industries will also contribute to increased fuelwood supply. The situation is less positive in Africa, where for a few countries declines are projected in plantation-based woodfuel production.

6

Ozone Layer and Ultraviolet-b Radiation

OZONE

In the TAR, separate estimates for RF due to changes in tropospheric and stratospheric ozone were given. Stratospheric ozone RF was derived from observations of ozone change from roughly 1979 to 1998. Tropospheric ozone RF was based on chemical model results employing changes in precursor hydrocarbons, CO and nitrogen oxides. Over the satellite era, stratospheric ozone trends have been primarily caused by the Montreal Protocol gases, and in Ramaswamy the stratospheric ozone RF was implicitly attributed to these gases. Studies since then have investigated a number of possible causes of ozone change in the stratosphere and troposphere and the attribution of ozone trends to a given precursor is less clear. Nevertheless, stratospheric ozone and tropospheric ozone RFs are still treated separately in this report. However, the RFs are more associated with the vertical location of the ozone change than they are with the agent(s) responsible for the change.

STRATOSPHERIC OZONE

The TAR reported that ozone depletion in the stratosphere had caused a negative RF of –0.15 W m^{-2} as a best estimate over the period since 1750. A number of recent reports have assessed changes in stratospheric ozone and the research into its causes. This part summarises the material from these reports and updates the key results using more recent research. Global ozone amounts decreased between the late 1970s and early 1990s, with the lowest values occurring during 1992 to 1993, and slightly increasing values thereafter. Global ozone for the period 2000 to 2003 was approximately 4% below the 1964 to 1980 average values. Whether or not recently observed changes in ozone trends are already indicative of recovery of the global ozone layer is not yet clear and requires more detailed attribution of the drivers of the changes. The largest ozone changes since 1980 have occurred during the late winter and spring over Antarctica where average total column ozone in September and October is about 40 to 50% below pre- 1980 values.

Ozone decreases over the Arctic have been less severe than have those over the Antarctic, due to higher temperature in the lower stratosphere and thus fewer polar stratospheric clouds to cause the chemical destruction. Arctic stratospheric ozone levels are more variable due to interannual variability in chemical loss and transport. The temporally and seasonally non-uniform nature of stratospheric ozone trends has important implications for the resulting RF. Global ozone decreases result primarily from changes in the lower stratospheric extratropics. Total column ozone changes over the mid-latitudes of the SH are significantly larger than over the mid-latitudes of the NH. Averaged over the period 2000 to 2003, SH values are 6% below pre-1980 values, while NH values are 3% lower. There is also significant seasonality in the NH ozone changes, with 4% decreases in winter to spring and 2% decreases in summer, while long-term SH changes are roughly 6% year round. Southern Hemisphere mid-latitude ozone shows significant decreases during the mid-1980s and essentially no response to the effects of the Mt. Pinatubo volcanic eruption in June 1991; both of these features remain unexplained. Pyle and Chipperfield assessed several studies that show that a substantial fraction of NH mid-latitude ozone trends are not directly attributable to anthropogenic chemistry, but are related to dynamical effects, such as tropopause height changes.

These dynamical effects are likely to have contributed a larger fraction of the ozone RF in the NH mid-latitudes. The only study to assess this found that 50% of the RF related to stratospheric ozone changes between 20°N to 60°N over the period 1970 to 1997 is attributable to dynamics. These dynamical changes may well have an anthropogenic origin and could even be partly caused by stratospheric ozone changes themselves through lower stratospheric temperature changes but are not directly related to chemical ozone loss. At the time of writing, no study has utilised ozone trend observations after 1998 to update the RF values presented in Ramaswamy.

However, Hansen repeated the RF calculation based on the same trend data set employed by studies assessed in Ramaswamy and found an RF of roughly –0.06 W m^{-2}. A considerably stronger RF of –0.2 ± 0.1 W m^{-2} previously estimated by the same group affected the Ramaswamy assessment. The two other studies assessed in Ramaswamy, using similar trend data sets, found RFs of –0.01 W m^{-2} and –0.10 W m^{-2}. Using the three estimates gives a revision of the observationally based RF for 1979 to 1998 to about –0.05 ± 0.05 W m^{-2}. Gauss compared results from six chemical transport models that included changes in ozone precursors to simulate both the increase in the ozone in the troposphere and the ozone reduction in the stratosphere over the industrial era. The 1850 to 2000 annually averaged global mean stratospheric ozone column reduction for these models ranged between 14 and 29 Dobson units. The overall pattern of the ozone changes from the models were similar but the magnitude of the ozone changes differed.

The models showed a reduction in the ozone at high latitudes, ranging from around 20 to 40% in the SH and smaller changes in the NH. All models have a maximum ozone reduction around 15 km at high latitudes in the SH. Differences between the models were also found in the tropics, with some models simulating about a 10% increase in the lower stratosphere and other models simulating decreases. These differences were especially related to the altitude where the ozone trend switched from an increase in the troposphere to a decrease in the stratosphere, which ranged from close to the tropopause to around 27 km. Several studies have shown that ozone changes in the tropical lower stratosphere are very important for the magnitude and sign of the ozone RF. The resulting stratospheric ozone RF ranged between –0.12 and +0.07 W m^{-2}. Note that the models with either a small negative or a positive RF also had a small increase in tropical lower stratospheric ozone, resulting from increases in tropospheric ozone precursors; most of this increase would have occurred before the time of stratospheric ozone destruction by the Montreal Protocol gases. These RF calculations also did not include any negative RF that may have resulted from stratospheric water vapour increases.

It has been suggested that stratospheric ozone during 1957 to 1975 was lower by about 7 DU relative to the first half of the 20th century as a result of possible stratospheric water vapour increases; however, these long-term increases in stratospheric water vapour are uncertain. The stratospheric ozone RF is assessed to be –0.05 ± 0.10 W m^{-2} between pre-industrial times and 2005. The best estimate is from the observationally based 1979 to 1998 RF of –0.05 ± 0.05 W m^{-2}, with the uncertainty range increased to take into account ozone change prior to 1979, using the model results of Gauss as a guide. Note that this estimate takes into account causes of stratospheric ozone change in addition to those due to the Montreal Protocol gases. The level of scientific understanding is medium, unchanged from the TAR.

Tropospheric Ozone

The TAR identified large regional differences in observed trends in tropospheric ozone from ozonesondes and surface observations. The TAR estimate of RF from tropospheric ozone was +0.35 ± 0.15 W m^{-2}. Due to limited spatial and temporal coverage of observations of tropospheric ozone, the RF estimate is based on model simulations. In the TAR, the models considered only changes in the tropospheric photochemical system, driven by estimated emission changes since pre-industrial times. Since the TAR, there have been major improvements in models. The new generation models include several Chemical Transport Models that couple stratospheric and tropospheric chemistry, as well as GCMs with on-line chemistry.

While the TAR simulations did not consider changes in ozone within the troposphere caused by reduced influx of ozone from the stratosphere, the new models include this process. This advancement in modelling capabilities and the need to be consistent with how the RF due to changes in stratospheric ozone is derived have led to a change in the definition of RF due to tropospheric ozone compared with that in the TAR. Changes in tropospheric ozone due to changes in transport of ozone across the tropopause, which are in turn caused by changes in stratospheric ozone, are now included. Trends in anthropogenic emissions of ozone precursors for the period 1990 to 2000 have been compiled by the Emission Database for Global Atmospheric Research consortium. For specific regions, there is significant variability over the period due to variations in the emissions from open biomass burning sources. For all components industrialised regions like the USA and Organisation for Economic Co-operation and Development Europe show reductions in emissions, while regions dominated by developing countries show significant growth in emissions. Recently, the tropospheric burdens of CO and NO_2 were estimated from satellite observations providing much needed data for model evaluation and very valuable constraints for emission estimates. Assessment of long-term trends in tropospheric ozone is difficult due to the scarcity of representative observing sites with long records. The long-term tropospheric ozone trends vary both in terms of sign and magnitude and in the possible causes for the change. Trends in tropospheric ozone at northern middle and high latitudes have been estimated based on ozonesonde data by WMO, Naja, Naja and Akimoto, Tarasick and Oltmans. Over Europe, ozone in the free troposphere increased from the early 20th century until the late 1980s; since then the trend has levelled off or been slightly negative.

Naja and Akimoto analysed 33 years of ozonesonde data from Japanese stations, and showed an increase in ozone in the lower troposphere between the periods 1970 to 1985 and 1986 to 2002 of 12 to 15% at Sapporo and Tsukuba and 35% at Kagoshima. Trajectory analysis indicates that the more southerly station, Kagoshima, is significantly more influenced by air originating over China, while Sapporo and Tsukuba are more influenced by air from Eurasia. At Naha a positive trend is found between 700 and 300 hPa while between the surface and 700 hPa a slightly negative trend is observed. Ozonesondes from Canadian stations show negative trends in tropospheric ozone between 1980 and 1990, and a rebound with positive trends during 1991 to 2001.

Analysis of stratosphere-troposphere exchange processes indicates that the rebound during the 1990s may be partly a result of small changes in atmospheric circulation. Trends are also derived from surface observations. Jaffe derived a positive trend of 1.4% yr^{-1} between 1988 and 2003 using measurements from Lassen Volcanic Park in California consistent with the trend derived by comparing two aircraft campaigns. However, a number of other sites show insignificant changes over the USA over the last 15 years. Over Europe and North America, observations from Whiteface Mountain, Wallops Island,

Hohenpeisenberg, Zugspitze and Mace Head show small trends or reductions during summer, while there is an increase during winter. These observations are consistent with reduced NOx emissions. North Atlantic stations indicate increased ozone. Over the North Atlantic measurements from ships show insignificant trends in ozone, however, at Mace Head a positive trend of 0.49 $\pm$ 0.19 ppb yr^{-1} for the period 1987 to 2003 is found, with the largest contribution from air coming from the Atlantic sector. In the tropics, very few long-term ozonesonde measurements are available.

At Irene in South Africa, Diab found an increase between the 1990 to 1994 and 1998 to 2002 periods of about 10 ppb close to the surface and in the upper troposphere during winter. Thompson found no significant trend during 1979 to 1992, based on Total Ozone Mapping Spectrometer satellite data. More recent observations show significant trends in free-tropospheric ozone in the tropics: 1.12 $\pm$ 0.05 ppb yr^{-1} and 1.03 $\pm$ 0.08 ppb yr^{-1} in the NH tropics and SH tropics, respectively. Ozonesonde measurements over the southwest Pacific indicate an increased frequency of nearzero ozone in the upper troposphere, suggesting a link to an increased frequency of deep convection there since the 1980s. At southern mid-latitudes, surface observations from Cape Point, Cape Grim, the Atlantic Ocean and from sondes at Lauder show positive trends in ozone concentrations, in particular during the biomass burning season in the SH. However, the trend is not accompanied by a similar trend in CO, as expected if biomass burning had increased. The increase is largest at Cape Point, reaching 20% per decade.

At Lauder, the increase is confined to the lower troposphere. Changes in tropospheric ozone and the corresponding RF have been estimated in a number of recent model studies. In addition, a multi-model experiment including 10 global models was organised through the Atmospheric Composition Change: an European Network. Four of the ten ACCENT models have detailed stratospheric chemistry. The adjusted RF for all models was calculated by the same radiative transfer model.

The normalised adjusted RF for the ACCENT models was +0.032 $\pm$ 0.006 W m^{-2} DU^{-1}, which is significantly lower than the TAR estimate of +0.042 W m^{-2} DU^{-1}. The simulated RFs for tropospheric ozone increases since 1750. Most of the calculations used the same set of assumptions about pre-industrial emissions. Emissions of NOx from soils and biogenic hydrocarbons were generally assumed to be natural and were thus not changed. In one study, pre-industrial NOx emissions from soils were reduced based on changes in the use of fertilizers. Six of the ACCENT models also made coupled climate-chemistry simulations including climate change since pre-industrial times. The difference between the RFs in the coupled climate-chemistry and the chemistry-only simulations, which indicate the possible climate feedback to tropospheric ozone, was positive in all models but generally small. A general feature of the models is their inability to reproduce the low ozone concentrations indicated by the very uncertain semi-quantitative observations during the late 19th century. Mickley tuned their model by reducing pre-industrial lightning and soil sources of NOx and increasing natural NMVOC emissions to obtain close agreement with the observations. The ozone RF then increased by 50 to 80% compared to their standard calculations. However, there are still several aspects of the early observations that the tuned model did not capture.

The best estimate for the RF of tropospheric ozone increases is +0.35 W m^{-2}, taken as the median of the RF values. The best estimate is unchanged from the TAR. The uncertainties in the estimated RF by tropospheric ozone originate from two factors: the models used and the potential overestimation of pre-industrial ozone levels in the models. The 5 to 95% confidence interval. A medium level of scientific understanding is adopted, also unchanged from the TAR.

THE OZONE LAYER

The ozone layer is a layer in Earth's atmosphere which contains relatively high concentrations of ozone (O_3). This layer absorbs 97–99% of the Sun's high frequency ultraviolet light, which is damaging to life on Earth. It is mainly located in the lower portion of the stratosphere from approximately 13 km to 40 km above Earth, though the thickness varies seasonally and geographically. The ozone layer was discovered in 1913 by the French physicists Charles Fabry and Henri Buisson. Its properties were explored in detail by the British meteorologist G. M. B. Dobson, who developed a simple spectrophotometer (the Dobsonmeter) that could be used to measure stratospheric ozone from the ground. Between 1928 and 1958 Dobson established a worldwide network of ozone monitoring stations which continues to operate today. The "Dobson unit", a convenient measure of the columnar density of ozone overhead, is named in his honour.

ORIGIN OF OZONE

The photochemical mechanisms that give rise to the ozone layer were discovered by the British physicist Sidney Chapman in 1930. Ozone in the Earth's stratosphere is created by ultraviolet light striking oxygen molecules containing two oxygen atoms (O_2), splitting them into individual oxygen atoms (atomic oxygen); the atomic oxygen then combines with unbroken O_2 to create ozone, O_3. The ozone molecule is also unstable (although, in the stratosphere, long-lived) and when ultraviolet light hits ozone it splits into a molecule of O_2 and an atom of atomic oxygen, a continuing process called the ozone-oxygen cycle, thus creating an ozone layer in the stratosphere, the region from about 10 to 50 km (32,000 to 164,000 feet) above Earth's surface. About 90% of the ozone in our atmosphere is contained in the stratosphere. Ozone concentrations are greatest between about 20 and 40 km, where they range from about 2 to 8 parts per million. If all of the ozone were compressed to the pressure of the air at sea level, it would be only a few millimeters thick.

ULTRAVIOLET LIGHT AND OZONE

Although the concentration of the ozone in the ozone layer is very small, it is vitally important to life because it absorbs biologically harmful ultraviolet (UV) radiation coming from the Sun. UV radiation is divided into three categories, based on its wavelength; these are referred to as UV-A (400–315 nm), UV-B (315–280 nm), and UV-C (280–100 nm). UV-C, which would be very harmful to all living things, is entirely screened out by ozone at around 35 km altitude. UV-B radiation can be harmful to the skin and is the main cause of sunburn; excessive exposure can also cause genetic damage, resulting in problems such as skin cancer. The ozone layer is very effective at screening out UV-B; for radiation with a wavelength of 290 nm, the intensity at the top of the atmosphere is 350 million times stronger than at the Earth's surface. Nevertheless, some UV-B reaches the surface. Most UV-A reaches the surface; this radiation is significantly less harmful, although it can potentially cause genetic damage.

DISTRIBUTION OF OZONE IN THE STRATOSPHERE

The thickness of the ozone layer—that is, the total amount of ozone in a column overhead—varies by a large factor worldwide, being in general smaller near the equator and larger towards the poles. It also varies with season, being in general thicker during the spring and thinner during the autumn in the northern hemisphere. The reasons for this latitude and seasonal dependence are complicated, involving atmospheric circulation patterns as well as solar intensity.

Since stratospheric ozone is produced by solar UV radiation, one might expect to find the highest ozone levels over the tropics and the lowest over polar regions. The same argument would lead one to expect the highest ozone levels in the summer and the lowest in the winter. The observed behaviour is very different: most of the ozone is found in the mid-to-high latitudes of the northern and southern hemispheres, and the highest levels are found in the spring, not summer, and the lowest in the autumn, not winter in the northern hemisphere. During winter, the ozone layer actually increases in depth. This puzzle is explained by the prevailing stratospheric wind patterns, known as the Brewer-Dobson circulation. While most of the ozone is indeed created over the tropics, the stratospheric circulation then transports it poleward and downward to the lower stratosphere of the high latitudes. However in the southern hemisphere, owing to the ozone hole phenomenon, the lowest amounts of column ozone found anywhere in the world are over the Antarctic in the southern spring period of September and October. The ozone layer is higher in altitude in the tropics, and lower in altitude in the extratropics, especially in the polar regions. This altitude variation of ozone results from the slow circulation that lifts the ozone-poor air out of the troposphere into the stratosphere. As this air slowly rises in the tropics, ozone is produced by the overhead sun which photolyzes oxygen molecules. As this slow circulation bends towards the mid-latitudes, it carries the ozone-rich air from the tropical middle stratosphere to the mid-and-high latitudes lower stratosphere. The high ozone concentrations at high latitudes are due to the accumulation of ozone at lower altitudes.

The Brewer-Dobson circulation moves very slowly. The time needed to lift an air parcel from the tropical tropopause near 16 km (50,000 ft) to 20 km is about 4–5 months (about 30 feet (9.1 m) per day). Even though ozone in the lower tropical stratosphere is produced at a very slow rate, the lifting circulation is so slow that ozone can build up to relatively high levels by the time it reaches 26 km.

Ozone amounts over the continental United States (25°N to 49°N) are highest in the northern spring (April and May). These ozone amounts fall over the course of the summer to their lowest amounts in October, and then rise again over the course of the winter. Again, wind transport of ozone is principally responsible for the seasonal evolution of these higher latitude ozone patterns. The total column amount of ozone generally increases as we move from the tropics to higher latitudes in both hemispheres. However, the overall column amounts are greater in the northern hemisphere high latitudes than in the southern hemisphere high latitudes. In addition, while the highest amounts of column ozone over the Arctic occur in the northern spring (March–April), the opposite is true over the Antarctic, where the lowest amounts of column ozone occur in the southern spring (September–October).

OBSERVED CHANGES IN STRATOSPHERIC OZONE

Global and hemispheric-scale variations in stratospheric ozone can be quantified from extensive observational records covering the past 20 to 30 years. There are numerous ways to measure ozone in the atmosphere, but they fall broadly into two categories: measurements of column ozone and measurements of the vertical profile of ozone. Approximately 90% of the vertically integrated ozone column resides in the stratosphere. There are more independent measurements, longer time-series, and better global coverage for column ozone. Regular measurements of column ozone are available from a network of surface stations, mostly in the mid-latitude NH, with reasonable coverage extending back to the 1960s. Near-global, continuous column ozone data are available from satellite measurements beginning in 1979. The different observational data sets can be used to estimate past ozone changes, and the differences between data sets provide a lower bound of overall uncertainty. The differences indicate good overall agreement between different data sources for changes in column ozone, and thus we have reasonable confidence in describing the spatial and temporal characteristics of past changes. Five data sets of zonal and monthly mean column ozone values developed by different scientific teams were used to quantify past ozone changes; they include ground-based measurements covering 1964–2001, and several different satellite data sets extending in time over 1979–2001. The analyses first remove the seasonal cycle from each data set and the deviations are area weighted and expressed as anomalies with respect to the period 1964–1980. The global ozone amount shows decreasing values between the late 1970s and the early 1990s, a relative minimum during 1992–1994, and slightly increasing values during the late 1990s.

Global ozone for the period 1997–2004 was approximately 3% below the 1964–1980 average values. Since systematic global observations began in the mid-1960s, the lowest annually averaged global ozone occurred during 1992–1993. These changes are evident in each of the available global data sets. No significant long-term changes in column ozone have been observed in the tropics. Column ozone changes averaged over mid-latitudes are significantly larger in the SH than in the NH; averaged for the period 1997–2001, SH values are 6% below pre-1980 values, whereas NH values are 3% lower. Also, there is significant seasonality to the NH mid-latitude losses whereas long-term SH losses are about 6% year round. The most dramatic changes in ozone have occurred during the spring season over Antarctica, with the development during the 1980s of a phenomenon known as the ozone hole.

The ozone hole now recurs every spring, with some interannual variability and occasional extreme behaviour. In most years the ozone concentration is reduced to nearly zero over a layer several kilometres deep within the lower stratosphere in the Antarctic polar vortex. Since the early 1990s, the average October column ozone poleward of 63°S has been more than 100 DU below pre-ozone-hole values with up to a 70% local decrease for periods of a week or so. Compared with the Antarctic, Arctic ozone abundance in the winter and spring is highly variable because of interannual variability in chemical loss and in dynamical transport. Dynamical variability within the winter stratosphere leads to changes in ozone transport to high latitudes, and these transport changes are correlated with polar temperature variability–with less ozone transport being associated with lower temperatures. Low temperatures favour halogen-induced chemical ozone loss.

Thus, in recent decades, halogen-induced polar ozone chemistry has acted in concert with the dynamically induced ozone variability, and has led to Arctic column ozone losses of up to 30% in particularly cold winters. In particularly dynamically active, warm winters, however, the estimated chemical ozone loss has been very small. Changes in the vertical profile of ozone are derived primarily from satellites, ground-based measurements and balloon-borne ozonesondes. Long records from ground-based and balloon data are available mainly for stations over NH mid-latitudes. Ozone profile changes over NH mid-latitudes exhibit two maxima, with decreasing trends in the upper stratosphere and in the lower stratosphere during the period 1979–2000. Ozone profile trends show a minimum near 30 km. The vertically integrated profile trends are in agreement with the measured changes in column ozone.

OBSERVED CHANGES IN ODSS

As a result of reduced emissions because of the Montreal Protocol and its Amendments and Adjustments, mixing ratios for most ODSs have stopped increasing near the Earth's surface. The response to the Protocol, however, is reflected in quite different observed behaviour for different substances. By 2003, the mixing ratios for CFC-12 were close to their peak, CFC-11 had clearly decreased, while methyl chloroform had dropped by 80% from its maximum. Halons and HCFCs are among the few ODSs whose mixing ratios were still increasing in 2000. Halons contain bromine, which is on average 40 to 50 times more efficient on a per-atom basis at destroying stratospheric ozone than chlorine.

However, growth rates for most halons have steadily decreased during recent years. Furthermore, increases in tropospheric bromine from halons have been offset by the decline observed for methyl bromide since 1998. Atmospheric amounts of HCFCs continue to increase because of their use as CFC substitutes. Chlorine from HCFCs has increased at a fairly constant rate of 10 ppt Cl yr–1 since 1996, although HCFCs accounted for only 5% of all chlorine from long-lived gases in the atmosphere by 2000. The ODPs of the most abundant HCFCs are only about 5–10% of those of the CFCs. The behaviour of CFC-11 is representative of the behaviour of CFCs in general, and the behaviour of HCFC-22 and HFC-134a is representative of HCFCs in general. The most rapid growth rates of CFC-11 occurred in the 1970s and 1980s. The largest emissions were in the NH, and concentrations in the SH lagged behind those in the NH, consistent with an inter-hemispheric mixing time scale of about 1 to 2 years.

In recent years, following the implementation of the Montreal Protocol, the observed growth rate has declined, concentrations appear to be at their peak and, as emissions have declined, the inter-hemispheric gradient has almost disappeared. In contrast, HCFC-22 and HFC-134a concentrations are still growing rapidly and there is a marked inter-hemispheric gradient. Ground-based observations suggest that by 2003 the cumulative totals of both chlorine- and bromine-containing gases regulated by the Montreal Protocol were decreasing in the lower atmosphere. Although tropospheric chlorine levels peaked in the early 1990s and have since declined, atmospheric bromine began decreasing in 1998. The net effect of changes in the abundance of both chlorine and bromine on the total ozone-depleting halogens in the stratosphere is estimated roughly by calculating the equivalent effective stratospheric chlorine. The calculation of EESC includes consideration of the total amount of chlorine and bromine accounted for by long-lived halocarbons, how rapidly these halocarbons degrade and release their halogen in the stratosphere and a nominal lag time of three years to allow for transport from the troposphere into the stratosphere. The tropospheric observational data suggest that EESC peaked in the mid-1990s and has been decreasing at a mean rate of 22 ppt yr^{-1} over the past eight years. Direct stratospheric measurements show that stratospheric chlorine reached a broad plateau after 1996, characterized by variability.

OBSERVED CHANGES IN STRATOSPHERIC AEROSOLS, WATER VAPOUR, METHANE AND NITROUS OXIDE

In addition to ODSs, stratospheric ozone is influenced by the abundance of stratospheric aerosols, water vapour, methane and nitrous oxide. During the past three decades, aerosol loading in the stratosphere has primarily reflected the effects of a few volcanic eruptions that inject aerosol and its gaseous precursors into the stratosphere. The most noteworthy of these eruptions are El Chichón and Pinatubo. The 1991 Pinatubo eruption likely had the largest impact of any event in the 20th century, producing about 30 Tg of aerosol that persisted into at least the late 1990s. Current aerosol loading, which is at the lowest observed levels, is less than 0.5 Tg, so the Pinatubo event represents nearly a factor of 100 enhancement relative to non-volcanic levels. The source of the non-volcanic stratospheric aerosol is primarily carbonyl sulfide and there is general agreement between the aerosols estimated by modelling the transformation of observed OCS to sulphate aerosols and observed aerosols.

However, there is a significant dearth of SO_2 measurements, and the role of tropospheric SO_2 in the stratospheric aerosol budget–while significant–remains a matter of some uncertainty. Because of the high variability of stratospheric aerosol loading it is difficult to detect trends in the non-volcanic aerosol component. Trends derived from the late 1970s to the current period are likely to encompass a value of zero. The recent Stratospheric Processes And their Role in Climate Assessment of Upper Tropospheric and Stratospheric Water Vapour provided an extensive review of data sources and quality for stratospheric water vapour, together with detailed analyses of observed seasonal and interannual variability. The longest continuous reliable data set is at a single location is based on balloon-borne frost-point hygrometer measurements and dates back to 1980. Over the period 1980–2000, a statistically significant positive trend of approximately 1% yr–1 is observed at all levels between about 15 and 26 km in altitude. However, although a linear trend can be fitted to these data, there is a high degree of variability in the infrequent sampling, and the increases are neither continuous nor steady. Long-term increases in stratospheric water vapour are also inferred from a number of other ground-based, balloon, aircraft and satellite data sets spanning approximately 1980–2000, although the time records are short and the sampling uncertainty is high in many cases.

Global stratospheric water vapour measurements have been made by the Halogen Occultation Experiment satellite instrument for more than a decade. Interannual changes in water vapour derived from HALOE data show excellent agreement with Polar Ozone and Aerosol Measurement satellite data, and also exhibit strong coherence with tropical tropopause temperature changes. The Boulder and HALOE data show reasonable agreement for the early part of the record but there is an offset after 1997, with the Boulder data showing higher values than HALOE measurements. As a result, linear trends derived from these two data sets for the period 1992–2004 show very different results. The reason for the differences between the balloon and satellite data is unclear at present, but the discrepancy calls into question interpretation of water vapour trends derived from short or infrequently sampled data records.

It will be important to reconcile these differences because these data sets are the two longest and most continuous data records available for stratospheric water vapour. It is a challenge to explain the magnitude of the water vapour increases seen in the Boulder frost-point data. Somewhat less than half the observed increase of about 10% per decade can be explained as a result of increasing tropospheric methane. The remaining increase could be reconciled with a warming of the tropical tropopause of approximately 1 K per decade, assuming that air entered the stratosphere with water vapour in equilibrium with ice. However, observations suggest that the tropical tropopause has cooled slightly for this time period, by approximately 0.5 K per decade, and risen slightly in altitude by about 20 m per decade. Although regional-scale processes may also influence stratospheric water vapour, there is no evidence for increases in tropopause temperature in these regions either. From this perspective, the extent of the decadal water vapour increases inferred from the Boulder measurements is inconsistent with the observed tropical tropopause cooling, and this inconsistency limits confidence in predicting the future evolution of stratospheric water vapour. The atmospheric abundance of methane has increased by a factor of about 2.5 since the pre-industrial era. Measurements of methane from a global monitoring network showed increasing values through the 1990s, but approximately constant values during 1999–2002. Changes in stratospheric methane have been monitored on a global scale using HALOE satellite measurements since 1991. The HALOE data show increases in lower stratospheric methane during 1992–1997 that are in reasonable agreement with tropospheric increases during this time.

In the upper stratosphere the HALOE data show an overall decrease in methane since 1991, which is likely attributable to a combination of chemical and dynamical influences. Measurements of tropospheric N_2O show a consistent increase of about 3% per decade. Because tropospheric air is transported into the stratosphere, these positive N_2O trends produce increases in stratospheric reactive nitrogen which plays a key role in ozone photochemistry. Measurements of stratospheric column NO_2 from the SH and the NH mid-latitudes show long-term increases of approximately 6% per decade and these are reconciled with the N_2O changes by considering effects of changing levels of stratospheric ozone, water vapour and halogens.

OBSERVED TEMPERATURE CHANGES IN THE STRATOSPHERE

There is strong evidence of a large and significant cooling in most of the stratosphere since 1980. Recent updates to the observed changes in stratospheric temperature and to the understanding of those changes have been presented of WMO. Current long-term monitoring of stratospheric temperature relies on satellite instruments and radiosonde analyses. The Microwave Sounding Unit and Stratospheric Sounding Unit instruments record temperatures in several 10–15 km thick layers between 17 and 50 km in altitude. Radiosonde trend analyses are available up to altitudes of roughly 25 km. Determining accurate trends with these data sets is difficult. In particular the radiosonde coverage is not global and suffers from data quality concerns, whereas the satellite trend data is a result of merging data sets from several different instruments. It reveals a strong imprint of 1 to 2 years of warming following the volcanic eruptions of Agung, El Chichón and Mt. Pinatubo. When these years are excluded from long-term trend analyses, a significant global cooling is seen in both the radiosonde and the satellite records over the last few decades. This cooling is significant at all levels of the stratosphere except the 30 km level in the SSU record. The largest global cooling is found in the upper stratosphere, where it is fairly uniform in time at a rate of about 2 K per decade. In the lower stratosphere this long-term global cooling manifests itself as more of a step-like change following the volcanic warming events. The cooling also varies with latitude.

The lower stratosphere extratropics show a cooling of 0.4–0.8 K per decade in both hemispheres, which remains roughly constant throughout the year. At high latitudes most of the cooling, up to 2 K per decade for both hemispheres, occurs during spring. Much recent progress has been made in modelling these temperature trends. Models range from one-dimensional fixed dynamical heating rate calculations to three-dimensional coupled chemistry-climate models; many of their findings are presented later in this stage. FDH is a simple way of determining the radiative response to an imposed change whilst 'fixing' the background dynamical heating to its climatological value; this makes the calculation of temperature change simpler, as only a radiation model needs to be used. Changes in dynamical circulations simulated by models can lead to effects over latitudinal bands, which vary between models.

However, dynamics cannot easily produce a global mean temperature change. Because of this, global mean temperature is radiatively controlled and provides an important focus for attribution. For the global mean in the upper stratosphere the models suggest roughly equal contributions to the cooling from ozone decreases and carbon dioxide increases. In the global mean mid-stratosphere there appears to be some discrepancy between the SSU and modelled trends near 30 km: models predict a definite radiative cooling from carbon dioxide at these altitudes, which is not evident in the SSU record.

In the lower stratosphere the cooling from carbon dioxide is much smaller than higher in the stratosphere. Although ozone depletion probably accounts for up to half of the observed cooling trend, it appears that a significant cooling from another mechanism may be needed to account for the rest of the observed cooling.

One possible cause of this extra cooling could be stratospheric water vapour increases. However, stratospheric water vapour changes are currently too uncertain to pinpoint their precise role. Tropospheric ozone increases have probably contributed slightly to lower stratospheric cooling by reducing upwelling thermal radiation; one estimate suggests a cooling of 0.05 K per decade at 50 hPa and a total cooling of up to 0.5 K over the last century. The springtime cooling in the Antarctic lower stratosphere is almost certainly nearly all caused by stratospheric ozone depletion.

However, the similar magnitude of cooling in the Arctic spring does not seem to be solely caused by ozone changes, which are much smaller than in the Antarctic; interannual variability may contribute substantially to the cooling observed in the Arctic wintertime. One mechanism for altering temperatures in the upper troposphere and lower stratosphere comes from the direct radiative effects of halocarbons. In contrast with the role of carbon dioxide, halocarbons can actually warm the upper tropospheric and lower stratospheric region.

In summary, stratospheric temperature changes over the past few decades are significant and there are clear quantifiable features of the contributions from ozone, carbon dioxide and volcanism in the past stratospheric temperature record. More definite attribution of the causes of these trends is limited by the short timeseries. Future increases of carbon dioxide can be expected to substantially cool the upper stratosphere.

However, this cooling could be partially offset by any future ozone increase. In the lower stratosphere, both ozone recovery and halocarbon increases would warm this region compared with the present. Any changes in stratospheric water vapour or changes in tropospheric conditions, such as high-cloud properties and tropospheric ozone, would also affect future temperatures in the lower stratosphere. Furthermore, circulation changes can affect temperatures over sub-global scales, especially at midand polar latitudes. Several studies have modelled parts of these expected temperature changes.

OZONE DEPLETION

The ozone layer can be depleted by free radical catalysts, including nitric oxide (NO), nitrous oxide (N_2O), hydroxyl (OH), atomic chlorine (Cl), and atomic bromine (Br). While there are natural sources for all of these species, the concentrations of chlorine and bromine have increased markedly in recent years due to the release of large quantities of man-made organohalogen compounds, especially chlorof-luorocarbons (CFCs) and bromofluorocarbons. These highly stable compounds are capable of surviving the rise to the stratosphere, where Cl and Br radicals are liberated by the action of ultraviolet light.

Each radical is then free to initiate and catalyze a chain reaction capable of breaking down over 100,000 ozone molecules. The breakdown of ozone in the stratosphere results in the ozone molecules being unable to absorb ultraviolet radiation. Consequently, unabsorbed and dangerous ultraviolet-B radiation is able to reach the Earth's surface. Ozone levels, over the northern hemisphere, have been dropping by 4% per decade. Over approximately 5% of the Earth's surface, around the north and south poles, much larger (but seasonal) declines have been seen; these are the ozone holes. In 2009, nitrous oxide (N_2O) was the largest ozone-depleting substance emitted through human activities.

Regulation

In 1978, the United States, Canada and Norway enacted bans on CFC-containing aerosol sprays that are thought to damage the ozone layer. The European Community rejected an analogous proposal to do the same. In the U.S., chlorofluorocarbons continued to be used in other applications, such as refrigeration and industrial cleaning, until after the discovery of the Antarctic ozone hole in 1985. After negotiation of an international treaty (the Montreal Protocol), CFC production was sharply limited beginning in 1987 and phased out completely by 1996. On August 2, 2003, scientists announced that the depletion of the ozone layer may be slowing down due to the international ban on CFCs. Three satellites and three ground stations confirmed that the upper atmosphere ozone depletion rate has slowed down significantly during the past decade. The study was organized by the American Geophysical Union. Some breakdown can be expected to continue due to CFCs used by nations which have not banned them, and due to gases which are already in the stratosphere. CFCs have very long atmospheric lifetimes, ranging from 50 to over 100 years, so the final recovery of the ozone layer is expected to require several lifetimes.

Compounds containing C–H bonds have been designed to replace the function of CFC's (such as HCFC), since these compounds are more reactive and less likely to survive long enough in the atmosphere to reach the stratosphere where they could affect the ozone layer. While being less damaging than CFC's, HCFC's also have a significant negative impact on the ozone layer. HCFC's are also being phased out.

WHY IS THE OZONE LAYER IMPORTANT?

Ozone's unique physical properties allow the ozone layer to act as our planet's sunscreen, providing an invisible filter to help protect all life forms from the Sun's damaging ultraviolet (UV) rays. Most incoming UV radiation is absorbed by ozone and prevented from reaching the Earth's surface. Without the protective effect of ozone, life on Earth would not have evolved in the way it has.

WHAT CAUSES OZONE DEPLETION?

It was first suggested, by Drs. M. Molina and S. Rowland in 1974, that a man-made group of compounds known as the chloro-fluorocarbons (CFCs) were likely to be the main source of ozone depletion. However, this idea was not taken seriously until the discovery of the ozone hole over Antarctica in 1985. Chlorofluorocarbons are not "washed" back to Earth by rain or destroyed in reactions with other chemicals. They simply do not break down in the lower atmosphere and they can remain in the atmosphere from 20 to 120 years or more. As a consequence of their relative stability, CFCs are instead transported into the stratosphere where they are eventually broken down by ultraviolet radiation, releasing free chlorine.

The chlorine becomes actively involved in the process of destruction of ozone. The net result is that two molecules of ozone are replaced by three of molecular oxygen, leaving the chlorine free to repeat the process:

$$Cl + O_3 \rightarrow ClO + O_2$$

$$ClO + O \rightarrow Cl + O_2$$

Ozone is converted to oxygen, leaving the chlorine atom free to repeat the process up to 100,000 times, resulting in a reduced level of ozone. Bromine compounds, or halons, can also destroy stratospheric ozone. Compounds containing chlorine and bromine from manmade synthetic compounds are known as industrial halocarbons.

HOW LONG HAS OZONE DEPLETION BEEN OCCURRING?

Based on data collected since the 1950s, scientists have determined that ozone levels were relatively stable until the late 1970s. Severe depletion over the Antarctic has been occurring since 1979 and a general downturn in global ozone levels has been observed since the early 1980s.

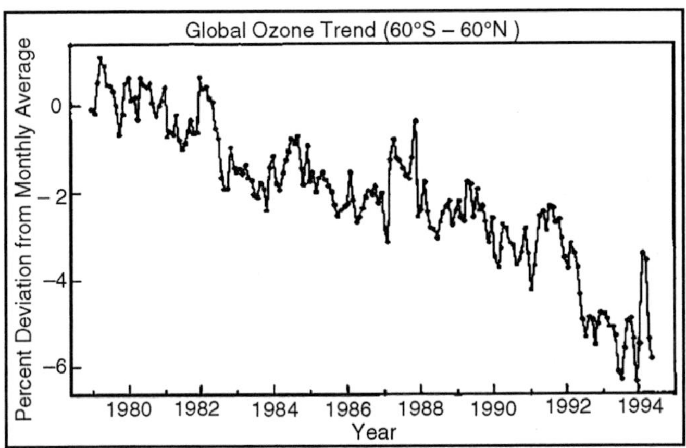

HOW MUCH OF THE OZONE LAYER HAS BEEN DEPLETED AROUND THE WORLD?

Global ozone levels have declined an average of about 3% between 1979 and 1991. This rate of decline is about three times faster than that recorded in the 1970s. In addition to Antarctica, ozone depletion now affects almost all of North America, Europe, Russia, Australia, New Zealand, and a sizeable part of South America. Short term losses of ozone can be much greater than the long term average. In Canada, ozone depletion is usually greatest in the late winter and early spring. In 1993, for example, average ozone values over Canada were 14% below normal from January to April.

OZONE IN THE ATMOSPHERE AND ITS ROLE IN CLIMATE

A number of important concepts concerning the ozone layer and its role in the stratosphere. Ozone, like water vapour and carbon dioxide, is an important and naturally occurring greenhouse gas; that is, it absorbs and emits radiation in the thermal infrared, trapping heat to warm the Earth's surface. In contrast to the so-called wellmixed greenhouse gases, stratospheric ozone has two distinguishing properties. First, its relatively short chemical lifetime means that it is not uniformly mixed throughout the atmosphere and therefore its distribution is controlled by both dynamical and chemical processes. In fact, unlike the WMGHGs, ozone is produced entirely within the atmosphere rather than being emitted into it. Second, it is a very strong absorber of short wavelength UV radiation.

The ozone layer's absorption of this UV radiation leads to the characteristic increase of temperature with altitude in the stratosphere and, in consequence, to a strong resistance to vertical motion. As well as ozone's role in climate it also has more direct links to humans: its absorption of UV radiation protects much of Earth's biota from this potentially damaging short wavelength radiation. In contrast to the benefits of stratospheric ozone, high surface ozone values are detrimental to human health. The distribution of ozone in the atmosphere is maintained by a balance between photochemical production and loss, and by transport between regions of net production and net loss. A number of different chemical regimes can be identified for ozone. In the upper stratosphere, the ozone distribution arises from a balance between production following photolysis of molecular oxygen and destruction via a number of catalytic cycles involving hydrogen, nitrogen and halogen radical species. The halogens arise mainly from anthropogenic ODSs. In the upper stratosphere, the rates of ozone destruction depend on temperature and on the concentrations of the radical species. In the lower stratosphere, reactions on aerosols become important. The distribution of the radicals can be affected by heterogeneous and multiphase chemistry acting on condensed matter. At the low temperature of the wintertime polar lower stratosphere, this is the chemistry that leads to the ozone hole.

The large-scale circulation of the stratosphere, known as the Brewer-Dobson circulation, systematically transports ozone poleward and downward. Because ozone photochemical reactions proceed quickly in the sunlit upper stratosphere, this transport has little effect on the ozone distribution there as ozone removal by transport is quickly replenished by photochemical production. However, this transport leads to significant variations of ozone in the extra-tropical lower stratosphere, where the photochemical relaxation time is very long and ozone can accumulate on seasonal time scales. Due to the seasonality of the Brewer- Dobson circulation, ozone builds up in the extra-tropical lower stratosphere during winter and spring through transport, and then decays photochemically during the summer when transport is weaker. The columnozone distribution is dominated by its distribution in the lower stratosphere and reflects this seasonality. Furthermore, planetary waves are stronger in the Northern Hemisphere than in the Southern Hemisphere, because of the asymmetric distribution of the surface features that, in combination with surface winds, force the waves.

The stratospheric Brewer- Dobson circulation is stronger during the NH winter, and the resulting extra-tropical build-up of ozone during the winter and spring is greater in the NH than in the SH. Variations in the Brewer-Dobson circulation also influence polar temperatures in the lower stratosphere; stronger wave forcing coincides with enhanced circulation and higher polar temperatures. Since temperature affects ozone chemistry, dynamical and chemical effects on column ozone thus tend to act in concert and are coupled. Human activities have led to changes in the atmospheric concentrations of several greenhouse gases, including tropospheric and stratospheric ozone and ODSs and their substitutes. Changes to the concentrations of these gases alter the radiative balance of the Earth's atmosphere by changing the balance between incoming solar radiation and outgoing infrared radiation. Such an alteration in the Earth's radiative balance is called a radiative forcing. This report, past IPCC reports and climate change protocols have universally adopted the concept of radiative forcing as a tool to gauge and contrast surface climate change caused by different mechanisms. Positive radiative forcings are expected to warm the Earth's surface and negative radiative forcings are expected to cool it. Changes in carbon dioxide provide the largest radiative forcing term and are expected to be the largest overall contributor to climate change. In contrast with the positive radiative forcings due to increases in other greenhouse gases, the radiative forcing due to stratospheric ozone depletion is negative. Halocarbons are particularly effective greenhouse gases in part because they absorb the Earth's outgoing infrared radiation in a spectral range where energy is not removed by carbon dioxide or water vapour.

Halocarbon molecules can be many thousands of times more efficient at absorbing the radiant energy emitted from the Earth than a molecule of carbon dioxide, which explains why relatively small amounts of these gases can contribute significantly to radiative forcing of the climate system. Because halocarbons have low concentrations and absorb in the atmospheric window, the magnitude of the direct radiative forcing from a halocarbon is given by the product of its tropospheric mixing ratio and its radiative efficiency. In contrast, for the more abundant greenhouse gases there is a nonlinear relationship between the mixing ratio and the radiative forcing. Since 1970 the growth in halocarbon concentrations and the changes in ozone concentrations have been very significant contributors to the total radiative forcing of the Earth's atmosphere.

Because halocarbons have likely caused most of the stratospheric ozone loss, there is the possibility of a partial offset between the positive forcing of halocarbon that are ODSs and the negative forcing from stratospheric ozone loss. The climate impacts of ozone changes are not confined to the surface: stratospheric ozone changes are probably responsible for a significant fraction of the observed cooling in the lower stratosphere over the last two decades and may alter atmospheric dynamics and chemistry. Further, it was predicted that depletion of stratospheric ozone would lead to a global increase in erythemal UV radiation at the surface of about 3%, with much larger increases at high latitudes; these predicted high latitude increases in surface UV dose have indeed been observed.

CURRENT UNDERSTANDING OF PAST OZONE CHANGES

A number of processes were described that can influence the distribution of and changes in stratospheric ozone. Here we discuss the current understanding of past ozone changes, based in part on numerical models that attempt to include these processes. A hierarchy of models of increasing sophistication is used. Some models include relatively few processes whereas others attempt to include many more. The models have their different uses, and strengths. The material here relies heavily on that report, which deals with these topics in much greater detail.

MID-LATITUDE OZONE DEPLETION

In the upper stratosphere the observed ozone depletion over the last 25 years is statistically robust in the sense that its value is not sensitive to small differences in the choice of the time period analysed. This behaviour accords with the fact that in this region of the atmosphere, ozone is under photochemical control and thus only weakly affected by dynamical variability. We therefore expect upper-stratospheric ozone abundance to reflect long-term changes in temperature and in the abundance of species that react chemically with ozone. Over the past 25 years there have been large changes in the abundance of halogen compounds, and the extent of the observed ozone decrease is consistent with the observed increase in anthropogenic chlorine, as originally predicted by Molina and Rowland and Crutzen.

In particular, the vertical and latitudinal profiles of ozone trends in the upper stratosphere are reproduced by 2-D photochemistry models. In the upper stratosphere the attribution of ozone loss in the last couple of decades to anthropogenic halogens is clear-cut. The 2-D models indicate that changes in halogens make the largest contribution to the observed loss of about 7% per decade. The observed cooling of the upper stratosphere will have reduced the rate of ozone destruction in this region. If we take the observed variation of ozone with temperature in the upper stratosphere, then a cooling of 2 K per decade should have led to an increase in ozone of about 2% per decade, partially compensating the loss caused by halogens. Note, however, that in the upper stratosphere radiative- transfer models suggest roughly equal contributions to the cooling from ozone decreases and carbon dioxide increases; the changes in the ozone-temperature system are nonlinear. Changes in CH_4 and N_2O will have also played a small role here. Note also that although 2-D models have been widely used in many ozone assessments, they have important limitations in terms of, for example, their ability to include the full range of feedbacks between chemistry and dynamics, or their ability to reproduce the polar vortex.

They are expected to be most accurate in the upper stratosphere, where dynamical effects on ozone are weakest. However, the changes in upper-stratospheric ozone represent only a small contribution to the total changes in column ozone observed over the last 25 years, except in the tropics where there is no statistically significant trend in column ozone. Most of the column ozone depletion in the extratropics occurs in the lower stratosphere, where the photochemical time scale for ozone becomes long and the ozone distribution is sensitive to dynamical variability as well as to chemical processes. This complicates the problem of attribution of the observed ozone decline. Strong ozone variability and complex coupled interactions between dynamics and chemistry, which are not separable in a simple manner, make attribution particularly difficult.

A number of 2-D models contributed of WMO and a schematic showing their simulations of column ozone between 60°S and 60°N from 1980 to 2050. As there are no significant trends in tropical ozone, the observed and modelled changes are attributable to mid-latitudes. The shaded area shows the results of 2-D photochemistry models forced by observed changes in halocarbons, other source gases and aerosols from 1980 to 2000. Overall, the models broadly reproduce the long-term changes in mid-latitude column ozone for 1980–2000, within the range of uncertainties of the observations and the model range. The spread in the model results comes mainly from their large spread over the SH mid-latitudes, and is at least partly a result of their treatment of the Antarctic ozone hole. In addition, the agreement between models and observations over 60°S to 60°N hides some important disagreements within each hemisphere. In particular, models suggest that the chemical signal of ozone loss following the major eruption of the Mt. Pinatubo volcano in the early 1990s should have been symmetric between the hemispheres, but observations show a large degree of inter-hemispheric asymmetry in mid-latitudes. Changes in atmospheric dynamics can also have a significant influence on NH mid-latitude column ozone on decadal time scales. Natural variability, changes in greenhouse gases and changes in column ozone itself are all likely to contribute to these dynamical changes. Furthermore, because chemical and dynamical processes are coupled, their contributions to ozone changes cannot be considered in isolation. This coupling is especially complex and difficult to understand with regard to dynamical changes in the tropopause region.

There is an observed relationship between column ozone and several tropospheric circulation indices, including tropopause height. Over time scales of up to about one month, it is the dynamical changes that cause the ozone changes, whereas on longer time scales feedbacks occur and the causality in the relationship becomes unclear. Thus, although various tropospheric circulation indices have changed over the last 20 years in the NH in such a way as to imply a decrease in column ozone, this inference is based on an extrapolation of short-time-scale correlations to longer time scales, which may not be valid. The tropopause, stratospheric PWD drives the seasonal winter-spring ozone build-up in the extratropics, and has essentially no interannual memory. It follows that the observed decrease in NH PWD in the late winter and spring has likely contributed to the observed decrease in NH column ozone over the last 20 to 25 years.

The effect of changes in PWD on the ozone distribution is understood in general terms, but its quantification in observations is relatively crude. The seasonality of the long-term changes in mid-latitude column ozone differs between hemispheres. In the NH, the maximum decrease is found in spring, and it decays through to late autumn. Fioletov and Shepherd have shown that in the NH the ozone decreases in summer and early autumn are the photochemically damped signal of winter-spring losses, and thus arise from the winter-spring losses without any need for perturbed chemistry in the summertime. The same seasonality is not enough for explaining the summer and early autumn ozone decreases observed in the SH mid-latitudes because they are comparable with the winter-spring losses and therefore point to the influence of transport of ozone-depleted air into mid-latitudes following the break-up of the ozone hole.

In summary, the vertical, latitudinal and seasonal characteristics of past changes in mid-latitude ozone are broadly consistent with the understanding that halogens are the primary cause of these changes. However, to account for decadal variations it is necessary to include consideration of the interplay between dynamical and chemical effects as well as the impact of variations in aerosol loading; our inadequate quantitative understanding of these processes limits our predictive capability.

Winter-Spring Polar Depletion

Polar ozone depletion in the winter-spring period is generally considered separately from mid-latitude depletion, because of the extremely severe depletion that can occur in polar regions from heterogeneous chemistry on PSCs. We consider first the Antarctic, and then the Arctic. The Antarctic ozone hole represents the most striking example of ozone depletion in the atmosphere. It developed through the 1980s as chlorine loading increased, and has recurred every year since then. The Antarctic ozone hole has been clearly attributed to anthropogenic chlorine through field campaigns and photochemical modelling.

An ozone hole now occurs every year because, in addition to the availability of anthropogenic chlorine, wintertime temperatures in the Antarctic lower stratosphere are always low enough for chlorine activation to occur on PSCs prior to the return of sunlight to the vortex in the spring, and the air is always sufficiently isolated within the vortex for it to become strongly depleted of ozone. There is nevertheless dynamically induced variability in the extent and severity of the ozone hole. The most dramatic instance of such variability occurred in 2002, when the Antarctic stratosphere experienced its first observed sudden warming. The sudden warming split the vortex in two and halted the development of the ozone hole that year. Such events occur commonly in the NH, as a result of the stronger planetary-wave forcing in the NH, but had never before been seen in the SH, although there had been previous instances of disturbed winters.

Although unprecedented, this event resulted from dynamical variability and was not indicative of ozone recovery; indeed, the ozone hole in 2003 was back to a severity characteristic of the 1990s. Given the fairly predictable nature of the Antarctic ozone hole, its simulation constitutes a basic test for models. Twodimensional models are not expected to represent the ozone hole well because they cannot properly represent isolation of polar vortex air and their estimates of Antarctic ozone loss vary greatly. However, an emerging tool for attribution is the 3-D chemistry-climate model. These models include an on-line feedback of chemical composition to the dynamical and radiative components of the model. The CCMs currently available have been developed with an emphasis on the troposphere and stratosphere. They consider a wide range of chemical, dynamical and radiative feedback processes and can be used to address the question of why changes in stratospheric dynamics and chemistry may have occurred as a result of anthropogenic forcing. The models specify WMGHGs, source gases, aerosols and often SSTs, but otherwise run freely and exhibit considerable interannual variability. Thus in addition to issues of model accuracy, it is necessary to consider how representative are the model simulations. In recent years a number of CCMs have been employed to examine the effects of climate change on ozone. They have been used for long-term simulations to try to reproduce observed past changes, such as ozone and temperature trends. The models have been run either with fixed forcings to investigate the subsequent 'equilibrium climate' or with time-varying forcings.

The current CCMs can reproduce qualitatively the most important atmospheric features with respect to the mean conditions and the seasonal and interannual variability; the long-term changes in the dynamical and chemical composition of the upper troposphere and the stratosphere are also in reasonable agreement with observations. But a reasonable reproduction of the timing of the ozone hole and an adequate description of total column ozone do not necessarily mean that all processes in the CCMs are correctly captured. There are obvious discrepancies between the models themselves, and between model results and detailed observations. Some of the differences among the models are caused by the fact that the specific model systems employed differ considerably, not only in complexity but also in the vertical extent of the model domain and in the horizontal and vertical resolutions. An important difference between models and observations originates to a large extent from the cold bias that is found in the high latitudes of many CCMs, particularly near the tropopause and in the lower stratosphere.

The low-temperature bias of the models is generally largest in winter over the South Pole, where it is of the order of 5–10 K; this magnitude of bias could be significant in controlling planetary- wave dynamics and restricting the interannual variability of the models. A low-temperature bias in the lower stratosphere of a model has a significant impact not only on model heterogeneous chemistry but also on the transport of chemical species and its potential change due to changes in circulation. Moreover, changes in stratospheric dynamics alter the conditions for wave forcing and wave propagation which in turn influence the seasonal and interannual variability of the atmosphere. The reasons for the cold bias in the models are still unknown. This bias has been reduced in some models, for example by considering non-orographic gravity-wave drag schemes but the problem is not yet solved. Without its resolution, the reliability of these models for attribution and prediction is reduced. Despite these potential problems, CCMs simulate the development of the Antarctic ozone hole reasonably well. The models examined in the intercomparison of Austin et al. agree with observations of minimum Antarctic springtime column ozone over 1980–2000 within the model variability, confirming that the Antarctic ozone hole is indeed a robust response to anthropogenic chlorine. The cold-pole biases of some models seem not to affect minimum column ozone too much, although some modelled minima are significantly lower than observed.

None of the models exhibited a sudden warming as seen in 2002, although several exhibit years with a disturbed vortex, as seen in their interannual variability in minimum column ozone. However, minimum total ozone is not the only diagnostic of the ozone hole. Many models significantly underestimate the area of the ozone hole. In contrast with the Antarctic, winter-spring ozone abundance in the Arctic exhibits significant year-to-year variability. This variability arises from the highly disturbed nature of Arctic stratospheric dynamics, including relatively frequent sudden warmings. Because dynamical variability affects both transport and temperature, dynamical and chemical effects on ozone are coupled.

The dynamical variability is controlled by stratospheric PWD; years with strong PWD, compared with years with weak PWD, have stronger downwelling over the pole, and thus higher temperatures, a weaker vortex and more ozone transport. Because low temperatures and a stronger vortex tend to favour chemical ozone loss in the presence of elevated halogen levels, it follows that dynamical and chemical effects tend to act in concert. In the warmest years there is essentially no chemical ozone loss and ozone levels are similar to those seen pre-1980. The chemical ozone loss was calculated for cold years using various methods of WMO, and the agreement between the methods provided confidence in the estimates.

The calculations showed a roughly linear relationship between chemical ozone loss and temperature. Furthermore, the in situ chemical ozone loss was found to account for roughly one half of the observed ozone decrease between cold years and warm years. Given the high confidence in the estimates of chemical ozone loss, the remaining half can be attributed to reduced ozone transport, as is expected in cold years with weak PWD, a strong vortex, and weak downwelling over the pole. From this it can be concluded that in cold years, chlorine chemistry doubled the ozone decrease that would have occurred from transport alone. Compared with other recent decades, the 1990s had an unusually high number of cold years, and these led to low values of Arctic ozone. In more recent years, Arctic ozone has been generally higher, although still apparently below pre- 1980 values. This behaviour does not reflect the time evolution of stratospheric chlorine loading. Rather, it reflects the meteorological variability in the presence of chlorine loading. In this respect the Arctic ozone record needs to be interpreted in the context of the meteorology of a given year, far more than is generally the case in the Antarctic.

This interpretation of Arctic ozone changes is consistent with the fact that the observed decrease in Arctic stratospheric temperature over 1980–2000 cannot be explained from direct radiative forcing due to ozone depletion or changes in green- house gases alone, although they do make a contribution. The inference is that the observed springtime cooling was mainly the result of decadal meteorological variability in PWD. This is in contrast with the Antarctic, where the observed cooling in November and the prolonged persistence of the vortex have been shown to be the result of the ozone hole. There is seen to be little impact from CO_2 at these altitudes over this time period. In the Arctic, the cooling induced by ozone loss is too small and too late in the season to account for the observed cooling. Rather, the observed cooling is required in order to initiate severe Arctic ozone depletion. It is not known what has caused the recent decadal variations in Arctic temperature. Rex et al. have argued that the value of VPSC in cold years has systematically increased since the 1960s.

However, the Arctic exhibits significant decadal variability and it is not possible to exclude natural variability as the cause of these changes. These considerations have implications for the attribution of past changes. No matter how good a CCM is and how well its climate-change experiments are characterized, it cannot exactly reproduce the real atmosphere because the real atmosphere is only one possible realization of a chaotic system. The best one can expect, even for a perfect model is that the observations fall within the range of model-predicted behaviours, just as to appropriate statistical criteria. Whether this permits a meaningful prediction depends on the relative magnitudes and time scales of the forced signal and the natural noise. Whereas the evolution of Antarctic ozone is expected to be fairly predictable over decadal time scales, it is not at all clear whether this is the case in the Arctic.

Thus, it is not a priori obvious that even a perfect CCM would reproduce the decreases in Arctic ozone observed over the past 20 years, for example. Bearing these caveats in mind, the simulations of past Arctic minimum ozone from the CCMs considered in Austin et al. The models seem generally to have a positive bias with respect to the observations over the same period. None of the CCMs achieve column ozone values as low as those observed, and the modelled ozone trends are generally smaller than the observed trend. However, the range of variability exhibited by each of the models is considerable, and it is therefore difficult to say that the models are definitely deficient on the basis of the ozone behaviour alone; more detailed diagnostics are required and with observations of other chemical species. In summary, the development of the Antarctic ozone hole through the 1980s was a direct response to increasing chlorine loading, and the severity of the ozone decrease has not changed since the early 1990s, although there are year-to-year variations.

In the Arctic, the extent of chemical ozone loss due to chlorine depends on the meteorology of a given winter. In winters that are cold enough for the existence of PSCs, chemical ozone loss has been identified and acts in concert with reduced transport to give Arctic ozone depletion. There was particularly severe Arctic ozone depletion in many years of the 1990s as a result of a series of particularly cold winters. It is not known whether this period of low Arctic winter temperatures is just natural variability or a response to changes in greenhouse gases.

THE MONTREAL PROTOCOL, FUTURE OZONE CHANGES AND THEIR LINKS TO CLIMATE

The halogen loading of the stratosphere increased rapidly in the 1970s and 1980s. As a result of the Montreal Protocol and its Amendments and Adjustments, the stratospheric loading of chlorine and bromine is expected to decrease slowly in the coming decades, reaching pre-1980 levels some time around 2050. If chlorine and bromine were the only factors affecting stratosphere ozone, we would then expect stratospheric ozone to 'recover' at about the same time. Over this long time scale, the state of the stratosphere may well change because of other anthropogenic effects, in ways that affect ozone abundance. For example, increasing concentrations of CO_2 are expected to further cool the stratosphere, and therefore to influence the rates of ozone destruction.

Any changes in stratospheric water vapour, CH_4 and N_2O, all of which are difficult to predict quantitatively, will also affect stratospheric chemistry and radiation. In addition, natural climate variability including, volcanic eruptions, can affect ozone on decadal time scales. For these reasons, 'recovery' of stratospheric ozone is a complicated issue. A number of model calculations to investigate recovery were reported in WMO, using both 2-D and 3-D models. The main results from that assessment, and new studies reported since then, are summarized here. As with the discussion of past ozone changes, we first discuss mid-latitude changes and then polar changes.

Mid-Latitude Ozone

Predictions of future mid-latitude ozone change in response to expected decreasing halogen levels have been extensively discussed in WMO based primarily on simulations with eight separate 2-D photochemistry models that incorporated predicted future changes in halogen loading. Several scenarios for future changes in trace climate gases were also incorporated in these simulations, although the ozone results were not particularly sensitive to which scenario was employed. The models generally predict a minimum in global column ozone in 1992–1993 following the eruption of Mt. Pinatubo, followed by steady increases. The latter evolution is primarily determined by changes in atmospheric chlorine and bromine loading, which reaches a maximum in approximately 1995 and then slowly decreases.

There is, however, a large spread in the times the models predict that global ozone will return to 1980 levels, ranging from 2025 to after 2050. The spread of results arises in part from the differences between the models used. Two of the 2-D models used in the simulations included interactive temperature changes caused by increasing greenhouse gases, which result in a long-term cooling of the stratosphere. Cooling in the upper stratosphere results in slowing of the gas-phase chemical cycles that destroy ozone there and consequently these two models show a greater increase with time of total ozone than the other models. An important caveat is that most of the past column ozone change has occurred in the lower stratosphere, and future temperature feedback effects are more complicated and uncertain in the lower stratosphere.

Furthermore, these effects involve changes in transport, heterogeneous chemistry and polar processes, which are not accurately simulated in 2-D models. At present there is considerable uncertainty regarding the details of temperature feedbacks on future midlatitude column ozone changes. Because 2-D models can be expected to capture the main chemical processes involved in ozone recovery, reliable predictions of future changes in the coupled ozone-climate system require the use of CCMs, since they consider possible changes in climate feedback mechanisms. Despite some present limitations, these models have been employed in sensitivity studies to assess the global future development of the chemical composition of the stratosphere and climate. The results have been compared with each other to assess the uncertainties of such predictions and have been documented in international assessment reports. However most of the attention has been focused on polar ozone. Two dynamical influences appear to be related to NH mid-latitude ozone decreases over 1980–2000: a decrease in PWD and an increase in tropopause height.

These mechanisms likewise have the potential to influence future ozone. If the past dynamical changes represent natural variability, then one cannot extrapolate past dynamical trends, and ozone recovery could be either hastened or delayed by dynamical variability. If, on the other hand, the past dynamical changes represent the dynamical response to WMGHG-induced climate change, then one might expect these changes to increase in magnitude in the future. This would decrease future ozone levels and delay ozone recovery.

In the case of tropopause height changes, to the extent that they are themselves caused by ozone depletion, the changes should reverse in the future.

Another potential dynamical influence on ozone arises via water vapour. Stratospheric water vapour is controlled by two processes: methane oxidation within the stratosphere and the transport of water vapour into the stratosphere. The latter depends in large part on the temperature of the tropical tropopause, although the precise details of this relationship remain unclear. In the future, changes in tropical tropopause temperature could conceivably affect the water vapour content of the stratosphere. Unfortunately, our ability to predict future dynamical influences on ozone is very poor, for two reasons. First, dynamical processes affecting ozone exhibit significant temporal variability and are highly sensitive to other aspects of the atmospheric circulation.

This means that the WMGHG-induced signal is inherently difficult to isolate or to represent accurately in CCMs, whereas the climate noise is relatively high. Second, there are still uncertainties in the performance of the CCMs, which are the tools needed to address this question. Predictions of WMGHG-induced dynamical changes by CCMs do not even agree on the sign of the changes: some models predict a weakened PWD, which would decrease mid-latitude ozone by weakening transport, whereas others predict a strengthened PWD. Note that these dynamical changes would also affect the lifetime of stratospheric pollutants, with stronger PWD leading to shorter lifetimes Thus, the direct effect of PWD change on ozone transport would be reinforced over decadal time scales by altered chemical ozone loss arising from PWD-induced changes in stratospheric chlorine loading. To summarize the results from 2-D models, upper-stratospheric ozone in the mid-latitudes will recover to pre-1980 levels well before stratospheric chlorine loading returns to pre- 1980 levels, and could even overshoot pre-1980 levels because of CO_2-induced upper-stratospheric cooling. However, details of this future evolution are sensitive to many factors and there is significant divergence between different models. In terms of mid-latitude column ozone, where changes are expected to be dominated by ozone in the lower stratosphere, 2-D models all predict a steadily increasing ozone abundance as halogen levels decrease.

While these models usually include a detailed treatment of polar chemistry, they do not include a detailed treatment of polar dynamics and mixing to mid-latitudes, or of dynamical aspects of climate change. Any future changes in stratospheric circulation and transport, or a large Pinatubo-like volcanic eruption, could have the potential to affect global column ozone, both directly and indirectly via chemical processes. Quantitative prediction is clearly difficult. At this stage, CCM simulations provide sensitivity calculations, which allow for the exploration of some examples of possible future evolution.

Polar Ozone

As noted earlier, 2-D models do not provide a realistic treatment of polar processes, so predictions of polar ozone rely principally on CCMs. An intensive inter-comparison of CCM predictions, the first of its kind, was performed of WMO. It is important to recognize that, apart from model deficiencies, CCM predictions are themselves subject to model variability and thus must be viewed in a statistical sense. Because of computer limitations, a large number of simulations with a single CCM cannot be carried out, so the models cannot yet be generally employed for ensemble runs. The approach taken in Austin et al. was to regard the collection of different CCMs as representing an ensemble. The collection included, moreover, a mixture of transient and time-slice runs.

For the Antarctic, where the CCMs all reproduce the development of the ozone hole in a reasonably realistic manner. Considering all the models, there is an overall consensus that the recovery of the Antarctic ozone hole will essentially follow the stratospheric chlorine loading. There is a hint in of a slight delay in the recovery compared with the peak in chlorine loading, presumably because of a cooling arising from the specified increase in WMGHG concentrations.

Austin et al. estimated that the recovery of the Antarctic ozone layer can be expected to begin any year within the range 2001 to 2008. This would mean that Antarctic ozone depletion could slightly increase within the next few years despite a decrease in stratospheric chlorine loading, because lower temperatures would increase chlorine activation. However there is considerable natural variability, so it is also quite possible that the most severe Antarctic ozone hole has already occurred. In the Arctic, where ozone depletion is more sensitive to meteorological conditions, the picture drawn by the CCMs is not consistent. Whereas most CCMs also predict a delayed start of Arctic ozone recovery, others indicate a different development: they simulate an enhanced PWD that produces a more disturbed and warmer NH stratospheric vortex in the future. This 'dynamical heating' more than compensates for the radiative cooling due to enhanced greenhouse-gas concentrations.

Under these circumstances, and in combination with reduced stratospheric chlorine concentrations, polar ozone in these CCMs recovers within the next decade to values measured before the start of stratospheric ozone depletion. However, none of the current models suggest that an 'ozone hole', similar to that observed in the Antarctic, will occur over the Arctic. Analyses of model results show that the choice of the prescribed SSTs seems to play a critical role in the planetary-wave forcing. Predicted SSTs vary in response to the same forcings that are already included in CCMs. Realistic predictions of future SSTs are therefore a necessary prerequisite for stratospheric circulation predictions, at least in the NH extratropics. The differences between the currently available CCM results clearly indicate the uncertainties in the assessment of the future development of polar ozone. In the NH the differences are more pronounced than in the SH. Part of these differences arises from the highly variable nature of the NH circulation, as is reflected in the past record.

Until now, the different transient model predictions have been based on single realizations, not ensembles, so it is not yet possible to determine whether the differences between the models. The time-slice simulations give some indication of the range of natural variability that is possible, and allow for a wide range of future possibilities. Nevertheless there are also significant uncertainties arising from the performance of the underlying dynamical models. Cold biases have been found in the stratosphere of many of the models, with obvious direct consequences for chemistry. In consequence, at the current stage of model development, uncertainties in the details of PSC formation and sedimentation might be less important than model temperature biases for simulating accurate ozone amounts, although it is possible that the models underestimate the sensitivity of chemical ozone loss to temperature changes. However, because it has been shown that denitrification does contribute significantly to Arctic ozone loss and is very sensitive to temperature, both problems have to be solved before a reliable estimate of future Arctic ozone losses is possible. Further investigations and model developments, combined with data analysis, are needed in order to reduce or eliminate these various model deficiencies and quantify the natural variability. In summary, the Antarctic ozone hole is expected to recover more or less following the decrease in chlorine loading, and return to 1980 levels in the 2045 to 2055 time frame. There may be a slight delay arising from WMGHG-induced cooling, but natural variability in the extent of the ozone hole is sufficiently large that the most severe ozone hole may have already occurred, or may occur in the next five years or so.

With regard to the Arctic, the future evolution of ozone is potentially sensitive to climate change and to natural variability, and will not necessarily follow strictly the chlorine loading. There is uncertainty in even the sign of the dynamical feedback to WMGHG changes. Numerical models like CCMs are needed to make sensitivity studies to estimate possible future changes, although they are currently not fully evaluated and their deficiencies are obvious. Therefore, the interpretation of such 'predictions' must be performed with care. Progress will result from further development of CCMs and from comparisons of results between models and with observations. This will help to get a better understanding of potential feedbacks in the atmosphere, thereby leading to more reliable estimates of future changes.

ULTRAVIOLET RADIATION

WHAT IS ULTRAVIOLET RADIATION

Ultraviolet radiation is one form of radiant energy coming from the sun. The various forms of energy, or radiation, are classified just as to wavelength, measured in nanometres (one nm is a millionth of a millimetre). The shorter the wavelength, the more energetic the radiation. In order of decreasing energy, the principal forms of radiation are gamma rays, X rays, UV (ultraviolet radiation), visible light, infrared radiation, microwaves, and radio waves.

There are three categories of UV radiation:

1. UV-A, between 320 and 400 nm
2. UV-B, between 280 and 320 nm
3. UV-C, between 200 and 280 nm

HOW HARMFUL IS UV?

Generally, the shorter the wavelength, the more biologically damaging UV radiation can be if it reaches the Earth in sufficient quantity. UV-A is the least damaging (longest wavelength) form of UV radiation and reaches the Earth in greatest quantity. Most UV-A rays pass right through the ozone layer in the stratosphere. UV-B radiation can be very harmful. Fortunately, most of the sun's UV-B radiation is absorbed by ozone in the stratosphere. UV-C radiation is potentially the most damaging because it is very energetic. Fortunately, all UV-C is absorbed by oxygen and ozone in the stratosphere and never reaches the Earth's surface. In summary, the danger from ultraviolet radiation comes mainly from the UV-B range of the spectrum, although UV-A poses some risk if exposure is long enough, or the sunshine is particularly strong.

INFLUENCES ON UV RADIATION REACHING THE EARTH?

Although the ozone layer is the one constant defence against UV penetration, several other factors can have an effect:

- *Latitude*: The sun's rays are the most intense near the equator where they impact the Earth's surface at the most direct angle. Season: During winter months, the sun's rays strike at a more oblique angle than they do in the summer. This means that all solar radiation travels a longer path through the atmosphere to reach the Earth, and is therefore less intense.
- *Time of day*: Daily changes in the angle of the sun influence the amount of UV radiation that passes through the atmosphere. When the sun is low in the sky, its rays must travel a greater distance through the atmosphere and may be scattered and absorbed by water vapour and other atmospheric components. The greatest amount of UV reaches the Earth around midday when the sun is at its highest point.
- *Altitude*: The air is thinner and cleaner on a mountain top - more UV reaches there than at lower elevations.
- *Cloud cover*: Clouds can have a marked impact on the amount of UV radiation that reaches the Earth's surface; generally, thick clouds block more UV than thin cloud cover.
- *Rain*: Rainy conditions reduce the amount of UV transmission.

- *Air pollution*: Like clouds, urban smog can reduce the amount of UV radiation reaching the Earth.
- *Land cover*: Incoming UV radiation is reflected from most surfaces. Snow reflects up to 85 per cent, dry sand and concrete can reflect up to 12%. Water reflects only five per cent. Reflected UV can damage people, plants, and animals just as direct UV does.

BASIC EFFECTS OF UV-B RADIATION ON ORGANISMS AND THEIR PROTECTIVE RESPONSES

Enhanced UV-B radiation can have many direct and indirect effects on organisms. However, organisms have developed mechanisms of protection and mitigation of UV-B radiation damage. General deleterious effects include production of active oxygen species and free radicals, DNA damage and, for plants, partial inhibition of photosynthesis. Protective responses include radiation shielding due to structural or pigment changes and specific damage repair systems. Although photochemical lesions of DNA and proteins and damage as a result of active oxygen species and free radicals may occur, many of the effects of UV-B radiation may be expressed through increased regulation rather than sustained damage, In order for UV radiation to be effective in most organisms, it must effectively penetrate into the tissues and be absorbed. Structural and biochemical changes induced by enhanced levels of UV-B radiation ultimately modify the penetration of UV radiation into plants and other organisms.

The UV shielding in most animals is thought to be quite effective in minimizing UV-B damage, but this should be further examined. For example, different stages of insect larvae may be less well protected by UV-absorbing pigments. In plants, a certain amount of UV-screening pigments may be constitutive, and additional UV-absorbing compounds can be synthesized when plants are exposed to increased levels of UV radiation. This will naturally be important in reducing the penetration of UV-B radiation to underlying tissues. Experimental mutant plants that lack these pigments are very sensitive to natural sunlight UV-B. Other adjustments in plant leaves after exposure to increased UV-B radiation may also contribute to a heightened UV defence. At the structural level, increased leaf thickness is often induced by UV-B radiation that reduces UV-B penetration to internal leaf tissues.

Ultraviolet radiation penetration varies among different plant species and this may be reflected in the sensitivity of these species. Penetration of UV-B was found to be greatest in herbaceous dicotyledons and was progressively less in woody dicotyledons, grasses and conifers. The UV penetration also changes with leaf age; younger leaves attenuate UV-B radiation less than do the more mature leaves, as was shown for some conifers. Of the different kinds of molecular damage, radiation damage to DNA is potentially dangerous to cells, because a single photon hit in a single molecule may have dramatic, sometimes even lethal effects. Many different types of DNA damage are known that result from free radicals and reactive oxygen species formed in various photochemical processes. The two most common UV-B induced DNA lesions are the cyclobutane pyrimidine dimers and photoproducts which are pyrimidine adducts. These two types of lesions differ from other DNA lesions in that many organisms living in sunlit habitats possess special enzymes that can effectively repair many of these lesions in the presence of visible light and favourable temperatures. Some DNA repair systems can also operate without light. Much of the research in this area has been conducted under laboratory conditions, but the level of DNA lesions in intact plants has also been measured under field conditions. While these studies indicate effective repair of DNA damage, the UV component of sunlight is still sufficient to result in some level of persistent damage. Low temperature can slow this enzymatic repair of DNA damage.

Therefore, plants, cold-blooded animals and microbes in cold environments may suffer from a less favourable balance between damage and repair than others. Unfortunately, these environments overlap with those exposed to the greatest ozone depletion. When exposure to increased UV radiation leads to stimulation of UV-absorbing compounds in plant tissues, another protective effect can result from the antioxidant properties that certain of the compounds confer. Enhanced levels of UV-B radiation appear to selectively stimulate those flavonoids with potential antioxidant properties. This selective enhancement can be up to 500%. At present, it is not known how extensive this selective induction is within the plant kingdom. Many genes in plants, animals and microorganisms are regulated by UV-B, and changes in UV-B may have important consequences by altered gene action. The mechanisms of how the organism perceives UV-B radiation and how signals are transduced are not yet well understood. Active oxygen can be one trigger for altered gene activity.

No matter what the triggering agent, altered gene activity is important, since UV-B radiation is involved in changes of gene expression which are reflected in many aspects of plant function. For example, an increased amount of UV-B radiation results in enhanced synthesis of UVscreening pigments and is due to the expression of particular genes. It appears that the effects of UV-B radiation on photosynthesis, growth and development of plants are caused by altered gene action. This is currently a topic of intensive research. Decreased elongation also may be due to UV-induced destruction of the plant hormone auxin, that absorbs in the UV-B range and could be photodegraded by high levels of UV-B radiation. Oxidative enzymes, such as the peroxidases, the activity of which is increased by enhanced UV-B radiation, also may be involved in plant hormone-regulated growth responses, as shown in sunflower and rice plants. The levels of another plant hormone, ethylene, which causes greater radial growth and less elongation, are increased after UV-B irradiation in sunflower seedlings and cultured shoots of pear seedlings. Changes in hormone levels ultimately may be due to UV-B-induced gene expression, but this remains to be demonstrated.

THE BIOLOGICAL EFFECTIVENESS OF CHANGES IN SUNLIGHT

The biological effectiveness of solar UV-B radiation needs to be taken into account in assessing what ozone reduction, and the resulting changes in solar radiation, may mean for biological systems and processes. The biological weighting functions used for this purpose often come from action spectra. Action spectra assumed to be relevant for organisms, especially plants all indicate that the shorter UV-B wavelengths are the most important. However, the relative importance of shorter vs longer UV-B wavelengths varies considerably. Depending on these slopes and the tails of the spectra extending into UV-A, the Radiation Amplification Factors vary enormously.

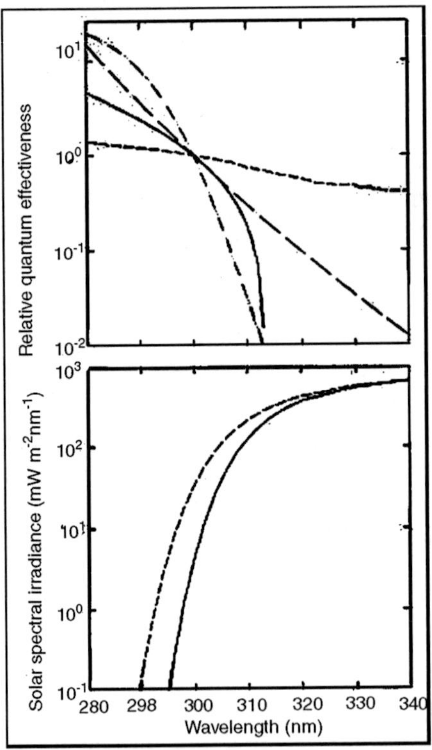

Fig. Action Spectra for "Naked" DNA Damage, DNA dimer Formation in intact Alfalfa seedlings, a Generalized plant action Spectrum Compiled from Various plant Spectra and a Spectrum for Putative lipid damage based on a Luminescence Indicator.

Only the weighting functions with steep slopes result in RAF values suggesting that ozone reduction is potentially important. Thus, the evaluation of weighting functions is critical. Although there is evidence that action spectra for some plant functions are steep, indicating that ozone reduction translates into large increases in effective solar UV-B, some more recent spectra developed specifically for evaluating the ozone reduction problem show somewhat flatter slopes than the earlier work. Still, many of these spectra are sufficiently steep so that ozone reduction must be taken seriously. Biological weighting functions also are needed to relate solar UV to UV from lamps used in many experiments.

The lower panel shows solar spectral irradiance at 360 and 180 Dobson Units of total atmospheric ozone. The solar irradiation is calculated for latitude 49 N at solar noon at the summer solstice using the model of Green.

PLANT GROWTH RESPONSES

In many plant species reduced leaf area and/or stem growth have been found in studies carried out in growth chambers, greenhouses and in the field. These studies have traditionally been conducted with specially filtered UV lamps. It is important in such experiments to maintain a realistic balance between different spectral regions since both UV-A and visible radiation can have strong ameliorating effects on responses of plants to UV-B. In growth chambers and greenhouses, the radiation conditions are usually quite different from those in nature. For example, the visible radiation which is used in photosynthesis and the UV-B/UV-A/PAR ratios are different from those in the field. If UV-A and PAR are low, the effects of UV-B may be much more severe. Thus, even if realistic levels of UV-B are used in simulating ozone reduction, the plant response may be exaggerated relative to that in the field.

In addition, other factors, such as temperature, water and nutrients differ from conditions in the field and this can alter response to UV-B radiation. It is, however, important that these studies conducted under controlled conditions be verified as much as possible under field conditions.

Even under field conditions, if applied UV-B is not adjusted downward during cloudy periods, the UV-B sensitivity may be unduly pronounced. Unfortunately, the most expensive and difficult experiments, *i.e.,* those conducted in the field with UV-B supplements adjusted for cloudiness and other atmospheric conditions, are seldom undertaken.

In the last few years more field experiments have been conducted and many of these employ lamp systems with controls to make continual adjustments just as to prevailing sunlight conditions. Also, there are several studies in which the UV component of existing sunlight has been altered by special filters in the field or in special small greenhouses or growth chambers located outdoors.

The filters have involved special glass, plastics, or in one series of studies, ozone gas in a UV-transparent plexiglass envelope. Many of these studies involving filtered sunlight have shown that normal ambient solar UV-B can cause somewhat reduced leaf area, smaller seedlings, etc. Plant species vary considerably in their response to UV-B in both controlled-environment and field studies. Also, varieties of the same species can vary in their response. For example, in the field, sizeable differences in response to UV-B were found among varieties of soybean and rice. Experiments in greenhouses covered by different materials that transmitted different amounts of UV indicated that varieties of bean from lower latitudes were less affected than those from higher latitudes under higher UV-B radiation.

PLANT REPRODUCTIVE PROCESSES

Ultraviolet-B radiation can alter both the timing of flowering as well as the number of flowers in certain species. Differences in timing of flowering may have important consequences for the availability of pollinators. Such effects may be due to regulatory alterations in the plant rather than damage *per se*. Poorly protected reproductive organs might, however, be susceptible to damaging effects. Most of the reproductive parts of plants, such as pollen and ovules, are rather well shielded from solar UV-B radiation. For example, anther walls can absorb more than 98% of incident UV-B radiation.

In addition, the pollen wall contains UV-B absorbing compounds affording protection during pollination as do the other flower parts such as sepals, petals and walls of the ovaries. Only after transfer to the stigma might pollen be susceptible to solar UV-B radiation. In vitro experiments have shown that germinating pollen can be sensitive at this time to UV-B radiation in some cases. However, often pollen germination itself is not affected, but pollen tube growth of many species can be retarded as shown in a survey of 34 plant species or varieties.

CARRY-OVER EFFECTS OF UV-B IRRADIATION IN SUBSEQUENT GENERATIONS

In sexually reproducing populations of an annual desert plant, effects of UV-B irradiation on growth and allocation of biomass appeared to accumulate as subsequent generations were exposed to UV-B irradiation. Furthermore, after four generations of UV-B irradiation, the effects persisted in a fifth generation that was not exposed to UV-B treatment. If this phenomenon is common, it could amplify the effects of UV-B radiation changes. This is somewhat analogous to apparent accumulated effects of UV-B irradiation over several growing seasons in longlived woody plants discussed later.

UV-B AND OTHER FACTORS

Plants and other organisms in nature seldom are affected by only a single stress factor, such as UV-B radiation. Instead, they typically respond to several factors acting in concert, such as water stress, increased atmospheric CO_2, mineral nutrient availability, heavy metals, tropospheric air pollutants and temperature. Therefore, it is important to keep in mind that the effectiveness of UV-B radiation can be greatly increased or decreased by such factors. Visible radiation is an important ameliorating factor and, thus, as natural levels as possible should be applied in laboratory experiments for attaining more realistic results, as discussed earlier.

Among the most common of factors in nature is water stress. In a field study, Sullivan and Teramura demonstrated that UV-B mediated reductions in photosynthesis and growth were observed only in well-watered soybeans. When soybeans were water stressed, there was no significant effect of the UV-B radiation on either photosynthesis or growth. The interpretation was that water stress resulted in a large reduction in photosynthesis and growth that masked the UV-B effect. Furthermore, water stressed plants resulted in a higher concentration of leaf flavonoids, which in turn, provided greater UV-B protection. Other interactions between UV-B radiation and water status of plants also occur. Elevated UV-B radiation in field experiments tended to alleviate drought symptoms in two Mediterranean pine species. In a moss species, UV-B radiation inhibited growth when the moss was under water stress, but stimulated growth when the moss was well hydrated. Increases of atmospheric CO_2 are a certain element of global climate change and atmospheric CO_2 concentration will likely double by the middle of the next century. Many experiments with elevated CO_2 employ a twice-ambient CO_2 concentration as a treatment condition.

Such a doubling often results in more pronounced plant responses than are evident in many elevated UV-B radiation lamp experiments designed to simulate up to $_2$0% ozone column reduction under field conditions. However, responses to CO_2 are small in semi-natural ecosystems where nutrient or water availability may strongly constrain plant growth. For example, Gwynn-Jones showed that growth responses to elevated CO_2 and enhanced UV-B were small during the first three years of experimentation in a sub-arctic heath. Also, most ecosystem-level effects of elevated CO_2 are mediated through changes in plant tissues. When studied independently, plant growth responses to changes in UV-B radiation and atmospheric CO_2 concentration generally are thought to be in opposite directions. Usually, however, in most experiments employing both elevated CO_2 and UV-B radiation, these factors do not yield interactions, with some exceptions.

Elevated CO_2 sometimes appears to provide some protection against elevated UV-B radiation for some species; yet, elevated UV-B radiation can limit the ability of some species to take advantage of elevated CO_2 in photosynthesis. Allocation of biomass in plants can also change in a complicated fashion with the combination of CO_2 and UV-B radiation treatments. Increased temperature is also a predicted element of global climate change. In a study combining two levels of UV-B radiation with two levels of CO_2 and two temperatures, the results indicated that either elevated CO_2 or somewhat higher temperature had similar effects in reducing the growth-inhibiting effects of elevated UV-B radiation on sunflower and maize seedlings. Plant uptake and translocation of mineral nutrients within the plant can be affected by elevated UV-B radiation, but the mineral nutrient status of plants also can affect plant responsiveness to UV-B radiation. Nitrogen concentration in plant tissues can increase under elevated UV-B which has been linked with reduced insect herbivory.

The uptake of certain nutrients may also be modified by UV-B radiation and cadmium. In oilseed rape plants grown under additional enhanced UV-B radiation and simultaneously exposed to different concentrations of cadmium, the manganese content in the shoots decreased in plants exposed to cadmium and UV-B radiation, while significant increases in magnesium, calcium, phosphate, copper and potassium occurred only in those plants exposed to cadmium and UV-B radiation. Cadmium uptake was not affected by UV-B radiation. The UV-B had no additional influence on the nutrient content of the roots. An earlier study showed that both cadmium and UV-B radiation negatively influenced photosynthetic efficiency in spruce seedlings.

Interaction of UV-B radiation with tropospheric air pollutants is also of concern although little work thus far has been conducted in this area. One field study of soybean plants showed them to be sensitive to ozone in the air, but not sensitive to UV-B supplements from lamps under the particular test conditions. There were no significant interactions of supplemental UV-B and ozone. However, in pine seedlings grown in a growth cabinet with a simulated solar UV radiation, increasing the ozone concentration increased the sensitivity of the pine seedlings to UV-B radiation since the ozone reduced the levels of UV-B-absorbing pigments in the plant tissues. In another experiment with tobacco, UV-B radiation increased the level of ozone-induced foliage lesions.

7

Oceans and the Climate

OCEAN DENSITY CHANGES

Sea level will rise if the ocean warms and fall if it cools, since the density of the water column will change. If the thermal expansivity were constant, global sea level change would parallel the global ocean heat content. However, since warm water expands more than cold water, and water at higher pressure expands more than at lower pressure, the global sea level change depends on the three-dimensional distribution of ocean temperature change.

Analysis of the last half century of temperature observations indicates that the ocean has warmed in all basins. The average rate of thermosteric sea level rise caused by heating of the global ocean is estimated to be 0.40 ± 0.09 mm yr^{-1} over 1955 to 1995 based on five-year mean temperature data down to 3,000 m. For the 0 to 700 m layer and the 1955 to 2003 period, the averaged thermosteric trend, based on annual mean temperature data from Levitus is 0.33 ± 0.07 mm yr^{-1}. For the same period and depth range, the mean thermosteric rate based on monthly ocean temperature data from Ishii is 0.36 ± 0.12 mm yr^{-1}. The thermosteric sea level curve over 1955 to 2003 for both the Levitus and Ishii data sets.

The rate of thermosteric sea level rise is clearly not constant in time and shows considerable fluctuations. A rise of more than 20 mm occurred from the late 1960s to the late 1970s with a smaller drop afterwards. Another large rise began in the 1990s, but after 2003, the steric sea level is decreasing in both estimates. Overlapping 10-year rates from these two estimates have a very high temporal correlation and the standard deviation of the rates is 0.7 mm yr^{-1}. The Levitus and Ishii data sets both give 0.32 ± 0.09 mm yr^{-1} for the upper 700 m during 1961 to 2003, but the Levitus data set of temperature down to 3,000 m ends in 1998.

From the results of Antonov for thermal expansion, the difference between the trends in the upper 3,000 m and the upper 700 m for 1961 to 1998 is about 0.1 mm yr^{-1}. Assuming that the ocean below 700 m continues to contribute beyond 1998 at a similar rate, with an uncertainty similar to that of the upperocean contribution, we assess the thermal expansion of the ocean down to 3,000 m during 1961 to 2003 as 0.42 ± 0.12 mm yr^{-1}. For the recent period 1993 to 2003, a value of 1.2 ± 0.5 mm yr^{-1} for thermal expansion in the upper 700 m is estimated both by Antonov and Ishii. Willis et al. estimate thermal expansion to be 1.6 ± 0.5 mm yr^{-1}, based on combined in situ temperature profiles down to 750 m and satellite measurements of altimetric height. Including the satellite data reduces the error caused by the inadequate sampling of the profile data. Error bars were estimated to be about 2 mm for individual years in the time series, with most of the remaining error due to inadequate profile availability. A close result was recently obtained by Lombard based on a combined analysis of in situ hydrographic data and satellite sea surface height and SST data. It is presently unclear why the latter two estimates are significantly larger than the thermosteric rates based on temperature data alone.

It is possible that the in situ data underestimate thermal expansion because of poor coverage in Southern Oceans, and it is interesting to note that a model based on assimilation of hydrographic data yields a somewhat higher estimate of 2.3 mm yr^{-1}. Published estimates of the steric sea level rates for 1955 to 2003 and 1993 to 2003. We assess the thermal expansion of the upper 700 m during 1993 to 2003 as 1.5 ± 0.5 mm yr^{-1}, and that of the upper 3,000 m as 1.6 ± 0.5 mm yr^{-1}, allowing for the ocean below 700 m as for the earlier period. Antonov attributed about 10% of the global average steric sea level rise during recent decades to halosteric expansion. A similar result was obtained by Ishii et al. who estimated a halosteric contribution to 1955 to 2003 sea level rise of 0.04 ± 0.02 mm yr^{-1}. While it is of interest to quantify this effect, only about 1% of the halosteric expansion contributes to the global sea level rise budget. This is because the halosteric expansion is nearly compensated by a decrease in volume of the added freshwater when its salinity is raised to the mean ocean value; the compensation would be exact for a linear state equation. Hence, for global sums of sea level change, halosteric expansion cannot be counted separately from the volume of added land freshwater. However, for regional changes in sea level, thermosteric and halosteric contributions can be comparably important.

INTERPRETATION OF REGIONAL VARIATIONS IN THE RATE OF SEA LEVEL CHANGE

Sea level observations show that whatever the time span considered, rates of sea level change display considerable regional variability. A number of processes can cause regional sea level variations.

Steric Sea Level Changes

Like the sea level trends observed by satellite altimetry, the global distribution of thermosteric sea level trends is not spatially uniform. The geographical distribution of thermosteric sea level trends over two different periods, 1993 to 2003 and 1955 to 2003 respectively. Some regions experienced sea level rise while others experienced a fall, often with rates that are several times the global mean. However, the patterns of thermosteric sea level rise over the approximately 50-year period are different from those seen in the 1990s. This occurs because the spatial patterns, like the global average, are also subject to decadal variability. In other words, variability on different time scales may have different characteristic patterns.

An EOF analysis of gridded thermosteric sea level time series since 1955 displays a spatial pattern that is similar to the spatial distribution of thermosteric sea level trends over the same time span. In addition, the first principal component is negatively correlated with the Southern Oscillation Index. Thus, it appears that ENSO-related ocean variability accounts for the largest fraction of variance in spatial patterns of thermosteric sea level. Similarly, decadal thermosteric sea level in the North Pacific and North Atlantic appears strongly influenced by the PDO and NAO respectively. For the recent years, the geographic distribution of observed sea lèvel trends shows correlation with the spatial patterns of thermosteric sea level change. This suggests that at least part of the nonuniform pattern of sea level rise observed in the altimeter data over the past decade can be attributed to changes in the ocean's thermal structure, which is itself driven by surface heating effects and ocean circulation.

Note that the steric changes due to salinity changes have not been included in insufficient salinity data in parts of the World Ocean. Ocean salinity changes, while unimportant for sea level at the global scale, can have an effect on regional sea level. For example, in the subpolar gyre of the North Atlantic, especially in the Labrador Sea, the halosteric contribution nearly counteracts the thermosteric contribution. This observational result is supported by results from data assimilation into models. Since density changes can result not only from surface buoyancy fluxes but also from the wind, a simple attribution of density changes to buoyancy forcing is not possible. While much of the non-uniform pattern of sea level change can be attributed to thermosteric volume changes, the difference between observed and thermosteric spatial trends show a high residual signal in a number of regions, especially in the southern oceans. Part of these residuals is likely due to the lack of ocean temperature coverage in remote oceans as well as in deep layers and to regional salinity change.

Ocean Circulation Changes

The highly non-uniform geographical distribution of steric sea level trends is closely connected, through geostrophic balance, with changes in ocean surface circulation. Density and circulation changes result from changes in atmospheric forcing that is primarily by surface wind stress and buoyancy flux. The wind alone can therefore cause local changes in steric sea level. Ocean general circulation models based on the assimilation of ocean data satisfactorily reproduce the spatial structure of sea level trends for the past decade, and show in particular that the tropical Pacific pattern results from decadal fluctuations in the depth of the tropical thermocline and change in equatorial trade winds. The similarity of the patterns of steric and actual sea level change indicates that density changes are the dominant influence. Discrepancies may indicate a significant contribution from changes in the winddriven barotropic circulation, especially at high latitudes.

Surface Atmospheric Pressure Changes

Surface atmospheric pressure also causes regional sea level variations. Over time scales longer than a few days, the ocean adjusts nearly isostatically to changes in atmospheric pressure, that is, for each 1 hPa sea level pressure increase the ocean is depressed by approximately 10 mm, shifting the underlying mass sideways to other regions. For the temporal average, regional changes in sea level caused by atmospheric pressure loading reach about 0.2 m. Such effects are generally corrected for in tide gauge and altimetry-based sea level analyses. The inverted barometer effect has a negligible effect on global mean sea level, because water is nearly incompressible, but is significant when averaged over the area of T/P and Jason-1 altimetry, which does not cover the whole World Ocean. For that reason, the altimetry-based mean sea level curve is corrected for the inverted barometer effect.

Solid Earth and Geoid Changes

Geodynamical processes related to the solid Earth's elastic and viscoelastic response to spatially variable ice melt loading also cause non-uniform sea level change. The solid Earth and oceans continue to respond to the ice and complementary water loads associated with the late Pleistocene and early Holocene glacial cycles through GIA. This process not only drives large crustal uplift near the location of former ice complexes, but also produces a worldwide signature in sea level that results from gravitational, deformational and rotational effects: as the viscous mantle material flows to restore isostasy during and after the last deglaciation, uplift occurs under the former centres of the ice sheets while the surrounding peripheral bulges experience a subsidence. The return of the melt water to the oceans produces an ongoing geoid change resulting in subsidence of the ocean basins and an upward warping of the continents, while the flow of water into the subsiding peripheral bulges contributes a broad scale sea level fall in the far field of the ice complexes.

The combined gravitational and deformational effects also perturb the rotation vector of the planet, and this perturbation feeds back into variations in the position of the crust and the geoid. Corrections for GIA effects are made to both tide gauge and altimeter estimates of global sea level change. Self-gravitation and deformation of the Earth's surface in response to the ongoing change in loading by glaciers and ice sheets is another cause of regional sea level variations. Model predictions show quite different patterns of non-uniform sea level change depending on the source of the ice melt and associated regional sea level variations reach up to a few 0.1 mm yr^{-1}.

CHANGES IN SEA LEVEL

Present-day sea level change is of considerable interest because of its potential impact on human populations living in coastal regions and on islands. This part focuses on global and regional sea level variations, over time spans ranging from the last decade to the past century; a brief discussion of sea level change in previous centuries. Processes in several nonlinearly coupled components of the Earth system contribute to sea level change, and understanding these processes is therefore a highly interdisciplinary endeavour. On decadal and longer time scales, global mean sea level change results from two major processes, mostly related to recent climate change, that alter the volume of water in the global ocean:

1. Thermal expansion,
2. The exchange of water between oceans and other reservoirs.

All these processes cause geographically nonuniform sea level change as well as changes in the global mean; some oceanographic factors also affect sea level at the regional scale, while contributing negligibly to changes in the global mean. Vertical land movements such as resulting from glacial isostatic adjustment, tectonics, subsidence and sedimentation influence local sea level measurements but do not alter ocean water volume; nonetheless, they affect global mean sea level through their alteration of the shape and hence the volume of the ocean basins containing the water. Measurements of present-day sea level change rely on two different techniques: tide gauges and satellite altimetry. Tide gauges provide sea level variations with respect to the land on which they lie.

To extract the signal of sea level change due to ocean water volume and other oceanographic change, land motions need to be removed from the tide gauge measurement. Land motions related to GIA can be simulated in global geodynamic models. The estimation of other land motions is not generally possible unless there are adequate nearby geodetic or geological data, which is usually not the case. However, careful selection of tide gauge sites such that records reflecting major tectonic activity are rejected, and averaging over all selected gauges, results in a small uncertainty for global sea level estimates. Sea level change based on satellite altimetry is measured with respect to the Earth's centre of mass, and thus is not distorted by land motions, except for a small component due to large-scale deformation of ocean basins from GIA. The TAR stage on sea level change provided estimates of climate and other anthropogenic contributions to 20th-century sea level rise, based mostly on models. The sum of these contributions ranged from -0.8 to 2.2 mm yr^{-1}, with a mean value of 0.7 mm yr-1, and a large part of this uncertainty was due to the lack of information on anthropogenic land water change. For observed 20th-century sea level rise, based on tide gauge records, Church adopted as a best estimate a value in the range of 1 to 2 mm yr^{-1}, which was more than twice as large as the TAR's estimate of climate-related contributions. It thus appeared that either the processes causing sea level rise had been underestimated or the rate of sea level rise observed with tide gauges was biased towards higher values.

Since the TAR, a number of new results have been published. The global coverage of satellite altimetry since the early 1990s has improved the estimate of global sea level rise and has revealed the complex geographical patterns of sea level change in open oceans. Near-global ocean temperature data for the last 50 years have been recently made available, allowing the first observationally based estimate of the thermal expansion contribution to sea level rise in past decades. For recent years, better estimates of the land ice contribution to sea level are available from various observations of glaciers, ice caps and ice sheets. In this part, we summarise the current knowledge of present-day sea level rise. The observational results are assessed, followed by our current interpretation of these observations in terms of climate change and other processes, and ending with a discussion of the sea level budget.

OBSERVATIONS OF SEA LEVEL CHANGES

20th-Century Sea Level Rise from Tide Gauges

The TAR listed several estimates for global and regional 20th-century sea level trends based on the Permanent Service for Mean Sea Level data set. The concerns about geographical bias in the PSMSL data set remain, with most long sea level records stemming from the NH, and most from continental coastlines rather than ocean interiors. Based on a small number of high-quality tide gauge records from stable land regions, the rate of sea level rise has been estimated as 1.8 mm yr^{-1} for the past 70 years and Miller and Douglas find a range of 1.5 to 2.0 mm yr^{-1} for the 20th century from 9 stable tide gauge sites. Holgate and Woodworth estimated a rate of 1.7 ± 0.4 mm yr-1 sea level change averaged along the global coastline during the period 1948 to 2002, based on data from 177 stations divided into 13 regions. Church determined a global rise of 1.8 ± 0.3 mm yr-1 during 1950 to 2000, and Church and White determined a change of 1.7 ± 0.3 mm yr^{-1} for the 20th century. Changes in global sea level as derived from analyses of tide gauges are displayed.

Considering the results, and allowing for the ongoing higher trend in recent years shown by altimetry, we assess the rate for 1961 to 2003 as 1.8 ± 0.5 mm

yr^{-1} and for the 20th century as 1.7 ± 0.5 mm yr^{-1}. While the recently published estimates of sea level rise over the last decades remain within the range of the TAR values, there is an increasing opinion that the best estimate lies closer to 2 mm yr^{-1} than to 1 mm yr^{-1}. The lower bound reported in the TAR resulted from local and regional studies; local and regional rates may differ from the global mean. A critical issue concerns how the records are adjusted for vertical movements of the land upon which the tide gauges are located and of the oceans.

Trends in tide gauge records are corrected for GIA using models, but not for other land motions. The GIA correction ranges from about 1 mm yr^{-1} near to former ice sheets to a few tenths of a millimetre per year in the far field; the error in tide-gauge based global average sea level change resulting from GIA is assessed as 0.15 mm yr^{-1}. The TAR mentioned the developing geodetic technologies that hold the promise of measuring rates of vertical land movement at tide gauges, no matter if those movements are due to GIA or to other geological processes. Although there has been some model validation, especially for GIA models, systematic problems with such techniques, including short data spans, have yet to be fully resolved.

Sea Level Change during the Last Decade from Satellite Altimetry

Since 1992, global mean sea level can be computed at 10- day intervals by averaging the altimetric measurements from the TOPEX/Poseidon and Jason satellites over the area of coverage. Each 10-day estimate of global mean sea level has an accuracy of approximately 5 mm. Numerous documents on the altimetry results show a current rate of sea level rise of 3.1 ± 0.7 mm yr^{-1} over 1993 to 2003. A significant fraction of the 3 mm yr^{-1} rate of change has been shown to arise from changes in the Southern Ocean. The accuracy needed to compute mean sea level change pushes the altimeter measurement system to its performance limits, and thus care must be taken to ensure that the instrument is precisely calibrated.

The tide gauge calibration method provides diagnoses of problems in the altimeter instrument, the orbits, the measurement corrections and ultimately the final sea level data. Errors in determining the altimeter instrument drift using the tide gauge calibration, currently estimated to be about 0.4 mm yr^{-1}, are almost entirely driven by errors in knowledge of vertical land motion at the gauges. Altimetry-based sea level measurements include variations in the global ocean basin volume due to GIA. Averaged over the oceanic regions sampled by the altimeter satellites, this effect yields a value close to –0.3 mm yr^{-1} in sea level with possible uncertainty of 0.15 mm yr^{-1}. This number is subtracted from altimetry-derived global mean sea level in order to obtain the contribution due to ocean volume change. Altimetry from T/P allows the mapping of the geographical distribution of sea level change. Although regional variability in coastal sea level change had been reported from tide gauge analyses, the global coverage of satellite altimetry provides unambiguous evidence of non-uniform sea level change in open oceans, with some regions exhibiting rates of sea level change about five times the global mean. For the past decade, sea level rise shows the highest magnitude in the western Pacific and eastern Indian oceans, regions that exhibit large interannual variability associated with ENSO.

Except for the Gulf Stream region, most of the Atlantic Ocean shows sea level rise during the past decade. Despite the global mean rise, Sea level has been dropping in some regions. These spatial patterns likely reflect decadal fluctuations rather than long-term trends. Empirical Orthogonal Functions analyses of altimetry-based sea level maps over 1993 to 2003 show a strong influence of the 1997–1998 El Niño, with the geographical patterns of the dominant mode being very similar to those of the sea level trend map.

Reconstructions of Sea Level Change during the Last 50 Years Based on Satellite Altimetry and Tide Gauge

Attempts have been made to reconstruct historical sea level fields by combining the near-global coverage from satellite altimeter data with the longer but spatially sparse tide gauge records. These sea level reconstructions use the short altimeter record to determine the principal EOF of sea level variability, and the tide gauge data to estimate the evolution of the amplitude of the EOFs over time.

The method assumes that the geographical patterns of decadal sea level trends can be represented by a superposition of the patterns of variability that are manifest in interannual variability. As a caveat, note that variability on different time scales may have different characteristic patterns. The trends in the EOF amplitudes allow the reconstruction of a spatially variable rate of sea level rise.

The geographical distribution of linear sea level trends for 1955 to 2003 based on this reconstruction technique. Comparison with the altimetry-based trend map for the shorter period indicates quite different geographical patterns. These differences mainly arise from thermal expansion changes through time Changes in spatial sea level patterns through time may help reconcile apparently inconsistent estimates of regional variations in tide-gauge based sea level rise. For example, the minimum in rise along the northwest Australian coast is consistent with the results of Lambeck in having smaller rates of sea level rise and indeed sea level fall off north-western Australia over the last few decades. In addition, for the North Atlantic Ocean, the rate of rise reaches a maximum in a band running east-northeast from the US east coast. The trends are lower in the eastern than in the western Atlantic.

Interannual and Decadal Variability and Long-term Changes in Sea Level

Sea level records contain a considerable amount of interannual and decadal variability, the existence of which is coherent throughout extended parts of the ocean. For example, the global sea level curve shows an approximately 10 mm rise and fall of global mean sea level accompanying the 1997–1998 ENSO event. Over the past few decades, the time series of the first EOF of Church represents ENSO variability, as shown by a significant correlation with the Southern Oscillation Index. The signature of the 1997–1998 El Niño is also clear in the altimetric maps of sea level anomalies. Model results suggest that large volcanic eruptions produce interannual to decadal fluctuations in the global mean sea level.

Holgate and Woodworth concluded that the 1990s had one of the fastest recorded rates of sea level rise averaged along the global coastline, slightly higher than the altimetry-based open ocean sea level rise. However, their analysis also shows that some previous decades had comparably large rates of coastal sea level rise. White confirmed the larger sea level rise during the 1990s around coastlines compared to the open ocean but found that in some previous periods the coastal rate was smaller than the open ocean rate, and concluded that over the last 50 years the coastal and open ocean rates of change were the same on average.

The global reconstruction of Church and Church and White also exhibits large decadal variability in the rate of global mean sea level rise, and the 1993 to 2003 rate has been exceeded in some previous decades. The variability is smaller in the global reconstruction than in the Holgate and Woodworth coastal time series. The rather low temporal correlation between the two time series suggests that the statistical uncertainty in the linear trends calculated from either data set probably underestimates the systematic uncertainty in the results.

Interannual or longer variability is a major reason why no long-term acceleration of sea level has been identified using 20th-century data alone. Another possibility is that the sparse tide gauge network may have been inadequate to detect it if present. The longest records available from Europe and North America contain accelerations of the order of 0.4 mm yr^{-1} per century between the 19th and 20th century. Church and White found an acceleration of 1.3 ± 0.5 mm yr^{-1} per century over the period 1870 to 2000. These data support an inference that the onset of acceleration occurred during the 19th century. Geological observations indicate that during the last 2,000 years, sea level change was small, with an average rate of only 0.0 to 0.2 mm yr^{-1}. The use of proxy sea level data from archaeological sources is well established in the Mediterranean. Oscillations in sea level from 2,000 to 100 yr before present did not exceed ± 0.25 m, based on the Roman- Byzantine-Crusader well data.

Many Roman and Greek constructions are relatable to the level of the sea. Based on sea level data derived from Roman fish ponds, which are considered to be a particularly reliable source of such information, together with nearby tide gauge records, Lambeck concluded that the onset of the modern sea level rise occurred between 1850 and 1950. Donnelly and Gehrels, employing geological data from Connecticut, Maine and Nova Scotia salt-marshes together with nearby tide gauge records, demonstrated that the sea level rise observed during the 20th century was in excess of that averaged over the previous several centuries. The joint interpretation of the geological observations, the longest instrumental records and the current rate of sea level rise for the 20th century gives a clear indication that the rate of sea level rise has increased between the mid-19th and the mid- 20th centuries.

Regional Sea Level Change

Two regions are discussed here to give examples of local variability in sea level: the northeast Atlantic and small Pacific Islands. Interannual variability in northeast Atlantic sea level records exhibits a clear relationship to the air pressure and wind changes associated with the NAO, with the magnitude and sign of the response depending primarily upon latitude. The signal of the NAO can also be observed to some extent in ocean temperature records, suggesting a possible, smaller NAO influence on regional mean sea level via steric changes. In the Russian Arctic Ocean, sea level time series for recent decades also have pronounced decadal variability that correlates with the NAO index. In this region, wind stress and atmospheric pressure loading contribute nearly half of the observed sea level rise of 1.85 mm yr^{-1}. Small Pacific Islands are the subject of much concern in view of their vulnerability to sea level rise. The Pacific Ocean region is the centre of the strongest interannual variability of the climate system, the coupled ocean-atmosphere ENSO mode. There are only a few Pacific Island sea level records extending back to before 1950. Mitchell calculated rates of relative sea level rise for the stations in the Pacific region. Using their results and focusing on only the island stations with more than 50 years of data, the average rate of sea level rise is 1.6 mm yr^{-1}. For island stations with record lengths greater than 25 years, the average rate of relative sea level rise is 0.7 mm yr^{-1}. However, these data sets contain a large range of rates of relative sea level change, presumably as a result of poorly quantified vertical land motions. An example of the large interannual variability in sea level is Kwajalein.

The local tide gauge data, the sea level reconstructions of Church and Church and White and the shorter satellite altimeter record all agree and indicate that interannual variations associated with ENSO events are greater than 0.2 m. The Kwajalein data also suggest increased variability in sea level after the mid-1970s, consistent with the trend towards more frequent, persistent and intense ENSO events since the mid-1970s. For the Kwajalein record, the rate of sea level rise, after correction for GIA land motions and isostatic response to atmospheric pressure changes, is 1.9 ± 0.7 mm yr^{-1}. However, the uncertainties in rates of sea level change increase rapidly with decreasing record length and can be several mm yr^{-1} for decade-long records. Sea level change on the atolls of Tuvalu has been the subject of intense interest as a result of their low-lying nature and increasing incidence of flooding. There are two records available at Funafuti, Tuvalu; the first record commences in 1977 and the second in 1993. After allowing for subsidence affecting the first record, Church estimate sea level rise at Tuvalu to be 2.0 ± 1.7 mm yr^{-1}, in agreement with the reconstructed rate of sea level rise.

Changes in Extreme Sea Level

Societal impacts of sea level change primarily occur via the extreme levels rather than as a direct consequence of mean sea level changes. Apart from non-climatic events such as tsunamis, extreme sea levels occur mainly in the form of storm surges generated by tropical or extra tropical cyclones. Studies of variations in extreme sea levels during the 20th century based on tide gauge data are fewer than studies of changes in mean sea level for several reasons. A study on changes in extremes, which are caused by changes in mean sea level as well as changes in surges, is more complex than the study of mean sea level changes.

Moreover, the hourly sampling interval normally used in tide gauge records is not always sufficient to accurately capture the true extreme. Among the different parameters often used to describe extremes, annual maximum surge is a good indicator of climatic trends. For study of long records extending back to the 19th century or before, annual maximum surge-at-high-water is a better-suited parameter because during that period high waters and not the full tidal curve were recorded. Studies of the longest records of extremes are inevitably restricted to a small number of locations. From observed sea level extremes at Liverpool since 1768, Woodworth and Blackman concluded that the annual maximum surge-at-high-water was larger in the late 18th, late 19th and late 20th centuries than for most of the 20th century, qualitatively consistent with the long-term variability in storminess from meteorological data.

From the tide gauge record at Brest from 1860 to 1994, Bouligand and Pirazzoli found an increasing trend in annual maxima and 99th percentile of surges; however, a decreasing trend was found during the period 1953 to 1994.

From non-tidal residuals at San Francisco since 1858, Bromirski concluded that extreme winter residuals have exhibited a significant increasing trend since about 1950, a trend that is attributed to an increase in storminess during this period. Zhang concluded from records at 10 stations along the east coast of the USA since 1900 that the rise in extreme sea level closely followed the rise in mean sea level. A similar conclusion can be drawn from a recent study of Firing and Merrifield, who found long-term increases in the number and height of daily extremes at Honolulu but no evidence for an increase relative to the underlying upward mean sea level trend.

An analysis of 99th percentiles of hourly sea level at 141 stations over the globe for recent decades showed that there is evidence for an increase in extreme high sea level worldwide since 1975. In many cases, the secular changes in extremes were found to be similar to those in mean sea level. Likewise, interannual variability in extremes was found to be correlated with regional mean sea level, as well as to indices of regional climate patterns.

OCEAN BIOGEOCHEMICAL CHANGES

The observed increase in atmospheric carbon dioxide and the observed changes in the physical properties of the ocean reported in this stage can affect marine biogeochemical cycles. The increase in atmospheric CO_2 causes additional CO_2 to dissolve in the ocean. Changes in temperature and salinity affect the solubility and chemical equilibration of gases.

Changes in circulation affect the supply of carbon and nutrients from below, the ventilation of oxygen-depleted waters and the downward penetration of anthropogenic carbon. The combined physical and biogeochemical changes also affect biological activity, with further consequences for the biogeochemical cycles. The increase in surface ocean CO_2 has consequences for the chemical equilibrium of the ocean. As CO_2 increases, surface waters become more acidic and the concentration of carbonate ions decreases. This change in chemical equilibrium causes a reduction of the capacity of the ocean to take up additional CO_2.

However, the response of marine organisms to ocean acidification is poorly known and could cause further changes in the marine carbon cycle with consequences that are difficult to estimate. Dissolved oxygen in the ocean is affected by the same physical processes that affect CO_2, but in contrast to CO_2, O_2 is not affected by changes in its atmospheric concentration. Changes in oceanic O_2 concentration thus provide information on the changes in the physical or biological processes that occur within the ocean, such as ventilation mode water formation, upwelling or biological export and respiration. Furthermore, changes in the oceanic O_2 content are needed to estimate the CO_2 budget from atmospheric O_2/molecular nitrogen ratio measurements. However, the method currently estimates the change in air-sea fluxes of O_2 indirectly based on heat flux changes. This part reports observed changes in biogeochemical cycles and assesses their consistency with observed changes in physical properties. Changes in oceanic nitrous oxide and methane have not been assessed because of the lack of large-scale observations. Observations of the mean fluxes of N_2O and CH_4.

CARBON

Total Change in Dissolved Inorganic Carbon and Air-sea Carbon Dioxide Flux

Direct observations of oceanic dissolved inorganic carbon reflect changes in both the natural carbon cycle and the uptake of anthropogenic CO_2 from the atmosphere. Links between the main modes of climate variability and the marine carbon cycle have been observed on interannual time scales in several regions of the world. In the equatorial Pacific, the reduced upwelling associated with El Niño events decreases the regional outgas of natural CO_2 to the atmosphere. In the subtropical North Atlantic, reduced mode water formation and reduced deep winter mixing during the positive NAO phase increase the storage of carbon in the intermediate ocean.

These observations show that variability in the content of natural DIC in the ocean has occurred in association with climate variability. Longer observations exist for the partial pressure of CO_2 at the surface only. Over more than two decades, the oceanic pCO_2 increase has generally followed the atmospheric CO_2 within the given uncertainty, although regional differences have been observed. The three stations with the longest time series, all in the northern subtropics, show pCO_2 increases at a rate varying between 1.6 and 1.9 atm yr^{-1}, indistinguishable from the atmospheric increase of 1.5 to 1.9 atm yr^{-1}. Variability on the order of 20 atm over periods of five years was observed in the three time series, as well as in other data sets, and has been associated with regional changes in the natural carbon cycle driven by changes in ocean circulation and by climate variability or with variations in biological activity. Direct surface pCO_2 observations have been used to compute a global air-sea CO_2 flux of 1.6 ± 1 GtC yr^{-1} for the year 1995. It is not yet possible to detect large-scale changes in the global air-sea CO_2 flux from direct observations because of the large influence of climate variability. However, estimates from inverse methods of the air-sea CO_2 flux from the spatio-temporal distribution of atmospheric CO_2 suggest that the global air-sea CO_2 flux increased by 0.1 to 0.6 GtC yr^{-1} between the 1980s and 1990s, consistent with results from ocean models.

Anthropogenic Carbon Change

The recent uptake of anthropogenic carbon in the ocean is well constrained by observations to a decadal mean of 2.2 ± 0.4 GtC yr^{-1} for the 1990s. The uptake of anthropogenic carbon over longer time scales can be estimated from oceanic measurements. Changes in DIC between two time periods reflect the anthropogenic carbon uptake plus the changes in DIC concentration due to changes in water masses and biological activity. To estimate the contribution of anthropogenic carbon alone, several corrections must be applied. From observed DIC changes between surveys in the 1970s and the 1990s, an increase in anthropogenic carbon has been inferred down to depths of 1,100 m in the North Pacific, 200 to 1,200 m in the Indian Ocean and 1,900 m in the Southern Ocean. An indirect method was used to estimate anthropogenic carbon from observations made at a single time period based on well-known processes that control the distribution of natural DIC in the ocean.

The method corrects the observed DIC concentration for organic matter decomposition and dissolution of carbonate minerals, and removes an estimate of the DIC concentration of the water when it was last in contact with the atmosphere. With this method, a global DIC increase of 118 ± 19 GtC between pre-industrial times and 1994 has been estimated, using 9,618 profiles from the 1990s. The uncertainty of ±19 GtC in this estimate is based on uncertainties in the anthropogenic DIC estimates and mapping errors, which have characteristics of random error, and on an estimate of potential biases, which are not necessarily centred on the mean value. Potential biases of up to 7% in the technique have been identified, mostly caused by assumptions about the time evolution of CO_2, the age or the identification of water masses and the recent changes in surface warming and stratification.

Potential biases from assumptions of constant carbon and nutrient uptake ratios for biological activity have not been assessed. While the magnitude and direction of all potential biases are not yet clear, the given uncertainty of ±16% appears realistic compared to the biases already identified.

Because of the limited rate of vertical transport in the ocean, more than half of the anthropogenic carbon can still be found in the upper 400 m, and it is undetectable in most of the deep ocean.

The vertical penetration of anthropogenic carbon is consistent with the DIC changes observed between two cruises. Anthropogenic carbon has penetrated deeper in the North Atlantic and subantarctic Southern Ocean compared to other basins, due to a combination of: i) high surface alkalinity which favours the uptake of CO_2, and ii) more active vertical exchanges caused by intense winter mixing and by the formation of deep waters. The deeper penetration of anthropogenic carbon in these regions is consistent with similar features observed in the oceanic distribution of chlorofluorocarbons of atmospheric origin, confirming that it takes decades to many centuries to transport carbon from the surface into the thermocline and the deep ocean. Deeper penetration in the North Atlantic and subantarctic Southern Ocean is also observed in the changes in heat content.

The large storage of anthropogenic carbon observed in the subtropical gyres is caused by the lateral transport of carbon from the region of mode water formation towards the lower latitudes. The fraction of the net CO_2 emissions taken up by the ocean was possibly lower during 1980 to 2005 compared to 1750 to 1994; however the uncertainty in the estimates is larger than the difference between the estimates. The net CO_2 emissions include all emissions that have an influence on the atmospheric CO_2 concentration. It is equivalent to the sum of the atmospheric and oceanic CO_2 increase. Because the atmospheric CO_2 is well constrained by observations, the uncertainty in the net CO_2 emissions is nearly equal to the uncertainty in the oceanic CO_2 increase. The decrease in oceanic uptake fraction would be consistent with the understanding that the ocean CO_2 sink is limited by the transport rate of anthropogenic carbon from the surface to the deep ocean, and also with the nonlinearity in carbon chemistry that reduces the CO_2 uptake capacity of water as its CO_2 concentration increases.

Ocean Acidification by Carbon Dioxide

The uptake of anthropogenic carbon by the ocean changes the chemical equilibrium of the ocean. Dissolved CO_2 forms a weak acid.1 As CO_2 increases, pH decreases, that is, the ocean becomes more acidic. Ocean pH can be computed from measurements of DIC and alkalinity. A decrease in surface pH of 0.1 over the global ocean was calculated from the estimated uptake of anthropogenic carbon between 1750 and 1994 with the lowest decrease in the tropics and subtropics, and the highest decrease at high latitudes, consistent with the lower buffer capacity of the high latitudes compared to the low latitudes. The mean pH of surface waters ranges between 7.9 and 8.3 in the open ocean, so the ocean remains alkaline even after these decreases. For comparison, pH was higher by 0.1 unit during glaciations, and there is no evidence of pH values more than 0.6 units below the pre-industrial pH during the past 300 million years.

A decrease in ocean pH of 0.1 units corresponds to a 30% increase in the concentration of $H+$ in seawater, assuming that alkalinity and temperature remain constant. Changes in surface temperature may have induced an additional decrease in pH of <0.01. The calculated anthropogenic impact on pH is consistent with results from time series stations where a decrease in pH of 0.02 per decade was observed. Results from time series stations include not only the increase in anthropogenic carbon, but also other changes due to local physical and biological variability. The consequences of changes in pH on marine organisms are poorly known.

Change in Carbonate Species

The uptake of anthropogenic carbon occurs through the injection of CO_2 and causes a shift in the distribution of carbon species. The availability of carbonate is particularly important because it controls the maximum amount of CO_2 that the ocean is able to absorb. Marine organisms use carbonate to produce shells of calcite and aragonite. Currently, the surface ocean is supersaturated with respect to both calcite and aragonite, but undersaturated below a depth called the 'saturation horizon'. The undersaturation starts at a depth varying between 200 m in parts of the high-latitude and the Indian Ocean and 3,500 m in the Atlantic.

Calcium carbonate dissolves either when it sinks below the calcite or aragonite saturation horizons or under the action of biological activity. Shoaling of the aragonite saturation horizon has been observed in all ocean basins based on alkalinity, DIC and oxygen measurements. The amplitude and direction of the signal was everywhere consistent with the uptake of anthropogenic carbon, with potentially smaller contributions from changes in circulation, temperature and biology. Feely calculated that the uptake of anthropogenic carbon alone has caused a shoaling of the aragonite saturation horizon between 1750 and 1994 by 30 to 200 m in the eastern Atlantic, the North Pacific and the North Indian Ocean, and a shoaling of the calcite saturation horizon by 40 to 100 m in the Pacific.

This calculation is based on the anthropogenic DIC increase estimated by Sabine on a global compilation of biogeochemical data and on carbonate chemistry equations. Furthermore, an increase in total alkalinity at the depth of the aragonite saturation horizon between 1970 and 1990 has been reported. These results are consistent with the calculated increase in $CaCO_3$ dissolution as a result of the shoaling of the aragonite saturation horizon, but with large uncertainty. Carbonate decreases at high latitudes and particularly in the Southern Ocean may have consequences for marine ecosystems because the current saturation horizon is closer to the surface than in other basins.

OXYGEN

In the thermocline a decrease in the O_2 concentration has been observed between about the early 1970s and the late 1990s or later in several repeated hydrographic parts in the North and South Pacific, North Atlantic, and Southern Indian Oceans. A number of O_2 decreases that fit the overall message. The reported O_2 decreases range from 0.1 to 6 μmol kg^{-1} yr^{-1}, superposed on decadal variations of ± 2 μmol kg^{-1} yr^{-1}.

In all published studies, the observed O_2 decrease appeared to be driven primarily by changes in ocean circulation, and less by changes in the rate of O_2 demand from downward settling of organic matter. A few studies have quantified the contribution of the change in ocean circulation using estimates of changes in apparent CFC ages. In nearly all cases, the decrease in O_2 could entirely be accounted for by the increased apparent CFC age that resulted from reduced rate of renewal of intermediate waters. Changes in biological processes were only significant at the coast of California and may result from assumptions in the method.

It is unclear whether the recent changes in O_2 are indicative of trends or of variability. Recent data in the Indian Ocean have shown a reversal of the O_2

decrease between 1987 and 2002 in the South Indian Ocean of similar amplitude to the decrease observed during the previous decades. Variability has been observed on decadal time scales in the North Atlantic large enough to mask any potential trends. In the upper 100 m of the global ocean surface, decadal variations of ± 0.5 μmol kg^{-1} in O_2 concentration were observed for the period 1956 to 1998 based on a global analysis of 530,000 oxygen profiles, with no clear trends.

However, the near-surface changes in O_2 concentration are difficult to interpret. They can be caused by changes in biological activity, by changes in the physical transport of O_2 from intermediate waters or by changes in temperature and salinity. Because there is less confidence in the early measurements and the reported changes cannot be explained by known processes, it cannot be said whether the absence of a long-term trend in surface O_2 is realistic or not.

NUTRIENTS

Changes in nutrient concentrations can provide information on changes in the physical and biological processes that affect the carbon cycle and could potentially be used as indicators for large-scale changes in marine biology. However, only a few studies reported decadal changes in inorganic nutrient concentrations. In the North Pacific, the concentration of nitrate plus nitrite and phosphate decreased at the surface and increased below the surface in the past two decades. Nutrient changes were observed in the deep ocean of all basins but no clear pattern emerges from available observations. Pahlow and Riebesell found changes in the ratio of nutrients in the North Pacific and Atlantic Oceans, and no significant changes in the South Pacific.

In the North Pacific, Keller observed a decrease in N associated with the increase in O_2 between 1970 and 1990 at 1,050 m, opposite to the results of Pahlow and Riebesell's longer study. Using the same data set extended to the world, large regional changes in nutrient ratios were observed but no consistent basin-scale patterns. Uncertainties in deep ocean nutrient observations may be responsible for the lack of coherence in the nutrient changes. Sources of inaccuracy include the limited number of observations and the lack of compatibility between measurements from different laboratories at different times. In some cases, the observed trends in nutrients can be explained by either a change in thermocline ventilation or a change in biological activity but in other cases are mostly consistent with a reduction in thermocline ventilation.

Thus, all of the reported trends are consistent with a physical explanation of the observed changes, although changes in biological activity cannot be ruled out. The concentration of surface nutrients can also be influenced by surface mixing, as a reduction in mixing leads to a decreased concentration of surface nutrients. The observed changes in surface temperature and salinity are indicative of changes in the surface mixing. In most of the Pacific Ocean, surface warming and freshening act in the same direction and contribute to reduced mixing, consistent with regional observations. In the Atlantic and Indian Oceans, temperature and salinity trends generally act in opposite directions and changes in mixing have not been quantified regionally.

BIOLOGICAL CHANGES RELEVANT TO OCEAN BIOGEOCHEMISTRY

Changes in biological activity are an important part of the carbon cycle but are difficult to quantify at the global scale. Marine export production is the biological process that has the largest influence on element cycles. There are no global observations on changes in export production or respiration. However, estimates of changes in primary production provide partial information. A reduction in global oceanic primary production by about 6% between the early 1980s and the late 1990s was estimated based on the comparison of chlorophyll data from two satellites. The errors in this estimate are potentially large because it is based on the comparison of data from two different sensors. Nevertheless, a change in biological fluxes of this order of magnitude is plausible considering that biological production is controlled primarily by nutrient input from intermediate waters, and that a decrease in intermediate water renewal has been observed during that period as indicated by the decrease in O_2. Shifts and trends in plankton biomass have been observed for instance in the North Atlantic, the North Pacific and in the Southern Indian Ocean but the spatial and temporal coverage is limited.

CONSISTENCY WITH PHYSICAL CHANGES

It is clearly established that climate variability affects the oceanic content of natural and anthropogenic DIC and the airsea flux of CO_2, although the amplitude and physical processes responsible for the changes are less well known. Variability in the marine carbon cycle has been observed in response to physical changes associated with the dominant modes of climate variability such as El Niño events and the PDO and the NAO. The regional patterns of anthropogenic CO_2 storage are consistent with those of CFCs and with changes in heat content.

The observed trends in CO_2, DIC, pH and carbonate species can be primarily explained by the response of the ocean to the increase in atmospheric CO_2. Large-scale changes in the O_2 content of the thermocline have been observed between the 1970s and the late 1990s. These changes are everywhere consistent with the local changes in ocean ventilation as identified either by changes in density gradients or by changes in apparent CFC ages. Nevertheless, an influence of changes in marine biology cannot be ruled out. The available data are insufficient to say if the changes in O_2 are caused by natural variability or are trends that are likely to persist in the future, but they do indicate that large-scale changes in ocean physics influence natural biogeochemical cycles, and thus the cycles of O_2 and CO_2 are likely to undergo changes if ocean circulation changes persist in the future.

CLIMATE CHANGE AND THE GLOBAL CARBON CYCLE

In 2001 the IPCC concluded that most of the warming observed over the last half of the twentieth century can be attributed to human activities that have increased greenhouse gas concentrations in the atmosphere. They also warned that these changes will continue to drive rapid climate changes for several centuries to come. Chief amongst these greenhouse gases is CO_2, whose atmospheric concentrations have been dramatically altered by human perturbations to the global carbon cycle.

Over at least the 420 000 years prior to the twentieth century, the atmospheric concentration of CO_2 only varied between ~ 80 ppmv (parts per million volume) (during the glaciations when the global temperature was 8 - 9 ^oC colder than today) and ~ 180 ppmv (during the interglacial periods where the temperature was similar to present values). This range of variation in atmospheric CO_2 is remarkably narrow, given that its concentration is determined by a highly dynamic biogeochemical cycle.This suggests that the global carbon cycle was controlled by powerful biological feedback processes to maintain a close balance between net photosynthetic uptake of CO_2 by the biosphere and its total respiration-the net source and sink strengths of the biosphere was very close to zero over at least the last 420,000 years.

There is convincing evidence that the biosphere has played a major role in regulating Earth's climate. In particular, recent data shows that although there have been periods during which Earth's temperature changed abruptly without discernable accompanying changes in the atmospheric CO2 concentrations, the converse does not appear in the glacial-interglacial records. The warming from glacial to interglacial conditions was relatively rapid, while the cooling phase leading to glaciation was initially rapid (possibly suggesting perturbation by an external event), but eventually gradual (indicating strong feedbacks that act to counter the change). These patterns suggest a long-term asymmetry in the global rates of CO_2 uptake and release by the biosphere. The terrestrial and ocean ecosystems act as buffers to maintain the global temperature in a habitable range. In contrast to the long-term record, the atmospheric CO_2 concentration today is ~ 370 ppmv - nearly 35% higher than at any time in the past 420 000 years - as a result of human perturbations to the global carbon cycle. The concentration is also rising at a rate that is at least ten times and perhaps as much as one hundred times faster than ever before observed. Clearly the biosphere's regulation of the global carbon cycle - and hence the climate system - has changed. Although it is straightforward to quantify the direct anthropogenic inputs of CO_2 to the atmosphere, a quantitative explanation of the rates of atmospheric increase has proven immensely challenging, precisely because of the strong feedbacks exerted by terrestrial and ocean ecosystems to the changes. Understanding of the biospheric feedback - the response of the world's biota to the perturbations - is needed both to gauge the magnitude of future impacts and to design appropriate mitigation actions.

HUMAN PERTURBATIONS TO THE GLOBAL CARBON CYCLE

Human perturbations to the carbon cycle have been both direct and indirect. Obvious direct effects are the addition of new carbon to the active[2] global carbon cycle through the combustion of fossil fuels, and the modification of the vegetation structure and distribution through land-use change. Deforestation, the removal of forest vegetation and replacement by other surface cover, has the largest land-use change impact on the carbon cycle, both through the loss of photosynthetic capacity in forest vegetation and the simultaneous release of large carbon stocks accumulated in forest ecosystems over long periods of time. Indirect human impacts on the carbon cycle include changes in other major global biogeochemical cycles, alteration of the atmospheric composition through the additions of pollutants as well as CO_2, and changes in the biodiversity of landscapes and species.

Currently about three-quarters of the direct human perturbations to the global carbon cycle are due to fossil-fuel combustion, emissions of which currently exceed 6 Gt C/yr (gigatonnes of carbon per year) and are still increasing. To provide perspective, this emission is equivalent to the total incineration of half of all the trees in Canada - with no residues, charcoal or soot. Every year. Since the mid-nineteenth century, however, the cumulative addition of anthropogenic CO_2 to the atmosphere by land-use change has been nearly as large (~ 156 Gt C) as that from fossil fuel use (~ 280 Gt C), and continues to be an important anthropogenic emission (2.2 Gt C/yr). Of the 7.6 ± 0.8 Gt C/yr of CO_2 added to the atmosphere by human activities during the period 1980 to 1995, only 3.2 ± 1.0 Gt C/yr remains there, with the rest taken up about equally by the oceans and by terrestrial ecosystems. Earth's biosphere thus actively removes some of the new carbon that humans have added. Terrestrial ecosystems, in particular, sequestered (took up and retained) 2.3 ± 0.9 Gt C/yr, even after accounting for the loss of 2.0 - 2.2 Gt C/yr from deforestation.

Of pressing importance in the search for mitigation strategies - activities that slow or reverse the buildup of atmospheric CO_2 - is the understanding of the mechanisms responsible for the present net biospheric uptake. Will these mechanisms continue to offset direct anthropogenic emissions? Or will they decline in strength, or even fail entirely as the carbon cycle-climate system moves into a new mode of operation, as several new terrestrial and ocean model simulations alarmingly suggest?

THE ANATOMY OF SOURCES AND SINKS AT THE STAND AND LANDSCAPE SCALE

A forest ecosystem acts as a "sink" (a net removal of atmospheric CO_2) when there is an increase in the sum of all carbon stocks retained in the forest vegetation itself and the derived stocks of organic carbon in other reservoirs. The most important of these derived reservoirs are the detritus and soil organic matter pools of the forest ecosystem.

Figure shows the conceptual pools and transfers of carbon involved in forest ecosystems and in the forest sector. In addition to the ecosystem compartments (vegetation, detritus and soil pools) and the exported pools that are located off-site (including forest products and the waste created during their manufacture and abandonment in landfills), the use of fossil-fuel reserves by the forest sector is also shown.

The net carbon balance in a forest ecosystem (NEP) can be estimated by summing all the changes in ecosystem carbon stocks (the 'inventory' method), or by direct measurement of the net exchange of CO_2 with the atmosphere (e.g. eddy correlation techniques). Provided all stocks and fluxes are accounted for, the two approaches must give identical answers (conservation of mass). In practice, a combination of the two approaches is used. The net carbon balance of a given stand of trees (patch) varies with the prevailing conditions (affecting rates of CO_2 uptake and release) but also depends strongly on the stage of development of the stand and its past history. At the landscape (or biome) scale, a forest is a mosaic of many stands of trees (individual ecosystems) in various stages of development. The net carbon balance at this scale is the summation across all such ecosystems in the landscape.

Change in the net carbon accumulation at the landscape scale thus has two components: changes in the productivity of the individual ecosystems with environmental variations and the changes in the age-class distribution associated with landscape variation in mortality and recruitment. The age-class distribution is a histogram of the fraction of the forest in each age class (typically 10 or 20 years), and is a record of past mortality and recruitment events.

Landscape sources and sinks of atmospheric carbon can arise from changes in ecosystem productivity or from changes in the disturbance regime. If the disturbance rates increase, the age-class distribution shifts to the left (younger stands) and the total carbon retained in the ecosystems in the landscape decreases - the landscape is a transient net source of CO_2 to the atmosphere until a new stable age-class distribution is reached. (If carbon is transported out of the ecosystem landscape to decompose in off-site reservoirs, such as forest products, the landscape source is reduced by that amount - in essence this component of the source is exported.) Similarly, if disturbances are suppressed, the distributions shift to the right (older forest stands) and carbon stocks increase with a transient net removal of CO_2 from the atmosphere.

EFFECTS OF CONVERTING NATURAL FOREST TO OTHER LAND COVER TYPES ON WATER YIELD

Whilst streamflow totals are observed to eventually return to preclearing levels where regrowth is allowed, the conversion of native forest to other types of vegetation cover may produce permanent changes. For example, permanent increases in annual water yield are associated with the conversion of deciduous or evergreen native forest to agricultural cropping. Reported increases range from 60-125 mm year^{-1} under humid warm temperate conditions to 450mmyear–^{1} in the equatorial tropics. The diminished water use of annual crops compared with that of a full-grown forest reflects the diminished capacity of low vegetation not only to intercept rainfall but also to extract water from deeper soil layers during periods of drought.

The former relates primarily to the lesser aerodynamic roughness of short annual crops (and possibly to their smaller leaf area and thus storage capacity as well), whereas the reduced water uptake of crops reflects their more limited rooting depth. For the same reasons, conversion to pasture generally produces permanent increases in streamflow as well, although the magnitude of the increase depends on precipitation patterns and elevation. Interestingly, Hibbert (1969) noted that replacing deciduous hardwood forest in the south-eastern USA by vigorously growing *Festuca* grass did not lead to increased water yields, although increases of up to 125 mm year^{-1} were observed after the productivity of the grass declined.

Such results can be explained in part by the fact that serious water shortages that could have reduced water uptake by shallow-rooted plants do not develop in the rainy climate at Coweeta, whereas in addition the contrast between forest and grass was lessened further by the deciduous character of the forest. In contrast, the conversion of mixed deciduous hardwoods by eastern white pine *(Pinus strobus)* in the same area brought about a reduction of 250mmyear–^{1} after 25 years.

The decline in flow was ascribed both to a steady increase in intercepted rainfall as the pines grew older and to higher transpiration by the evergreen pine trees during the dormant season.

Similarly, the higher total water use *(ET)* of mature *Pinus radiata* plantations in south-eastern Australia compared with that of native eucalypt forest has been explained in terms of the distinctly higher rainfall interception by the pines. However, where pine plantations replaced non-deciduous native forests elsewhere in the world, streamflows have been reported to return to preconversion levels within 8 years, such as in Kenya and New Zealand.

The results presented thus far pertain mostly to small headwater catchment areas involving a unilateral change in cover. Although these experiments provide a clear and consistent picture of increased water yield after replacing tall vegetation by a shorter one, and *vice versa,* such effects are more difficult to discern in large river basins having a variety of land-use types and temporal changes therein.

In addition, there are the complications associated with strong spatial and temporal variability in rainfall, especially in tropical areas with largely convective rainfall, and the often large-scale withdrawals of water for municipal, agricultural and industrial purposes in populated areas. Qian (1983) was unable to detect any systematic changes in streamflow from catchments ranging in size from 7 to 727km^2 on the island of Hainan, southern China despite a 30% reduction in tall forest cover over three decades. Dyhr-Nielsen (1986) arrived at the same conclusion for the 14 500 km^2 Pasak River basin in northern Thailand, which lost 50% of its tall forest cover between 1955 and 1980.

On the other hand, Madduma Bandara and Kurupuarachchi (1988) observed an increase in averaged annual flow totals for the 1100 km^2 upper Mahaweli catchment in Sri Lanka over the period 1940-80, despite a weak negative trend in rainfall over the same period. Although both trends were not statistically significant at the 95% level, the associated increase in annual run-off ratios was highly significant.

The increased hydrological response was ascribed to the widespread conversion of tea plantations (not forest) to annual cropping and home gardens without appropriate soil conservation measures (Madduma Bandara and Kurupuarachchi 1988).

However, when analysing a longer time series (1940-97) for the 380 km^2 upper Nilwala catchment in Sri Lanka, which experienced a 35% reduction in forest cover during the observation period, a less consistent picture was obtained. Such contrasting findings illustrate the need for high-quality rainfall and streamflow data in historical hydrological data analyses, not only in the context of evaluating the effects of land-use transformations but also those associated with climatic change.

EFFECTS OF FOREST CLEARING ON STREAMFLOW REGIMES

In areas with seasonal rainfall, the distribution of streamflow throughout the year is often of greater importance than the total annual amount *per se.* Reports of greatly diminished streamflows during the dry season after forest clearance abound in the literature, particularly in the tropics. At first sight, this seems to contradict the evidence presented earlier, that forest removal leads to higher water yields, even more so because the bulk of the increase in flow after *experimental* clearing is observed during baseflow or dry-season conditions. However, the controlled conditions imposed during catchment experiments differ from those encountered in many real-world situations.

8

Forest Resources and Uses

ECOLOGY MOVEMENTS AND CONFLICTS OVER FOREST NATURAL RESOURCES

The recent period in human history contrasts with all the earlier ones in its strikingly high rate of resource utilization. Ever expanding and intensifying industrial and agricultural production has generated increasing demands on the world's total stock and flow of resources. These demands are mostly generated from the industrially advanced countries of the North and the industrial enclaves in the underdeveloped countries of the South. Paradoxically, the increasing dependence of the industrialised societies on natural resources, through the rapid spread of energy and resource-intensive production technologies, has been accompanied by the spread of the myth that increased dependence on modern technologies implies a decreased dependence on nature and natural resources This myth is supported by the introduction of a long and indirect chain of resource utilisation which leaves invisible the real material resource demands of the industrial processes.

Through this combination of resource intensity at the material level and resource indifference at the conceptual and political levels, conflicts over natural resources generated by the new pattern of resource utilisation are generally shrouded and overlooked. These conflicts become visible when resource and energy-intensive industrial technologies are challenged by communities whose survival depends on the conservation of resources threatened by destruction and overexploitation, or when the devastatingly destructive potential of some industrial technologies is demonstrated as in the Bhopal disaster.

For centuries, vital natural resources like land, water and forests had been controlled and used collectively by village communities thus ensuring a sustainable use of these renewable resources. The first radical change in resource control and the emergence of major conflicts over natural resources induced by non-local factors was associated with colonial domination of this part of the world. Colonial domination systematically transformed the common vital resources into commodities for generating profits and growth of revenues.

The first industrial revolution was to a large extent supported by this transformation of commons into commodities which permitted European industries access to the resources of South Asia.

With the collapse of the international colonial structure and the establishment of sovereign countries in the region, this international conflict over natural resources was expected to be reduced and replaced by resource policies guided by comprehensive national interests. However, resource use policies continued along the colonial pattern and, in the recent past, a second drastic change in resource use has been initiated to meet the international requirements and the demands of the elites in the Third World, leading to yet another acute conflict among the diverse interests. The most seriously threatened interest, in this conflict, appears to be that of the politically weak and socially disorganised group whose resource requirements are minimal and whose survival is primarily dependent directly on the products of nature outside the market system. Recent changes in resource utilisation have almost wholly by-passed the survival needs of these groups. These changes are primarily guided by the requirements of the countries of the North and of the elites of the South.

This book analyses environmental conflicts in contemporary human society. In general it relates to societies all over the world, but in particular it addresses the most intense and emerging social contradictions in India related to conflicts over natural resources. Science and technology are central to these conflicts because while scientific knowledge has been used by contemporary societies to considerably enlarge man's access to natural resources, it has also allowed the utilisation natural resources at extremely high rates.

The contemporary period is characterised by the emergence of ecology movements in all parts of the world which are attempting to redesign the pattern and extent of natural resource utilisation to ensure social equality and ecological sustainability. Ecology movements emerging from conflicts over natural resources and the people's right to survival are spreading in regions like the Indian subcontinent where most natural resources are already being utilised to fulfil the basic survival needs of a large majority of people.

The introduction of resource and energy-intensive production technologies under such conditions leads to economic growth for a small minority while, at the same time, undermines the material basis for the survival of the large majority.

In this way, ecology movements have questioned the validity of the dominant concepts and indicators of economic development. The ideology of economic development, which remained almost monolithic in the post World War II period, is thus faced with a major foundational challenge.

In this chapter an attempt has been made to provide a systematic conceptual framework for analysing the processes and structures of modern economic development from an ecological -perspective. It attempts to analyse the relationship between economic development and conflicts over natural resources to trace the roots of ecological movements.

Further, in the light of the ecological perspective, it examines the fundamental assumptions and categories of modern development economics that are used to determine the objectives of economic development as well as the criteria for the choice of technologies that are used to achieve these objectives.

ECONOMIC DEVELOPMENT AND ENVIRONMENTAL CONFLICTS IN INDIA

A characteristic of indian civilization has been its sensitivity to natural ecosystems. vital renewable natural resources like vegetation, soil and water were managed and utilised according to well defined social norms that respected the known ecological processes. The indigenous modes of natural resources utilisation were sensitive to the limits to which these resources could be used It is said that the codes of visiting important pilgrim centres Badrinath in the sensitive Himalayan ecosystem, included a maximum stay of one night so that the temple area would not put excess pressure on the local natural resources base. In the precolonial indigenous economic processes, the levels of utilisation of natural resources were not significant enough to result in drastic environmental problems. There were useful social norms for environmentally safe resource utilisation and people protested against the destructive use of resources even by kings. A major change in the utilisation of natural resources of India was introduced by the British who linked the resources of this country with the direct and large non-local demands of Western Europe. Natural resource utilisation by the East India Company, and later by the colonial rulers, replaced the indigenous organizations for the utilisation of natural resources, like water, forest and minerals, that were mainly managed as commons.

With the establishment of British colonial rule in India, the ever increasing resource demands of-the industrial revolution in England were largely met from colonies like India. Forced cultivation of indigo in Bengal and Bihar, cultivation of cotton in Gujarat and the Deccan led to large-scale commitment of land for the supply of raw materials for the British textile industry, the flagbearer of the industrial revolution. Forests in the sensitive mountain ecosystems like the Western Ghats or the Himalayas were felled to build battleships, or to meet the requirements of the expanding railway network.

Forests of the Bengal-Bihar-Orissa region were used for running wood fuel locomotives in the early stages of railway expansion. The latter stages of colonial resource utilisation and control included the monopolization. of water rights as in the Sambhar Lake of Rajasthan or the Damoda' Canal in Bengal. Colonial intervention in natural resource management in India led to conflicts over vital renewable natural resources like water or forests and induced new forms of poverty and deprivation. Changes in resource endowments and entitlements introduced by the British came into conflict with the local people's age old rights and practices related to natural resource utilisation As a result local responses were generated through which people tried to regain and retain control over local natural resources.

The indigo Movement in Eastern India, the Deccan Movement for land rights or the forest movement in all forest areas of the country, the Western Ghats, the Central Indian Hills or the Himalayas, were obvious expressions of protest generated by these newly created conflict's. Conflicts generated by the colonial modes of natural resource exploitation could not, however, grow with a local identity. With the progress of the anti-colonial people's movement at the national level, these local protests merged with the national struggle for independence. With the collapse of colonial rule internationally, and the emergence of sovereign independent countries in the Third World like India, resolution of these conflicts at the local level became a possibility. While political independence vested the control over natural resources with the Indian state, the colonial institutional framework for natural resource management did not change in essence. Where colonialism collapsed, the slogan of economic development stepped in. There was unfortunately no alternative institutional mechanism other than that of the classical model of development left by the British, with which the newly formed Indian state could respond to the accentuated aspirations of the Indian people for a better life.

The same institutions and concepts, nurtured and developed by the colonial rulers were applied to objectives which were exactly opposite to those of the colonial period. Concepts and categories relating to economic development and natural resource utilisation that had emerged in the specific context of capitalist growth and industrialization in the centres of colonial power were raised to the level of universal assumptions and applicability. The processes which led to deprivation were now entrusted with the responsibility of basic needs satisfaction. No serious thought was given to the fact that the historical specificity of early industrial development in Western Europe necessitated the permanent occupation of the colonies and the undermining of the local 'natural economy'.' This inexorable logic of resource exploitation, exhaustion and alienation integral to the classical model of economic development based on resource intensive technologies led Gandhi to seek an alternate path of development for India when he wrote:

God forbid that India should ever take to industrialism after the manner of the West. The economic imperialism of a single tiny island kingdom (England) is today keeping the world in chains. If an entire nation of 300 million took to similar economic exploitation, it would strip the world bare like locusts.

While Gandhi's critique was a forewarning against the problems likely to arise by following the classical path of resource-intensive development, at the time of India's independence, there was no clear and comprehensive work plan to realise the Gandhian dream of alternate development that would be resource prudent and would satisfy basic needs. The issues of resource constraints of economic development were, therefore, not highlighted at the theoretical level, partly due to the tremendous pressure of the enhanced developmental aspirations of a newly independent nation, and partly due to the lack of internalization of natural resource parameters within the framework of economics. As the scale of economic development activities escalated from one Five Year Plan to another, the disruption of ecological processes that maintain the productivity of the natural resource base started becoming increasingly apparent.

The classical model of economic development in the case of the newly independent nations resulted in the growth of urban-industrial enclaves where commodity production was concentrated, as well as rapid exhaustion of the internal colonies whose resources supported the enhanced demands of these enclaves. In the absence of ecologically enlightened resource management methods, the pressure of poverty enhanced the pace of economic development activities in the hope of a quick improvement in the standard of living for all, as in the case of Western Europe. For example, commercial forestry earned more revenue by making increasing amount of timber and pulpwood available in the market but in the process reduced the multipurpose biomass productivity or damaged the hydrology of the forests. People dependent on non-timber biomass outputs of forests like leaves, twigs, fruits, nuts, medicines and oils were unable to sustain themselves, in the face of the commercial exploitation of forests. The changed hydrological character of the forests affected both the micro-climate and the stream flows, disturbing the hydrological stability and affecting agricultural production.

There are similar examples from all parts of the country, related to almost all massive developmental interventions in India's natural resource system. Ecological degradation and economic deprivation generated by the resource insensitivity and intensity of the classical model of development have resulted in environmental conflicts, an understanding of which is imperative for the reorientation of our current development priorities and concepts. It is becoming in creasingly clear that these classical concepts and priorities are being used as an alibi to direct 'development' at the national level, while the educated minority elite is the main beneficiary of these 'development' processes.

The ecology movements that have emerged as major social movements in many parts of India are making visible many invisible externalities and pressing for their internalisation in the economic evaluation of the elite-oriented development process. In the context of a limited resource base and unlimited development aspirations, ecology movements have initiated a new political struggle for safeguarding the interests and survival of the poor, the marginalised, including women, tribals and poor peasants.

Ecology Movements and Survival

The intensity and range of ecology movements in independent india have continuously widened as predatory exploitation of natural resources to feed the process of development has increased in extent and intensity. This process has been characterised by the massive expansion of energy and resource-intensive industrial activity and major development projects like large dams, forest exploitation, mining and energy-intensive agriculture. The resource demand of development has led to the narrowing of the natural resource base for the survival of the economically poor and powerless, either by direct transfer of resources away from basic needs or by destruction of the essential ecological process that ensure renewability of the life-supportirig natural resources.

In the light of this background, ecology movements emerged as the people's response to this new threat to their survival and as a demand for the ecological conservation of vital life-support systems. The most significant life-support systems in addition to clean air are the common property resources of water, forests and land on which the majority of the poor people of India depend for survival. It is the threat to these resources that has been the focus of ecology movements in the last few decades. Among the various ecology movements in India, the Chipko movement (embrace the trees to oppose fellings) is the most well known.

It began as a movement of the hill people in the state of Uttar Pradesh to save the forest resources from exploitation by contractors from outside.' It later evolved into an ecological movement that was aimed at the maintenance of the ecological stability of the major upland watersheds in India. Spontaneous people's response to save vital forest resources was seen in Jharkhand area in Bihar-Orissa border region as well as in Bastar area of Madbya Pradesh where there were attempts to convert the mixed natural forests into plantations of commercial tree species, to the complete detriment of the tribal people. In the southern part of India the Appiko movement, which was inspired by the success of the Chipko movement in the Himalayas, is actively involved in stopping illegal over-felling of forests and in replanting forest lands with multipurpose broad leaved tree species. In Himachal Pradesh the Chipko activists have intensified their opposition to the expansion of monoculture plantation of the commercial Chir Pine (Pinus roxburghii). In the Aravalli Hills of Rajasthan there has been a massive programme of tree planting to give employment to those hands which were hitherto engaged in felling of trees.

The exploitation of mineral resources, in particular the opencast mining in the sensitive watersheds of the Himalayas, the Western Ghats and Central

India have also resulted in a great deal of environmental damage. As a consequence, environmental movements have come up in these regions to oppose the reckless mining operations. Most successful among them is the movement against limestone quarrying in the Doon Valley. Here, volunteers of the Chipko movement have led thousands of villagers, in peaceful resistance, to oppose the reckless functioning of limestone quarries that is seen by the people as a direct threat to their economic and physical survival.'

While the Doon Valley instance has a long history of popular opposition to the quarrying of limestone and a Supreme Court order has restricted the area of quarrying to a minimum, examples of such success' of ecology movements are rare People's ecology movements against mineral exploitation in the neighbouring areas of Almora and Pithoragarh still seem to be ignored, probably due to the relative isolation of these interior areas. Beyond the Himalayas, the ecology movement in the Gandhamardan Hills in Orissa against the ecological havoc of bauxite mining has gained momentum and it draws inspiration from the Chipko movement.

The mining project of the Bharat Aluminium Company (BALCO) in the Gandhamardan Hills is being opposed by local youth organisations and tribal people whose survival is directly under threat. The peaceful demonstrators have claimed that the project could be only continued 'over our dead bodies. The situation is more or less the same in large parts of Orissa-Madhya Pradesh region where rich mineral and coal deposits are being opened up for exploitation and thousands of people in these interior areas are being pushed to deprivation and destitution. This is also true of the coal mining areas around the energy capital of the country in Singrauli. In these interior areas of Central India, movements against both mining and forestry are becoming increasingly volatile and people's resistance is growing.

Large river valley projects, which are coming up in India at a very rapid pace, is another group of development projects against which people have organised ecology movements. The large-scale submersion of forest and agricultural lands, a prerequisite for the large river valley projects, always takes a heavy toll of dense forests and the best food growing lands. These have usually been the material basis for the survival of a large number of people in India, specially tribal people.

The Silent Valley project in Kerala was opposed by the ecology movement on the ground of its being a threat, not to the survival of the people directly, but to the gene pool of the Tropical Rainforests threatened by submersion. The ecological movement against the Tehri high dam in the UP Himalaya exposes the possible threat to people living both above and below the dam site through large-scale destabilization of land by seepage and strong seismic movements that could be induced by impoundment. The Tehri Dam Opposition Committee has appealed to the Supreme Court against the proposed dam by identifying it as a threat to the survival of all people living near the river Ganga up to West Bengal. Most notable among the people's movements against dams on the issue of direct threat to survival from submersion are Bedthi lcchampalli, Bhopalpatnam, Narmada Sagar, Koel-Karo, Bodhghat, etc.

In the context of the already overutilised land resources, the proper rehabilitation on a land-to-land basis of millions of people displaced through the construction of dams seems impossible. The cash compensation given instead is inadequate in all respects for providing an alternate livelihood for the majority of the displaced. Destitution is thus the first and foremost precondition for initiating large dam projects

While the process of construction of dams itself invites opposition from ecology movements, the functioning of water projects dependent on the constructed dams results in further ecological disasters and movements. People's movements against widespread water-logging, salinisation and the resulting desertification in the command areas of many dams have been registered. Among them are instances of protests against the Tawa, Kosi, Gandak, Tungabhadra, Malaprabha, Ghatprabha projects and the canal irrigated areas of Punjab and Haryana. While excess water led to ecological destruction in these cases, improper and unsustainable use of water in the arid and semi-arid regions generated ecology movements in a different way. The anti-drought and desertification movement is gaining momentum in the dry areas of Maharashtra, Karnataka, Rajasthan, Orissa, etc. Ecological water use for survival is being advocated by water based movements like Pani Chetana, Pani Panchayat, and Mukti Sangharsh. Another major movement originating from the ecological destruction of resources by growth based development is spreading all along the 7,000 km long coastline of India. It is the movement of the small fishing communities against the ecological destruction caused by mechanised fishing whose instant profit motive is destroying the coastal ecology and its long-term biological productivity in a big way.

No amount of threat to survival in India from environmental hazards can be complete without a reference to the Bhopal tragedy on 2 December 1984, in which several thousand people died and several lakhs faced serious health hazards following the leakage of poisonous Methyl Iso Cyanate from a pesticide plant of Union Carbide (India) Limited. People's movements for clean air and water are growing in ail parts of the country just as ecologically irresponsible industrialization is moving deeper into the hinterland in search of new resources.

NATURAL RESOURCE MANAGERS

Conservation efforts often fail to take into consideration that alternatives have to be available if local populations are to change their natural resource utilization patterns. Along the Cimanuk River in West Java, efforts to prohibit agriculture on the raised riverbanks to curb seasonal flooding irremediably failed until viable economic alternatives were introduced. Through participatory development communication, the dialogue between researchers, decision-makers and small and landless farmers made it possible for sheep raising to flourish as an alternative to the practice of agriculture on the riverbanks that was making villages more vulnerable to annual flooding.

Fifteen years ago, as we were preparing a baseline study to explore the development options that could improve the living conditions of small farmers and landless farm labourers living along the Cimanuk River in the district of Majalengka, West Java, we wondered whether those resource-poor farmers would be able to participate in managing a strip of public land stretched out along the river. We also wondered whether they would be able to derive benefits from such an undertaking.

At that time, the main concern of the local government and of the communities living along the river was flood control. Every rainy season, heavy downpours filled up the river and flooded the villages. To control this annual flooding, the Department of Public Works bought a strip of land along the river that goes through the district's villages and small towns, and raised the riverbanks to a certain height. Hence, the annual flooding was controlled and the communities were freed from the yearly threat of natural disaster.

However, this flood prevention technique created another problem for the small and landless farmers who had been farming on the riversides long before the banks were raised. According to the new government rules and regulations, farming activities were forbidden on the public land along the river due to the risks of destabilization of the land structure and hindrance of the waterways.

In spite of the government rules and regulations, the small and landless farmers pursued their farming activities as usual. On their part, officials of the Department of Public Works continued to enforce the regulations by eradicating the crops that could possibly weaken the raised riverbank's land structure.

Looking for Alternatives

After several years of playing cat and mouse and tired of conflicting with the local communities, the Department of Public Works finally proposed a win–win solution to the farmers. The latter were finally allowed to farm along the riversides, provided that they grew grass at least 1m away from the river's brim and on both sides of the raised bank. However, they were not allowed to grow bananas or any other type of land crops that could destabilize the land structure. To further motivate the farmers to grow grass, the government officials promised to supply them with sheep. However, this promise did not materialize very well. The farmers were disappointed with the few low-quality sheep delivered to them and the abundant grass was wasted.

At that critical time, a baseline study was undertaken in order to understand the availability of local natural resources such as land, crops, animals and water, as well as the limitations, problems and opportunities associated with their use and development. Moreover, the study also aimed at understanding the traditional systems and at assessing indigenous capacity. The study concluded that there was potential for involving the small farmers in managing the public land located along the river and the irrigation channels, and for deriving benefits from it, provided that a funding body be put in place.

A follow-up pilot study was then undertaken. Essentially, the challenge consisted in helping farmers to solve their problems and benefit from the abundant grass, while ensuring the conservation of the riverbank. After discussing the best-suited approach, our team decided to use participatory development communication. Thus, linkages were established between university researchers, local government, the Department of Public Works, the local livestock extension service, the village authorities and local farm communities.

USING APPROPRIATE COMMUNICATION TOOLS

Prior to the fieldwork, communication materials such as sound-slide shows, posters and even a comic book were prepared and tested with farmers and extension workers. Development communication graduate students were involved in producing and testing these materials, either through their message design course assignments or through their theses research.

Topics such as the social and economic potential of sheep raising in the district of Majalengka were specifically addressed with local decision-makers and livestock extension workers. Other topics concerning various technical aspects of sheep production, marketing and rural family budget management were developed for extension workers, farmers and their wives. The fieldwork started once the communication materials were ready. In the first quarter of 1992, a memorandum of understanding was officially signed between the government of the district of Majalengka and the Faculty of Animal Science at Bogor Agricultural University.

During the ceremony, a sound-slide show regarding the social and economic potentials of sheep raising in the district of Majalengka was shown to the government officials and to the local House of Representatives, prior to signing the agreement. This created enthusiasm among the officials attending the ceremony. Moreover, the local media covered the event, which was broadcast on the evening news by the regional television station. The next step consisted of promoting capacity-building for the local livestock extension personnel and farmer leaders. Once a month, researchers conducted an on-site training meeting for the livestock extension workers and other officials of the district livestock service. The technical aspects of sheep production, including housing, feeding, breeding and sheep healthcare, were addressed. Two sound-slide programmes about sheep production were shown and discussed with the extension workers. A pre-test and a post-test on the training subjects were always administered to the extension workers prior to and after the slide show in order to build up the agents' interest and to focus their attention on the subjects. Moreover, the tests were also useful in indicating whether the sound-slide shows had been effective in improving the agents' knowledge on the topics that were being discussed.

Following the agents' training, a short meeting was conducted for farmer leaders of the four pilot sites. On that occasion, they were informed that they were entitled to receiving an in-kind loan in the form of sheep. If they agreed, they would have to return a certain number of the sheep offspring to the project for a period of four years.

After the meeting, participants were invited to visit a senior farmer leader in the village of Balida to witness how he collected forage. One month later, they joined an excursion to the district of Garut to visit a sheep-raising initiative. The Garut sheep were particularly well known for their fast growth and prolificacy.

Trying out New Techniques

Upon their return, every farmer leader was asked to build a good sheep house capable of housing two adult rams and ten ewes, and to prepare a plot to grow improved varieties of grass and legumes to feed the sheep for demonstration purposes.

Intensive supervision and backstopping activities were provided to the farmer leaders in the following month to allow them to get first-hand experience in raising the new sheep breed in their own environment. After one month, the farmer leaders had enough experience in taking care of the sheep and were able to provide information about the Garut sheep raising to other farmers in the villages.

Upon completion of the trial run among the farmer leaders, monthly evening training meetings were conducted for the rest of the farmers in the four research sites. The first training meeting was about building a good low-cost and healthy sheep house using local materials, while the second one focused on the appropriate feeding for this specific breed of sheep. During the meetings, all farmers were given pre-tests and post-tests in order to measure their knowledge regarding the training subjects prior to and after the audio-visual presentations.

About two weeks after the training sessions, the farmer leaders were asked to check whether all the farmers had made or had improved their sheep houses for the incoming new sheep. Farmers who had not done so were encouraged to rebuild or improve their sheep houses.

Two weeks after completing the necessary preparations, a farmer leader with extensive experience in sheep breeding was asked to join the researchers to select rams and ewes for 100 farmers in the four research sites. The training meetings continued after the sheep had been distributed to the farmers. The topics addressed ranged from technical aspects of sheep raising to rural family budget management.

Meanwhile, the research assistant, together with the farmer leaders, visited every farmer once a week to check on the sheep's condition and health and to provide further backstopping activities to the farmers.

About one year after the distribution of the sheep, all farmer leaders were invited to a meeting in the district capital to share their experience in Garut sheep raising with farmers from other villages. This farmer-to-farmer communication turned out to be an excellent medium for raising other farmers' interest in Garut sheep production, better than anything that researchers or extension workers could have done.

Promising Results

The initiative had a first positive result when the head of the local livestock service, using the narrative progress report for the first year, was able to convince the local government to provide his office with extra funding

Towards the end of the second year, results began to be experienced at the grassroots level, when several farmers began to return the sheep offspring to the initiative. According to a previous agreement, the sheep offspring were to be split fifty–fifty. The returned offspring were to be revolved to other farmers interested in sheep raising. This was an indication that the initiative was having a positive impact on the small farmers involved. Indeed, between September and December 1993, the sheep population increased by 159 per cent. One year later, the increase reached 271 per cent, while at the end of 1995, it had risen to 306 per cent. The value of the participating farmers' assets also increased accordingly. In less than three years, the value of the sheep almost tripled – a positive contribution to the local economy. Moreover, the welfare of most farmers involved in the initiative improved significantly since the income generated by selling the sheep was worth more than 1.5 times the capital loaned to them.

In addition to the economic benefits, such as having more sheep in barns and gaining additional income, the initiative also brought about some social benefits to the villagers involved, who gained respect from their families and communities. One of the farmer leaders even managed to send his son to the Bogor Agricultural University, where he obtained his degree in Animal Science last year.

In May 1996, after about five years of continued fieldwork, the initiative officially came to an end. However, since the ewes continued to give birth and farmers kept on returning the sheep offspring, the revolving activities continued on their own.

A New Spur to the Initiative

In early 1999, following the severe Asian financial crises that badly hurt the Indonesian economy, the initiative obtained additional funding to buy 55 new sheep. These sheep were distributed to two other villages in the district of Majalengka. Meanwhile, around the same time, one former project group in the village of Kadipaten received a 300 million Indonesian rupiah loan through the World Bank's social security network funds. The purpose of this loan was to help the local communities cope with the economic crisis. The farmer group used it to further develop their sheep raising ventures.

Eight years after the initiative terminated, we are still serving the farmers, though at a slower pace and on a smaller scale. To cover the operational costs of supervision and backstopping activities, fieldworkers have been allowed to sell the culled sheep.

Today, livestock production, including sheep production, is growing at a good pace in the district of Majalengka. Both public and private investments have been increasing steadily, especially since the 1998 economic crisis. Between 2002 and 2004, the government of Majalengka invested over 9 million rupiahs in livestock production.

Public funds have been channelled to farmers as loans through local banks, mainly for beef and cattle production. In addition, a big livestock marketing infrastructure has recently been built in the region. Livestock traders from east and central Java bring and sell their sheep and cattle in the newly operated livestock market to local and other traders and buyers from Bandung, Bogor and Jakarta.In other words, the situation of livestock economics in the district of Majalengka has greatly evolved compared to the situation that prevailed in 1989, when the research team first arrived. We believe that over the last 15 years, our work has contributed to this situation by raising a flag to both public and private audiences, informing them about the existing potential in a remote district of West Java that was waiting to be explored and developed.

During recent times, when driving along the river and the irrigation channels in the former research sites, one can see that the irrigation channels are well maintained and that farmers periodically harvest the grass on the riverbanks. From a distance, one can see numerous small white plastic flags stuck orderly in the ground every 10m, on both sides of the irrigation channel. What do these flags mean? Are they a border line? Yes, they are. No one is now allowed to harvest the grass except for the farmer who sticks the flags on the ground. The grass now means something to the rural communities as it has played a key role in local economic development, while contributing to solving environmental problems.

In summary, through participatory development communication, researchers, extension workers, government officials and cooperating small farmers were able to work hand in hand to solve certain community problems.

Through the introduction of sheep raising as an alternative economic activity, the small and landless farmers living in the farming communities along the sides of the Cimanuk River have been able to ensure the management of the meagre public natural resources in their environment.

Problems Encountered

This story would not be complete without describing some of the problems that the researchers encountered in the course of the initiative.

The first problem occurred during the early phase of the fieldwork, when we were trying to gain acceptance and support from the village heads. Most of the village heads in the research sites had shown great interest in the benefits that they could obtain from the activities that were to occur in their areas, and they were keen on having a cut. Denying this opportunity to the village heads could have meant troubles ahead or, worse, it could have led to the failure of the initiative altogether. In contrast, letting them have a share of the benefits could possibly spare the researchers and the farmer leaders many feuds and tensions with the village officials.

For those reasons, the wisest choice at that time seemed to be to give them a small slice of the benefits. With that in mind, we included the village heads' relatives on the list of sheep recipients, provided that they agreed to abide by the project's policies and rules. It seemed to us that by doing this, we would save a lot of energy and avoid further conflict between farmer leaders and village heads. This compromise would also make it possible for us to use the village halls for training meetings.

As the activities unfolded, we found out that high mortality rates occurred among the project sheep raised by the village heads' relatives. Why? The answer was simple. They were not farmers and did not have the necessary skills to raise sheep. Moreover, they did not spend enough time on their tasks. Consequently, they failed in raising the sheep and also failed in returning the sheep offspring.

The second problem we encountered was the implementation of inappropriate feeding practices. Farmers were used to feeding natural grass to their animals and gave similar feeds to the new prolific sheep, without adding enough legumes in the rations. As a result, many animals suffered malnutrition.

Malnutrition brought about many miscarriages among the pregnant ewes and high mortality rates among newly born lambs. To solve these problems, the research team introduced urea molasses block as a protein and mineral supplement for the sheep.

The third problem was the lack of grass and forage supply during the dry season. This recurring problem was seasonal and continuously haunted the farmers year after year. At the beginning of the fieldwork, the research team, together with certain farmer leaders, initiated a demonstration plot to cultivate improved grass and legume varieties in the four research sites. However, very few farmers were willing to cultivate the grass and the legumes, arguing that the grass supplies had always been abundant. Why bother cultivating it?

Another difficulty encountered was that farmers did not respond well to the suggestion of silage-making as a way of preserving the grass, particularly during the rainy season, when there is a large supply of grass. They did not like its smell. Lastly, after the project's end, when we had to reduce the frequency of our visits to farmers due to limited funds, many farmers did not keep their promise to return good-quality sheep offspring to the initiative. This situation badly affected the efforts to revolve the sheep offspring to other potential farmers. The sustainability question then came up. How much longer would the project last?

Despite these shortcomings, this initiative provided a number of learning opportunities. Looking back on the work accomplished, we can draw lessons and identify the most outstanding features that contributed to making it a success story. First, conducting a comprehensive baseline study on the research sites was very useful in order to understand the situation, the problems and the opportunities related to local natural resource development, as well as the traditional systems and indigenous capacities of local communities. This also allowed for the development of a sound action plan to implement the initiative.

Second, the strong commitment of a funding partner to ensure adequate funding in order to implement the action plan and prepare the communication materials proved to be a key issue. Furthermore, support from local policy-makers, village authorities and farmer leaders also proved extremely important to ensure the successful implementation of the action plan. The direct involvement of local extension workers and farmer leaders in disseminating the materials, in distributing the sheep and in providing backstopping activities also seems to have greatly contributed to building the necessary trust with local farmers. Moreover, holding monthly meetings proved to be a good way of getting feedback and monitoring activities, while providing further information to farmer groups about sheep raising and marketing their products.

Finally, holding dissemination seminars on the project's success, with the participation of farmer leaders, the academic community and policy-makers, was very important for the project's replication and expansion.

RESOURCES OF THE NATIONAL FORESTS

The Forest Service began in earnest the development of all the resources of the national forests. Mature timber was sold wherever there was a demand for it and the permanent welfare of the forests and protection of the streams permitted, but always so as to prevent waste, guard against fire, protect young. growth and ensure reproduction. Regulations were adopted which allowed small sales to be made without formality or delay. secured for the government the full value of timber sold and eliminated unnecessary routine. Care was taken to safeguard the interests of the government and provide for the maintenance of good technical standards. The conduct of local business was entrusted to local officers. Large transactions with general policies were controlled from Washington, but with careful 2 = *District Boundaries and Numbers* provision for first-hand knowledge and close touch with the work in the field.

Business efficiency and the convenience of the public were carefully studied. In short, an organization was created capable of handling safely, speedily and satisfactorily the complex business of making useful a forest property of vast extent, scattered through sixteen different states of an aggregate area of over 1,50o,000 sq. m. and with a population of 9,000,000.

The growth since the 1st of July 1897 of the area of the national forests, of the expenditures of the government for forestry and of the receipts from the national forests, is shown by the statement which follows. Though the act of June 4, 1897, became effective immediately upon its passage, the fiscal year 1899 was the first of actual administration, because the first for which Congress made the appropriation necessary to carry out the law.

Forest is ripe for the axe, the demand is strong and control by trained men makes it safe to cut more freely. The increase is marked both in small and in large sales, but a score of sales for less than $5000 are made against one for more. The total cut is still far below the annual increment of the forests. As the demand grows restrictions must increase in order to husband the present supply until the next crop matures.

- The stumpage price would seem on the face of the figures to have risen from about one dollar to more than three dollars per thousand board-feet. The receipts, however, for any one year are not exclusively for the timber cut in that year, since payments are made in advance. In the year 1907 the average price obtained was something less than $2.50 per thousand. It is therefore true that stumpage prices have risen greatly, although conditions new to the American lumbermen are im *Area of National Forests, Annual Expenditures of the Federal Government for Forestry and National Forest Administration and Receipts from National Forests, 1898-1909* Until 1906, the sole source of receipts was the sale of timber. In the fiscal year 1907, however, timber sales furnished less than half the receipts. The following statement concerning the timber sales of the fiscal years 1904-1907 will serve to bring out the change that followed the transfer of control to the forest service in the midst of the fiscal year 1905:
  - A large excess each year in the amount of timber sold over that cut and paid for;
  - Nine times as much timber sold at the end of the four-year period as at the beginning and three times as much cut;

  - A much higher price obtained per thousand board-feet at the end of the period than at the beginning. Each of these matters calls for comment. The sales are of stumpage only; the government does no logging on its own account.
- More timber is sold each year than is cut and paid for, because many of the sales extend over several years. With increasing sales the amount sold each year for future removal has exceeded the amount to be removed during that year under sales of earlier years. Large sales covering a term of years are made because the national forests contain much overmature timber, which needs removal, but which is frequently too inaccessible to be saleable in small amounts. To prevent speculation the time allowed for cutting is never more than five years and cutting must begin at once and be continued steadily.
- The volume of sales has increased rapidly because much 1 The United States fiscal year ends June 30 and receives its designation from the calendar year in which it terminates. Thus, the fiscal year 1898 is the year July 1, 1897 - June 30, 1898.

Administration transferred to Bureau of Forestry, February I, 1905. Full utilization of all merchantable material, care of young growth in felling and logging and the piling of brush, to be subsequently burned by the forest officers if burning is necessary, are among these conditions.

Timber to be cut must first be marked by the forest officers. Sales of more than $100 in value are made only after public advertisement.

Only the simplest forms of silviculture have as yet been introduced. The vast area of the national forests, the comparatively sparse population of the West, the rough and broken character of the forests themselves and the newness of the problems which their management presents, make the general application of intensive methods for the present impracticable.

Natural reproduction is secured. The selection system is most used, often under the rough and ready method of an approximate diameter limit, with the reservation of seed trees where needed. The tendency, however, is strongly towards a more flexible and effective application of the selection principle, as a better trained field force is developed and as market conditions improve. One conspicuous achievement was the reduction of loss by fires on the national forests. During the unusually dry season of 1905 there were only eight fires of any importance and the area burned over amounted only to about 16 of 1% of the total area. In 1906 about 12 of 1% was burned. This was accomplished by efficient patrol, co-operation of the public and by preventive measures, such as piling and burning the brush on cut-over areas.

Since the beginning of 1906 the largest source of income from the national forests was their use for grazing. Stock-raising is one of the most important industries of the West. Formerly cattle and sheep grazed freely on all parts of the public domain. In the early days of the national forests the wisdom of permitting any grazing at all upon them was sharply questioned. Unrestricted grazing had led to friction between individuals, the deterioration of much of the range through overstocking and serious injury to the forests and stream flow. The forests of the West, however, are largely of open growth and contain many grassy parks, the results of old fires and many high mountain meadows.

Under proper regulations the grass and other forage plants which they produce in great quantity can be used without detriment to the forests themselves and with great benefit to the stock industry, which often can find summer pasturage nowhere else.

Except in southern California grazing is now permitted on all national forests unless the watersheds furnish water for domestic use; but the time of entering and leaving, the number of head to be grazed by each applicant and the part of the range to be occupied are carefully prescribed. Planted areas and cut-over areas are closed to stock until the young growth is safe from harm and goats are allowed only in the brushland of the foothills.

The results of regulation, in addition to the protection of forest growth and streams, are the prevention of disputes, improved range, better stock, stable conditions in the stock industry and the best use of the range in the interest of progress and development. The first right to graze stock on the forests is given to residents, small owners and those who have used the range before. Thus the crowding out of the weaker by the stronger and of the settler by the roving outsider has been stopped. In 1906 the forest service began to impose a moderate charge for the use of the national forest range.

The following statement shows the amount of stock grazed on the national forests 1904-09 and the receipts for the grazing charge: - A work of enormous magnitude which has now begun is planting on the national forests. At present, with low stumpage prices and incomplete utilization of forest products, clear cutting with subsequent planting is not practicable.

There are, however, many million acres of denuded land within the national forests which require planting. Such planting is still confined chiefly to watersheds which supply cities and towns with water. The first planting was done in 1892, in California. Since then similar work has been done on city watersheds in Colorado, Utah, Idaho and New Mexico. Other plantations are in the Black Hills national forest, where large areas of cut-over and burned-over land are entirely without seed trees and in the sandhill region of Nebraska. Up to 1908 about 2,000,000 seedlings had been planted, on over 2000 acres - a small beginning, but the work was entirely new and presented many hard problems. The nursery operations of the forest service are concentrated at seven stations, located in southern California, Nebraska, Colorado, New Mexico Utah and Idaho, where stock is raised for local planting and for shipment elsewhere. These nurseries are small. Their annual productive capacity is between 8,000,000 and seedlings. Each nursery is practically an experimental forest-planting station, at which a large variety of species are grown and various methods are tried.

The organization of the administrative work of the national forests is by single forests. On the 1st of January 1908 the total number of forests was 165 with a total area of 162,023,190 acres. In charge of each forest is a forest supervisor. Under the supervisors are forest rangers and forest guards, whose duties include patrol, marking timber and scaling logs, enforcing the regulations and conducting some of the minor business arising from the use of the forests. Guards are temporary employes; rangers are employed by the year. The supervisors report directly to and receive instructions from the central office at Washington.

In this office there are four branches - operation, grazing, silviculture and products - each of which directs that part of the work which belongs to it, dealing directly with the supervisor.

For inspection purposes, however, the forests are separated into six districts, in each of which is located a chief inspector with a corps of assistants. The inspectors are without administrative authority, but assist by their counsel the supervisors and through inspection reports keep the Washington office informed of the condition of all lines of administrative work in progress. Administrative officers alternate frequently between field and office duties.

The number of forest officers in the several grades on the 1st of January 1908 were: 6 chief inspectors, 26 inspectors, 106 forest supervisors, 41 deputy forest supervisors, 820 forest rangers and 283 forest guards. The total number of employes of the forest service on the same date, including the clerical force, was 2034.

Besides the administration of the national forests, the forest service conducts general investigations, carries on an extensive educational work and co-operates with private owners who contemplate forest management upon their own tracts. This last work is undertaken because of the need of bringing forestry into practice, the lack of trained foresters outside of the employ of the government and the lack of information as to how to apply forestry and what returns may be obtained. Co-operation takes the form of advice upon the ground and, on occasion, of the making of working plans.

The educational work of the service is performed chiefly through publications, the purpose of which is to spread very widely a knowledge of the importance of forestry to the nation and of the principles upon which its practice rests. The investigations which the service conducts extend from studies. of the natural distribution and classification of American forests. and of their varied silvicultural problems to statistics of lumber production and laboratory researches which bear upon the economical utilization of forest products. As examples of these researches may be mentioned tests of the strength of timber, studies of the preservative treatment of wood for various uses, wood-pulp investigations and studies in wood chemistry.

THE USES OF WOOD

The invention of metal working, first in bronze, then in iron, profoundly influenced man's relation to the forest. The operation of smelting created a technical use for fuel, hitherto needed chiefly for domestic purposes. Improved cutting tools speeded up the clearing of forests and added immeasurably to the ease of wood utilization, since better tools meant easier shaping of wood into useful articles. This in turn created a heavier draft than ever upon the forest, particularly for the construction of dwellings, public buildings and ships. The advent of framed structures to replace laboriously built stonemasonry and perishable huts of mud and wattle or rushes and thatch marked a most significant contribution of the forest to human advancement.

It should be noted that rapid progress of the ceramic arts— glass, pottery, brick and tile manufacture—and of such simple chemical arts as brewing and the manufacture of lye and dyestuffs all called for an expansion in the use of fuel. But it was the invention of steel, an alloy of iron and carbon, which was most devastating to the forests of Europe.

For this process charcoal was the essential source of carbon. Armament and empire went hand-in-hand and the age of Charlemagne is no more notable for its military exploits than for its wholesale destruction of forests for the manufacture of steel. An incidental, but important, by-product of such destruction has been the easier conquest and domestication of forest-dwelling tribes, thus deprived of their shelter.

ATTEMPTS TO COMPENSATE FOR THE STRONG INROADS ON THE FORESTS

There was, however, a conserving influence in the feudal military pattern. Forests were valued as hunting preserves for royalty and nobility. Thus the region of Avernus, which Virgil mentions as heavily forested, was later denuded except for the hunting preserves of the kings of Naples. Both on the Continent and in England the chief areas where forests were protected against the insatiable demand for charcoal and timber were those which, like the New Forest, were reserved for sport and recreation of the ruling class.

Meanwhile there was general ignorance of some of the less obvious functions of the forests, though Plato had linked the deforestation of the Grecian hills with the drying of the ancient springs. In 1609 Enrico Martinez stated that the washing of mud down into the Valley of Mexico and the danger of flood had been greatly increased by the destruction of forests and other plant cover on the highlands surrounding the basin. He referred to the hills as "descarnados," literally de-fleshed or stripped to the bone.

Scientific evidence as to the part played by forests in building and stabilizing the soil and regulating the flow of water is a product of fairly recent times and still incomplete. Although forested areas are obviously humid areas, their water economy becomes less efficient after clearing. It has been maintained by some authorities that the presence of forests actually increases rainfall.

This was at one time the subject of considerable debate between the United States Forest Service and the Weather Bureau. It was maintained by the former that evaporation from the forest makes an important contribution to rainfall farther inland. The weight of opinion seems to be now that the chief source of precipitation comes from the ocean. The relationship of forests to moisture, while of great importance, is less obvious and direct.

Delicate Relationship of Soils to Forests

The part played by forests in soil formation offers some peculiar problems, because forests, as we have said, generally occur in relatively moist regions. Here water supply and facilities for transportation may encourage heavy concentrations of population and rapid expansion of agriculture. However, the soil formed beneath forests tends to be much shallower than that in the grasslands and to be underlaid by leached mineral subsoil. It is generally acid as well. Such soil when cleared yields heavy crops for a few years, but unless it is handled with great skill, the rich, shallow top layer is easily destroyed by oxidation and erosion. When this happens, there is a rapid loss of fertility, leading to deterioration of farm economy or to costly programmes of fertilization.

On more level and sandy soils such cost may be offset by the nearness of good markets, but rougher lands become the site of submarginal farms or are abandoned to second-growth forest under conditions where forest management is often difficult. Closely connected with the fact that forest soils usually offer a narrower margin for agriculture is the circumstance that on many of the poorer soils trees constitute the most profitable crop under conditions of good management. An exception occurs on certain types of glacial soil, which are rich in minerals and can be restored by skilful management.

Soils in moist, tropical forest regions present an especially difficult problem. Here the thin humus layer is rapidly destroyed by exposure, while the heavy rains tend to dissolve out nutrient minerals. Owing to the high temperature, destructive bacterial action is very rapid once the protecting forest cover is removed. Some progress has been made in transforming these soils into well-managed tropical pasture lands.

However, because of the great variety of valuable products that can be obtained from tropical trees, it would seem that much more attention should be given to types of land use that depend upon the forest or upon skilful combination of tree crops with other crops, such as coffee, which can be grown in their shade.

The increasing pressure of world population in the temperate zones and the availability of extensive areas of tropical forest land make these problems especially important at this time.

Vital Relationship to Water

The value of forests in the protection of stream sources has long been recognized, whether tacitly or explicitly. The utility of the lower river basins in China and India is largely dependent upon the heavy forest cover in the vast mountain area that separates those two countries. The setting up of great public forest reserves in the United States has in view not only the future supply of timber but the protection of watersheds. This problem, however, has its complications. While it is generally agreed that forested headwaters do regulate and conserve stream flow, it has been shown that, where a large volume of flow is required for domestic uses, the presence of cleared areas, suitably protected by vegetation, may materially increase the amount of water available for runoff and immediate use.

The Complex Problem of Fire

There is a similar complication in the relation of fire to forests. While fire, along with clearing and pests, remains a great destroyer of forest and must be combatted effectively, fire can also be a useful tool for the forester. It has been demonstrated that the prudent use of fire is an important means of maintaining and improving the Southern forests of longleaf pine in the United States. Certain valuable species, such as the lodgepole pine of the Rocky Mountains, require the heat of forest fires to release their seeds from the cones.

While forest fires have always been occasioned by such natural causes as lightning or even volcanic eruptions, man causes most of the destructive fires of modern times. Much of this is due to carelessness. On the other hand, much is a direct heritage from the ancient practice of clearing the forests by fire. In many places, notably in Africa and in the southern United States, this practice has become crystallized into popular ritual, the woods are fired regularly and the most diverse reasons are given for the practice, which is asserted to improve the grass for cattle and to control noxious insects and even human diseases. That it does "green up" the woods there can be no doubt, but the best of woods pasture is very inferior to the best-managed grass and legume pasture.

Insect and Fungus Damage

Destructive as fire is at the present time, insects and fungus pests are more so. It is possible that danger from these sources has been intensified by breaking the continuity of the great forests that once covered the eastern part of the American continent. There is, however, abundant evidence that, even under natural conditions, forests were constantly subjected to violent disturbances by many natural forces, notably high wind as well as fire and pests.

Effect of Types of Ownership

The pattern of forest ownership has changed during the course of history and still varies according to the prevailing conception of property rights. Thus the American Indian had no sense of private property in the forests, although a general group priority for hunting and other purposes was recognized. In the tropical islands of the South Pacific, the ownership of coconut palms is recognized and allocated and during World War II military personnel who might become lost and forced to forage for food were advised to leave chits for the coconuts on certain islands, so that recompense could be made to the proper owner.

In the eastern Mediterranean area, the ownership of single olive trees may be shared, each owner claiming a certain sector, thus greatly complicating the business of improved orchard management. In Mexico the right of private exploitation of forests may be bestowed by concession and the laws against illegal felling of large trees are strict. However, it is the custom of the natives to cut out portions of the living tree for firewood bit by bit until it falls, when it becomes the property of anyone who can salvage it.

In parts of western Europe, in addition to public forests there are extensive private holdings. These, however, are subject to strict regulations as to management, which make government, in effect, a responsible party in the enterprise. In North America, in spite of the extensive reservation of government forests, the bulk of forest land is under private ownership. Until recently this has led to ruthless exploitation—a tendency now being reversed, especially in the case of large corporate owners who see the importance of a continuing supply of timber.

Because of the long period required for the maturity of harvestable timber and the frequent pressure upon private owners to raise immediate funds, private ownership of the bulk of woodlands poses difficult and challenging economic problems. The ratio of labour costs to the value of the wood and other forest products likewise must be considered. The European peasant does not hesitate to gather faggots or bundles of twigs for fuel. On the other hand, at the present time in the United States, it is often difficult to find anyone to cut cull trees, even though given the wood for sale as fuel.

Changing Uses of Forest Products

In the days of wooden sailing ships the possession of adequate forest supplies was a major factor in naval power. Oak and cypress and, in particular, tall, straight mast timbers were essential. So were framing timbers for houses, until structural steel became available. At first it would seem that with the development of metal and other structural materials the pressure on forests would be less. The contrary is the case. Not only does wood still have many technical functions in our civilization, but the variety of uses for which it is indispensable is increasing. The great field of plastics, instead of lessening the demand for wood, has augmented it, since cellulose is a necessary ingredient in many plastics. Likewise, the insatiable demand for paper of all kinds has enlarged rather than decreased the need for wood. Thanks to technological progress, many kinds of wood which were formerly worthless for that purpose can now be used for pulp.

An interesting if minor example of the irreplaceable properties of wood is found among the Eskimos. The willow wood, which is the chief kind available to them, cannot by itself be made into bows, but when laminated with other materials lends itself well to such use. Consequently the Eskimo, before the days of firearms, traveled long distances to get supplies of this otherwise not very useful wood. A significant phase of the use of wood in technology is afforded by the changing industrial picture due to the growing scarcity of wood. The ornate and tasteless architecture of the early post-Civil War period was closely related to the fact that at that time the white pine forests of Wisconsin and Michigan were being rapidly exploited. Under the compulsion to find markets the timber industry actually did a great deal, through the distribution of architectural plans, to promote the extravagant use of wood.

The notable severity of more modern designs, while partly traceable to an improvement in public taste, can be related as well to the increasing scarcity of a once abundant building material. Communities which supported great planing and cabinet mills in the days of abundant forests lost them, sometimes without understanding why, when raw materials became scarce. These mills were replaced by factories using smalldimension materials for boxes, handles, staves and hoops—until even the steady supply of small and inferior lumber was exhausted.

Exacting Demands for Wood on Forest Resources

So great is the continuing pressure on forest resources today that, wherever transportation is available, cutting rather generally exceeds the annual increment of merchantable timber. Reserves exist chiefly where distance or topography renders them difficult of access. Vast tropical areas of forest exist, but present their own peculiar problems.

Toward the equator, the number of species in the forest community increases rapidly, so that one encounters not merely the problems of dense growth and of difficult access, but mixtures of many kinds of trees within the same small area. These trees differ so much among themselves in physical properties that profitable mass harvesting is far less simple than in more temperate regions. Because of this, the tropical forests must be harvested, if at all, for widely differing purposes.

Among the more hopeful approaches that are being made to this problem are experiments in the manufacture of tropical plywood, in which woods of very different textures, densities and strengths can be satisfactorily combined. In any event, the harvesting of tropical timbers on a wide scale is likely to occur in the near future. Whether this will be done with more thought to a continuing supply than has been the case in temperate regions remains to be seen. A sensible world economy would certainly require it. Lumber is, of course, but one of many forest products. The heaviest use of wood, the world over, is still for fuel. In underdeveloped countries or regions where coal is lacking, much of this tree fuel is in the form of charcoal, a far more compact source of energy than the untreated wood and one which frequently affords twice the heat value. However, much of this charcoal is prepared by primitive and wasteful methods which do not conserve the various valuable distillates produced in the charring process.

In addition to lumber, fuel and pulp for paper and plastics, enormous amounts of wood are required for such special products as excelsior, matches, spools, tool handles and the like. Forests are also important sources of raw materials for the chemical industry.

Native rubber, resins, gums, oils, dyes, drugs and aromatics belong in this category. While many of them can be produced in artificial plantations, the cheapness of labour and land in tropical forest areas makes the wild product a significant factor in the world market. Anyone familiar with the food habits of primitive forest dwellers knows that there are considerable possibilities for sustenance from this source It is estimated that the coconut palm alone has several hundred technical uses among native peoples, apart from the obvious one of food.

Manifold Possibilities for Wood Utilization

One of the most intriguing phases of forest utilization lies in the rare and unique qualities possessed by certain woods for particular, highly specialized purposes. The instances that could be cited are legion, including, for example: tough, straight-grained ash for medieval lances and modern tool handles; butter-like pearwood for exquisite carvings such as those of Grinling Gibbons; end grain ivory-like box for wood engraving; unsplittable lignum vitae and gum for mallets, house rollers and bowling balls; bitter quassia, which imparts its flavour and medicinal qualities to water placed in bowls of this material; teak and cypress, so valuable for the decking of ships; dogwood for "glats" or wedges, used in splitting rails and firewood in the days when steel was scarcer than now—and so on.

Although such uses, like the uses of medicinal and food plants, were discovered before the days of modern science, it is of interest to note that the rich variety of tropical woods is now being re-explored in laboratories-as for example at Yale-in the hope of finding substitutes for the ever scarcer woods of familiar commerce, as well as of finding uses that are quite new.

Vast Implications of the Forest for Human Culture

The significance of forests to mankind goes far deeper than economic utility, whether direct, as in the case of usable forest products, or indirect, as in that of watershed protection. Forests have a powerful, though often intricate, relation to the more intangible aspects of human life.

That forests give aesthetic pleasure needs no particular emphasis. Poets and artists have made this clear enough. More is involved, however, than the simple fact that forests can delight the eye and gratify the body with their cooling shelter. At least one of Lincoln's biographers has intimated that the melancholy of that great man was partly a product of the gloom and shadow of the forest that helped condition his childhood. This may be pure guess, but there is no doubt that a prolonged stay in the tall twilight of the Engelmann spruce of the High Rockies induces a mood that is vastly relieved when one descends from them and emerges into the foothills to look out over the broad expanse of sunlit plain.

The violent deforestation of the Mexican uplands following the Spanish Conquest was accomplished by axe and fire and consolidated by goats and other cattle brought from Europe. But there is testimony that the destruction was swifter than it might have been because of the Spanish craving for a landscape resembling that of their own treeless homeland. Forgetting alike that much of the poverty of the Spanish peninsula was due to its having been deforested and overgrazed and that much of the wealth that remained was due to tree crops such as cork, olive and citrus fruits, the conquerors allowed themselves to repeat ancient and costly errors.

If the forest, by presence or absence, can be so deeply involved in the human emotions, it is not surprising to find it likewise involved in religious symbolism and practice. Worn, but still beautiful, is the line "the groves were God's first temples." Even though the Judaeo-Christian tradition stems from pastoral lands, its literature is rich in symbolism connected with the forest—the tree of life, the tree of the knowledge of good and evil, the wooden Cross itself. The use of wood in altar sacrifice, the hearth as a symbol of home and worship and ashes of wood in the ceremony of grief and repentance, are all familiar and ancient.

Among the Greeks not only forests, but also individual trees, were personified as the dwelling places of god and demigod. In the Orient, crowded for living space as it is, groves surround the temples and it is believed that it was in such groves that the living fossil, the ginkgo tree, now so much valued for its beauty in this country, was preserved from extinction.

In the Druidic worship of western Europe, as also in the robust Thor worship of the ancient Norsemen, the forest played an intrinsic part. The mistletoe was sacred in both-with a shaft of mistletoe the gentle Freya was killed by the mischievous Loki. Our use of Christmas greens and trees stems from the ancient Teutonic festivals that brightened the season of long night, during which we celebrate the birth of Christ.

Even the mood of gloom and depression that the deep shadow of the forest may induce is akin to the feeling of awe and reverence. This the primeval forest clearly inspired among the ancients-as it continues to do today—an effect not only evident in religious belief, but in the inseparable fields of poetry, music and art.

NATURAL RESOURCES: WATER, FORESTS AND LAND

Abundantly endowed by nature, Chhattisgarh is a land blessed with a pleasant climate as well as a priceless heritage that has sustained and nourished its people through the ages. The State has rich resources including land, forests and water. These abundant resources are of high quality and are spread across the State, allowing an exceptional degree of access and availability. In return, the people and communities have treated the gifts of nature with reverence.

They have evolved a way of life unique to this verdant land, a way that seeks to protect this legacy for future generations. Pivotal to life and livelihood are the trinity of water (jal), forests (jangal) and land (jameen). Each of these is important, but they are also dependent on each other – and on the people. Water nourishes the forests and the land.

The forests are repositories of diversity and natural wealth. From the land comes food and security, each year. For centuries, people have built their traditions, consumption, habitat patterns and livelihoods around these resources. Celebrated in song and folklore, deified and venerated, the people know and understand the importance of these resources. These themes aroused passions and a great deal of discussion. The intensity and fervour were inversely proportional to the distance from the resources.

The District and Village Jan Rapats present a variety of generic as well as specific submissions with respect to these natural resources with differing language. The Village Reports tend to be more specific and their tenor more impassioned. They speak of natural resources as a chronicle of existence, of survival, and of life itself. The District Reports express issues and relationships related to natural resources as a corollary to life and livelihood.

Their definitions and content reflect a more controlled livelihood pattern and a larger economic dimension, one that extends beyond the immediate space of a village, a settlement or a city. Even so, in all the Reports it is amply clear that the relationship between the people and natural resources transcends the confines of modern market and employment approaches. These relationships are immediate yet dynamic, and are a part of a complex web in which culture, societal values, environment, health, knowledge, and lifestyles are intertwined.

Table. Availability of Natural Resources

Region	Water	Forests	Land for common purposes
Northern region	42	77.5	57
Central plains	33	39	48
Southern region	57.46	79.1	64.93
State	44.2	65.2	56.6

There are differences too in perception, based on topography, location and region. In the plains of Chhattisgarh, land and water are seen as the primary resources. For the people of the hill tracts in the north and the south of the State, water and the forests are the critical resources, seen as the key to survival, sustenance and advancement.

The availability of water resources is better in the north and the south of the State. In either case, the relationship with natural resources is direct and immediate, the difference being the degree of dependence on one or the other. The exceptions to this are few, and are in the context of the tertiary sector, or in urbanised environs.

The Reports, in particular the Village Jan Rapats, have enumerated and quantified their natural resources, and elaborated the dependence of the village on these resources. They have identified issues related not only to access and control, quality, exploitation and conservation, but to the technical and legal dimensions as well. This chapter examines the three main resources – water, forests and land – separately and discusses some of the significant issues raised in the Jan Rapats. It then discusses the relationship between women and natural resources and the critical issue of common property and its management. This is followed by suggestions for intervention and concluding remarks.

WATER

Chhattisgarh abounds in water bodies – rivers and streams, lakes and tanks (dabrees). It also receives, in normal years, rainfall adequate to replenish water resources, and to meet the needs of the people. The annual average rainfall varies between 1200 mm to 1400 mm.

Despite the abundance of water, people have learnt to conserve water, and use it judiciously and equitably, through systems and practices that have evolved over hundreds of years. A combination of wisdom, intuition and experience enables the people to tide over situations of adversity – the preceding years of drought and poor monsoon provide an excellent example of how the people survive difficult times.

The recent drought-like conditions (in 2000 and 2002, the average annual rainfall was less than 1000 mm) and the resulting hardship find recurring mention. The Reports refer not just to the impact on agriculture and the consequent need for more irrigation. The Jan Rapats speak of declining water tables and of biotic pressure on the forests, the natural reservoirs of water and moisture. The need to ensure clean drinking water to all is cited as urgent. Looking ahead, many refer to the need to ensure that ground water is used wisely and sustainably, and that the forests are protected. Control and management of local water bodies are vested in the tiered system of Panchayats. The Gram Panchayats, Janpad Panchayats and Zila Panchayats are authorised to manage and lease out water bodies.

This has enabled public participation in their use and management, and in ensuring the rights of user groups – fishermen, cultivators and other users. At the same time, it has brought about visible changes in the perceptions of this resource. Building on age-old traditions of community management, equity and shared responsibility, the Panchayats have become effective instruments of dispute resolution.

WATER RESOURCES OF CHHATTISGARH

The State of Chhattisgarh forms part of the extended river basin of four major rivers – the Mahanadi, Godavari, Narmada and the Ganga. The combined river length flowing through the State is 1,885 kilometres. These rivers provide a large network of surface water and support the primary sources of irrigation in the State. There are also smaller rivers and tributaries, seasonal nallahs and natural springs. It is estimated that surface water available for use is 41,720 million cubic metres (mcm). The State has three major, 30 medium and 2,017 minor irrigation projects maintained by the Water Resources Department. Small tanks are maintained by the Panchayats. Ground water is an unregulated resource, one that land users have freedom to harness.

It is relatively under-utilised, and there is scope for increasing ground water based irrigation. According to the Central Ground Water Board, the ground water available for use in the State in 1995 was over 8,000 mcm per year, and the ground water exploitation could be significantly enhanced. In most districts, less than 10 percent of the potential is currently being utilised.

Drinking Water

In the past, drinking water was obtained from wells, natural springs, streams, rivers, tanks and lakes. In the plains, where drinking water has been generally insufficient, wells, tanks and small rivers have been the main sources. In hilly and undulating regions, springs, rivulets and wells provide drinking water. Most households in rural areas now rely on hand pumps for their supply of drinking water.

Despite their increasing density, there are still places where hand pumps are not available or functioning. In these locations drinking water is sourced from tube wells or even rivers. Piped and tap water is still not common. The Jan Rapats confirm the improved availability of drinking water and acknowledge the improved quality and access. They state that many more hamlets and households now have direct access to drinking water. Hand pumps are sometimes non functional.

This may be due to irregular or poor maintenance or due to the drying up of water sources. This is more common in remote habitations and small hamlets, particularly in hilly terrains. In such settlements, other water sources such as natural springs or streams are then used.

While the perception of water quality varies considerably, most Village Jan Rapats indicate that water from hand pumps is usually clean and suitable for domestic use. They have also commented positively on the practice of adding chlorine to drinking water. In some villages especially those located in the mining and industrial belt, the issue of water pollution due to industrial waste is of concern.

Sources of Water

Traditionally, water from open wells and tanks has been utilised for domestic and drinking purposes, while canal and river water has been used for irrigation.

Table: Adequate availability of water

Region	*Drinking water*	*Water for household needs*	*Irrigation*
Northern region	42	57	19.6
Central plains	26	28	36
Southern region	56	62	16.7
State	41.3	49.0	24.1

Long-established irrigation systems provide for the diversion of water from small rivers, nallahs, seasonal streams, and ancient water tanks. Small water storage tanks (dabrees) constructed in cultivated fields store rainwater for irrigation. These are supplemented by animal power operated water drawl systems. Traditional methods are suitable for small, compact areas but are inadequate to meet the needs of large scale, assured irrigation.

Since they are substantially dependent on rainfall, they tend to be most efficient during the monsoon and shortly thereafter, and are ideally suited for single crop based agriculture. The Village Reports show that the availability of water for drinking and household needs is best in the southern region. There is a shortage of drinking water in the central plains region, although irrigation is more prevalent in this region.

Most water bodies at the village level are managed by traditional and community based systems. Over the years these have begun to break down in the face of social and economic change and due to the emergence of alternate structures of authority. Water bodies and structures that were created or regenerated under Government programmes have not been very successful in aligning or integrating themselves with community based systems.

One reason for this is that the State has not recognised or supported the traditional systems while taking over and exercising its provisioning authority. Unfortunately, there is little to indicate that alternate forms of community based and community owned systems of managing water bodies are replacing the traditional systems.

There is a need to evolve new systems that support the efficient management of modern irrigation structures. These should involve communities and users in their operations, and provide a blend of old and new ways that combines the best of both.

Irrigation

Cropping intensities in the State are low, since agriculture continues to be largely dependent on the monsoon, and most cultivators still practise single-crop agriculture. The Jan Rapats speak of efforts made by Government, Panchayats and individuals to increase irrigation coverage end effectiveness.

Most of the Governmental effort has gone into surface water exploitation, and there is a perception that groundwater needs to be systematically harnessed, with the support of the Government. Small and marginal farmers, with low ability to invest the capital needed, are particularly in need of support. There are private tube wells in some villages but these usually belong to well-off farmers.

PROBLEMS ASSOCIATED WITH WATER

A number of problems and issues have been highlighted in the reports, ranging from the contamination of ground water sources as a result of mining, chemical fertilizers and industrial activity to the declining water table.

Irrigation initiatives undertaken by the State

A major programme has been taken up to motivate cultivators to set apart marginal plots under their ownership for construction of tanks (dabrees). These help in moisture retention, meet domestic needs, prevent soil erosion and provide water to fields. The construction of dabrees, (using public funds) also provides wage employment under drought relief operations.

Khet Ganga Yojana

This initiative seeks to provide irrigation to the rain-shadow areas of the State. It aims at tapping ground water potential and riparian run-off through tube-wells and lift irrigation. To safeguard against the excessive exploitation of groundwater, it is mandatory to maintain a minimum distance of 300 metres between two tube wells. Assistance (subsidy) ranges from Rs 10,000 to Rs 18,000 for drilling and Rs 10,000 to Rs 25,000 for installation and energizing pump sets. Failed tube wells are usually compensated for. Subsidy reimbursements for farmers from the Scheduled Castes and Tribes have a higher ceiling of Rs 45,000.

Gaon Ganga Yojana

This scheme aims to create at least one source of water in every village/ habitation.

This will be achieved through conservation of existing sources of water through community initiated maintenance and renewal with the involvement of the Panchayat. It also seeks to develop new sources of water through sustainable exploration of ground water potential and prevent waste and run-off through appropriate community level interventions.

FORESTS

The people of Chhattisgarh have a symbiotic relationship with forests. There is religious reverence and a grateful recognition of nature's benevolence. There is also an appreciation and understanding of the impact of the environment on the lives of the people. With its vast forest

The distribution of the other sources of irrigation (apart from rivers and canals) shows a distinct concentration of these sources in the central plains. Three out of four tube wells are installed in the central plains cover (135,224 square kilometres, 44 percent of the State's area), the State's economy, culture, tradition and livelihood are inextricably linked to the forests.

Water pollution due to economic activities: The Korea District Report refers to the contamination of groundwater due to mining activities. In Bilaspur, a paper mill is cited in the Report has been held responsible for toxins in the water. Many Village Reports also refer to water pollution due to the use of chemical fertilisers.

Improper maintenance of drinking water sources and disrepair: A recurring theme in the reports has been the irregular and unsatisfactory maintenance of hand pumps. Several reports have indicated that the Government appointed mechanics do not respond to complaints and that little local expertise is available to repair hand pumps.

Declining water table: While most parts of the State have good availability of ground water, there has been a decline in the water table. Digma village of Block Ambikapur of Surguja district, reports that in earlier times, their forefathers had to dig to a depth of 25 to 30 feet (4 to 5 porish5) to reach water but now they have to dig up to 60 to 70 feet (8 to 10 porish).

Similar reports have come from other villages, especially in the hill regions, from villages where ground water is being used for irrigation and from villages where industry and mining compete for limited water resources.The forests of the State are of two major types: tropical moist deciduous and tropical dry deciduous. Most of the dense forests are concentrated in the northern (Surguja, Korea, Jashpur and Korba districts) and the southern regions of the State (Bastar, Kanker and Dantewada districts).

These areas also have large tribal populations. The plains of the central region of the State have much less forest cover. In this region, the dependence on agriculture and therefore on land as a source of livelihood is much higher.

Direct benefits: Indirect benefits:

- Food items such as fruits, roots and shoots directly available from trees, and animal products such as honey and meat
- Raw material for the production of soap, oil and liquor
- Nistaar8 items such as fuel wood, fodder and timber
- Medicinal plants and herbs such as safed musli, brahmi and ashwagandha
- MFPs for the market such as tendu patta, sal seeds, gum, lac and wax
- Minor minerals and water.

Indirect benefits:

- Soil conservation
- Rainfall
- Climate control
- Biodiversity conservation

The Jan Rapats have documented the benefits from the forests, extensively. The forests provide food for the people and for their animals, raw materials for household based industry, firewood, medicinal plants and minor forest produce like tendu patta (tendu leaves) and lac. The Reports have also acknowledged that forests hold the key to the climatic conditions and bio diversity conservation in the area.

Dependence on Forests

The culture of the people of Chhattisgarh is linked to the forests and the people share an intense emotional bond with the 'jangal'. This is especially true of forest based and tribal communities.

FUEL WOOD

In rural areas non-commercial fuel wood and animal waste continue to be the main source of energy. Women are the main collectors of this resource, which is used for cooking and household activities. Women often have to traverse a large area in search of fuel wood. In the plains, where the forests may be far away, women depend on energy from animal waste.

Like cow dung, fuel wood has to be bought from the nistaar depots and this means an additional financial burden on poor households. The ownership of forests lies with the State. The Forest Department extends the privilege of extraction of forest products to the people within the stipulations of forest policies. These include the provision of nistaari rights to forest dependent communities.

- There are 797 nistaar depots in the State. Each family is eligible to get bamboo for domestic use at a subsidised rate from these depots.
- Forest dependent communities are entitled to access forests for grazing, limited by the carrying capacity of forests. They may collect (free of cost), dry and fallen fuel wood and fodder. Medicinal plants may also be collected (by non-destructive means) for sale.
- The Basod community is eligible to get 1,500 bamboos per family, per year, (subject to availability of bamboo) at subsidised rates, for bamboo-based income generating activities.
- Forest dependent communities are free to collect tendu patta, sal seed, harra and gum and sell this to notified outlets of Chhattisgarh Minor Forest Produce Cooperative Federation at pre-determined rates. Registered collectors of tendu patta are eligible for bonus and group insurance facilities.

FODDER

Most households in the State own livestock. Animals are used to till the land and they also provide energy. They are an investment and a valuable asset, especially in times of adversity like drought, or in an emergency. Most villages have common grazing and pasture lands for animals. In the plains, paddy straw is used as fodder for cattle. In the forested belts, animals too depend on the forest for fodder.

The most important issue related to fodder which has been elaborated in the Jan Rapats, is the degeneration and shrinking of grazing and pasture lands. Another issue is that of encroachment. Common lands are most susceptible to encroachments. This has directly affected the quantity of fodder available for the cattle. Over the years the availability of fuel wood and fodder from the forests has declined, measured in terms of availability as well as access. The factors responsible are sporadic clearing of forests and growing biotic pressure. Adding to the complexity of the situation is the fact that the traditional systems of people managing and selfgoverning common lands have been eroded.

MINOR FOREST PRODUCE

Forest produce is categorised in two main categories:

- Major forest produce (mainly wood or timber)
- Minor forest produce. Ownership of the major forest produce is with the Government.

People and communities may extract minor forest produce (MFP) for consumption or sale under certain conditions. Chhattisgarh accounts for about

20 percent of the total production of tendu patta in the country. Other major MFPs of the State include mahua flowers and seeds, harra, bahera, mehul leaves, tamarind, lac, gum and katha. These are mainly used to make brews, toys, disposable leaf plates, etc. Tamarind and katha are used in food items.

District Report, Uttar Bastar Kanker: Fuel wood, timber, MFPs (jamun, harra, bamboo, mahua flower and seed, tendu patta, lac, aavla, bahera, gond, madras, tikhur, ghaas, mile, aam guthli, dhaura, beeja palash, haldu, saja, sheshum, bel, ram dataun, mahul patta, etc) and other medicinal plants (bhui neem, kali hari, dev kanda, van pyaaz, chidchida, van haldi, charauta, peng beej, dudhi beej, hadsighadi, amaltaas, stavar, patalkumhada, kevti, safed musli, kali musli) are collected from the forest.

District Report, Korea: There are thick forests in Korea district. Other than timber and fuel wood, various kinds of forest produce are found in these forests. The forest dwellers collect MFPs like mahua fruit and flower, harra, bahera, aavla, chaar, chiraunji, mahul patta, tendu patta, sal beej, and other types of herbs and roots. The collection of forest produce is a major source of income for the people. These MFPs are the basis of many small and home industries.

District Report, Korba: Many medicinal plants grow in the forests and are regularly used in various treatments. These include chiraita, safed musli, kali musli, satavar, adusa, lat jeera, vaybirang, jungli pyaaz, hadjod and dhava.

In Chhattisgarh, the ownership of all minor forest produce in forest areas is now vested with tribal communities through Primary Cooperative Societies of actual collectors and Gram Van Samitis. This has become possible through the provisions of the Panchayats Extension to Scheduled Areas Act (PESA), 1996.

The collection of specified nationalised produce is done by the Societies and primary processing is also done by the Societies. The proceeds from the sale of MFP are transferred to the Societies. While elected representatives manage these Societies, the Forest Department continues to exercise control by holding key positions in the management.

The Village and District Jan Rapats have provided a comprehensive list and inventory of products acquired from the forest. The forest produce collected is sold at approved collection centres. These include sal seeds, tendu leaves, harra, bahera, mahua, char, tendu and imli, which are sold at pre-determined rates. In addition to the Joint Forest Management (JFM) programme, initiatives have been taken by the State Government to develop an efficient and people friendly system to manage minor forest produce as well as its marketing.

Recently, some efforts have been made to involve people in the marketing of major forest produce as well. Under the State sponsored scheme of Public Private Partnership (PPP), proceeds from the timber stock and bamboo in degraded areas, is made over to Gram Van Samitis (GVS). GVSs have been encouraged to enter into buy-back arrangements with private industry and the

Federation of Gram and Van Suraksha Samitis fix the selling rates. In degraded forest areas, where the Government spends on greening, 30 percent of the final harvest and all the intermediate yields (from thinning) go to the GVSs.

In well-stocked forests, 15 percent of the proceeds from the final harvest are distributed to Forest Protection Committees (Van Suraksha Samitis). The Forest Department estimates that about 2,00,000 tribal families are associated with the forest based economic activities of the department.

This number is expected to go up, as all economically significant minor forest produce, including medicinal plants, are brought under the PPP arrangement. However, caution will need to be exercised to ensure that the PPP programme does not become a means for enabling industry to gain access to forest resources.

FORESTS AND THEIR CHIEF PRODUCTS AND USES

In trying to see what lies ahead for world forestry it is first necessary to take cognizance of the dimensions and the complexities of the forest problem. It is also important to keep in mind the limitations of the forecasters, for the outlook depends very much on the lookout. In varying combinations, all men have hopes and fears, tendencies to look on the bright or the dark side, inclinations to hasty or slow judgment; some are dewy-eyed about the future and for that reason cannot see clearly, while others suffer from cataracts of vested prejudice that impair the vision. Furthermore, the lookouts are not all looking in the same direction.

Some can see only the truly remarkable achievements in technology and overlook the fact that for a long time to come the main wood problem of most of mankind will not be how to obtain rayon fabrics, cellophane wrappers, or molded monocoque airplane seats, but how to secure enough firewood for cooking purposes. Others are attracted by the improved methods of silviculture, developed particularly in northwestern Europe, but fail to keep in view the hard fact that much of what can be done in Europe cannot, for economic reasons, be done in North America and probably cannot, because of different physical conditions, be done in the tropical forests.

This personal factor is no doubt the main reason for the greatly varying reports of the prospect in view, whether they deal with trends in agriculture, forestry, or any other aspect of the man-earth relationship. Some say that a wood famine is inevitable, others assure us that no world shortage of timber is in sight and the reason for the difference in opinion obviously lies in men, not in the forest.

THE PROBLEM DEFINED

Forecasting in this field is not an exact science, partly because of personal factors that cannot be eliminated and partly because the information on which the forecast must be based is neither complete nor altogether accurate. Forest mensuration is excellent in a few countries, but the available data on forest area, volume of stand, annual growth and even yearly consumption of wood in the world as a whole are at best only rough estimates. These estimates are nevertheless important. They are, in the first place, the only quantitative data we have and they reveal the dimensions of the problem before us.

The forest inventory published in 1954 by the Food and Agriculture Organization of the United Nations (FAO) shows that the annual world consumption of wood for all purposes in recent years has been over 50 billion cubic feet and that the total area of the forests on the earth is estimated to be about 9½ billion acres.

It is not easy, indeed perhaps not possible, for the human mind to comprehend such large quantities. Many have seen the forests of Washington and Oregon and been impressed by their vastness, yet in area they constitute less than half of one per cent of the world's forests. It is not easy to envisage 9½ billion acres of forest land under management on a sustainedyield basis, but that seems to be the future goal implied by the statement in Unasylva.

"The inventory shows that the world's forests are potentially capable of furnishing a plentiful flow of forest products for a world population much higher than that of today" It is not possible to say exactly how much forest land is managed on a sustained-yield basis at the present time, because data are lacking and because good practices have been initiated so recently in many regions that we do not yet know whether yield will actually be sustained. Management of a type is practiced on some of the forests in India, Ceylon, Burma, Australia, New Zealand, in other countries and in some parts of Africa. In 1949, according to the American Forestry Association, about 55 million acres of public and private forest land in the United States were classed as under "intensive management," which is defined by the words: "high order of cutting and good fire protection" As of January 1955 nearly 34 million acres of private forests in the United States were in certified tree farms.

However, not all of these forests are truly managed as the term is understood in northwestern Europe. It is perhaps safe to say that, after more than a century of research in silviculture and experiments with forestry practices, some 160 to 180 million acres of forest land are now under management in northwestern Europe and, with an optimistic view, possibly an equal amount outside of Europe, making a total of about 350 million acres. That is, it may be conjectured that 4 or 5 per cent of the world's forest lands are at the present time under reasonably good management and 95 per cent are not. The sheer size of the task that remains to be done if the goal is to be achieved looms very large in the outlook.

The task is not only large but complex. The ultimate goal might be defined as the utilization of the world's forest lands in such a manner that they will permanently yield sufficient wood for all mankind at a price people can pay, while at the same time some of the forests will provide watershed, soil and wildlife protection, grazing opportunity and facilities for study and recreation. None of the terms in this definition can well be omitted.

The term "forest land" must be used, for in the long run it is the land that constitutes the resource and not the generation of trees that happens to occupy the land at a given time. Forestry is only one type of land use. In many regions it competes for site with other forms of land utilization: the demands of agriculture, for example, will have a decisive effect and the effect will be fatal to some of the forests. The term "permanently" must be included, for the problem is not merely to balance forest drain and growth in the immediate future, but to maintain the balance in perpetuity by increasing the growth to meet the rising demand of an increasing population. The phrase "all mankind" is required, for the goal cannot be regarded as having been reached so long as some forest lands produce species of trees that are considered useless because there is no market for them, while other regions are so short of wood that cow dung must be used for fuel.

"Price" is perhaps the most critical term. In one way or another and regardless of type of prevailing economic theory, every forestry operation and every forest product must be paid for, whether in energy, money, or barter. Good forestry practices are simply not possible unless they are economically feasible; for all we know, some of the forests may remain inaccessible forever because access may be too costly; and technological achievement, no matter how brilliant, will be of little use if people cannot pay for the products. The terms in the last part of the definition pertain to the concept of multiple use, an idea that may or may not turn out to be practicable in the long run.

In many parts of the world problems of soil and water are far more critical than those of timber supply and large sections of the forest, probably much larger than we commonly think, will have to be reserved for the primary function of watershed protection and will serve only secondarily, or perhaps not at all, as a source of wood. Other forest lands are withdrawn from ordinary economic utilization and reserved for recreation, study, wildlife sanctuaries, or for the purpose of saving some of the undisturbed forest for future generations. An example is found in the National Parks of the United States, which contain almost half as much timber land as there is in all of West Germany.

We may not always be able to afford this luxury, but as long as the world can keep its parks and wildlife reserves they will add up to a significantly large area which, like the protection forests, cannot be included in the area of the wood-yielding forest land. There are other complicating factors. The difficulties of silviculture and management are multiplied by the many different physical characteristics of the world's forests; people have different and deeply rooted attitudes toward trees; and the problem is not made easier by the fact that the forests are under about one hundred national jurisdictions. It is not a mere matter of balancing numbers. Unasylva suggests that the wood needed for a moderately good standard of living would be an annual per capita consumption of about 35 cubic feet, basing the suggestion on prewar European experience; and makes the assumption, which seems plausible enough, that with reasonably good management the forests of the world can produce an annual yield of 30 cubic feet per average acre.

But it would be very misleading to conclude that the balance is favorable, or that no particular problem exists because there are only 2½ billion people in the world and 9½ billion acres of forest, or almost 4 acres per person.

Expected Demands for Forest Products in 1975

The outlook for the distant future depends largely on what will happen in the near future, which may be defined conveniently as the next two decades, since 1975 is the common target year in several recently published forecasts. 1 Only the major conclusions of these reports need to be considered here.

The Paley Report (1952) expects that the cubic-foot requirement of wood for all purposes in the United States will increase by 17 per cent between 1950 and 1975 and the Stanford Report (1954) foresees an increase of 14 per cent between 1952 and 1975.

These forecasts are thus not very different from earlier estimates: for example, in 1948 the Forest Service looked forward to a rise in the over-all requirement of about 50 per cent between 1945 and the year 2020, or an increase of nearly 15 per cent for each 25 years of that period.

The reports are in substantial agreement in predicting a decline in the demand of wood for pilings, railroad ties, shingles and cooperage and a very marked decrease in the use of fuel wood, but they differ on some items.

The Paley Report expects that the need of pulpwood will increase by 50 per cent; the Stanford Report, by 60 per cent. The demand for peeler logs, mostly for plywood, will rise by 40 per cent according to the Paley study; by 90 per cent in the Stanford forecast. The most notable difference is in the expected board-foot requirement of saw timber for lumber, which is by far the bulkiest item in the over-all demand: the Paley Report foresees an increase of 10 per cent; the Stanford Report, one of only 3.4 per cent.

The principal reason for this difference is no doubt the fact that the Paley Report is based on the assumption of no significant change in price relationships, whereas the Stanford Report assumes that the price of lumber relative to competing materials will increase and that lumber will lose some of its markets.

The Stanford assumption is well worth noting. High wood prices, causing people to prefer other materials, might have the effect of alleviating to some extent the coming pressure on the forest. A world pulp survey of the same type as the Paley and Stanford studies is reported to be in preparation by the

FAO, but, to my knowledge, projections of the future world demand for all wood are not yet available. A partial world forecast, which does not include the countries in the Soviet zone, is made in the Paley Report.

The average total output and requirement of industrial wood during the years from 1947 to 1949 are compared with the average prospective output and requirement during the decade of the 1970's. Fuel wood and charcoal are not included.

The conclusion is that the total output will probably increase by 4 per cent and the total requirement by 40 per cent. It is very likely that the demand will rise, but the 4 per cent increase in output may be an overestimate. According to Egon Glesinger of the FAO, as reported in American Forests, the inventories have shown that the output of the world's forests is not rising, despite increasing demands.

The type of forest products demanded and regional differences between the United States and Latin America in the use of wood. The products are shown as percentages of the total cut of all timber, as measured in cubic feet.

An indication of regional differences in the demand and use of wood on a world-wide basis is presented in Unasylva, December 1954. Expressed in approximate percentages of world forest area, population and total production in cubic feet of roundwood in 1953, the data reveal the differences between the highly industrialized countries of the northern hemisphere and the rest of the world. By "North America" is meant Canada and the United States and "Europe" includes the Soviet Union.

The heaviest production and consumption of industrial wood products, such as lumber, plywood, pulp and paper, are clearly concentrated in North America and Europe and the demand for these products is steadily rising. Furthermore, a large part of the industrial timber cut elsewhere, chiefly in the tropical forests, is exported to Europe and North America. The industrial demand is preponderantly for lumber and pulp, but the most widespread and still the largest single requirement is fuel wood.

Almost half of the total output of some 50 billion cubic feet in 1953 was fuel wood; even in Europe this requirement accounted for slightly more than one third of the total cut; and in many countries besides Latin America 80 per cent or more of the entire production was for firewood. At the present time, then, the demand on the world's forests is mainly for fuel wood, lumber and pulp, in that order and this is likely to continue at least in the near future.

Whether or not the forecasts will come true naturally depends on how valid the underlying premises turn out to be. The Stanford Report is based on the general assumptions that:

- No major war will occur.
- No radical advance in technology will increase production at a Tate faster than in the past,

- Business cycles will become more stable, accompanied by high employment. Another assumption, based on data of the Bureau of Census, is that the population of the United States will increase by 35 per cent to reach 212 million in 1975. The premises of the Paley Report are generally similar; for example, the expectation of a 40 per cent rise in the world requirement of wood rests on the assumptions of a population increase of 47 per cent, continued high demand in the industrial countries and rising consumption in others.

The safest of the assumptions is that population will increase and this alone would mean a rising demand for forest products. If a major war should come, the drain on the forest would surely increase drastically, as it has in past ways and the over-all requirement might become much larger than the forecasts indicate. If peace lasts but good economic conditions do not, the demand for industrial forest products would probably be less, but the world requirement of fuel wood might remain unaffected.

The second assumption of the Stanford Report has already been questioned by some forestry people, who believe that the large sums of money currently invested in research will result in technological improvements which will have a marked effect not only on efficiency in production but will help the industry to hold and even expand its markets.

An editor of House Beautiful magazine, speaking before the annual convention of the National Lumber Manufacturers Association in November 1954, challenged some of the Stanford predictions, declaring that the concept of abundant living demanded more lumber for dwelling units and urged the manufacturers to gear their thinking to an expanding market and an expanding American home.

It may well be that a 15 per cent increase in the over-all requirement of wood is about what can be expected in the next quarter century. But thereafter will come other quarter centuries and no evidence is at hand showing that this or any other country intends to stop growing in 1975. On the contrary, like the editor who spoke to the lumber manufacturers, most people hope for a steadily expanding economy, which of necessity means a steadily increasing demand on the natural resources, including the forests. If the hope is fulfilled, the outlook is that before very long the requirement will increase not by a mere 15 per cent, but by 100 or 200 per cent and perhaps more.

The question is how the forest will stand up under the rising pressure. The forecasters do not provide clear answers, only clear affirmations of their faith that the demand can be met. The conclusion of most of the reports, whether they deal with the United States or the world, is that the area of the forest is large enough to provide the wood needed in the future, even if the need should become twice or three times as large as today, but only on two conditions: much of the now inaccessible forests must be opened to utilization and all forests must be placed under greatly improved sustained-yield management. These are not small conditions and in them, it might be said, lies the whole forest problem.

The problem can be resolved into three major parts. The first is the question of supply: of forest area, forest soil, stand, annual growth and natural factors affecting the trees, such as fire, insects and diseases. The second is the problem of how best to utilize the supply: of silviculture, forest genetics, methods of cutting, logging, manufacturing-in short, technological knowledge. The third can be called cultural or institutional in nature; it is the question of our social-economic-political ability to apply the technological skill to best advantage.

More particularly, then, the outlook depends on trends of development in all of these fields; and the direction of the trends cannot be determined by looking only toward the future: we must also be guided by the past. Perhaps the most important thing to grasp is that in nearly all of these fields we are dealing with processes that have the tremendous momentum of history behind them.

THE MANAGEMENT OF THE FOREST RESOURCES AND FOREST LAND

Laws of Brunei mandated the Forestry Department to be the primary government agency responsible in the management and administration of the forest resources of the country.

The Act provided the basic law for administration of the forests, reservation of forest lands, harvesting of forest produce, grant of customary gratuitous rights to forest-dependent inhabitants, stipulating penalties for violations, and prescribing forest royalty. In addition to this legal instrument, the management and administration of the forest resources is also guided by the 1989 National Forest Policy and the long-term forestry sector strategic plan [20 years plan for the period of 2004-2023]. Relevant legislations that strengthen the management of forest lands are the Land Code & Land Acquisition Act; Wildlife Act; Town & Country Planning Act; Antiquities & Treasure Trove Act; and Wild Flora & Fauna Order 2007.

The provisions in the Forest Act provides the Forestry Department with the power to manage and administer the forest resources of the country, however, as far as the overall planning, development and management of the land with existing forests remains under the jurisdiction of the relevant government agencies & individuals, where the land are gazetted or allocated to. The Forestry Department has the sole right to the overall planning, development and management of forest land which has been gazetted as Forest Reserve. However, while the overall responsibility of managing the forested land on the State Land, Reserved Land and Alienated Land falls under the jurisdiction of the respective government agencies or individuals. The role of the Forestry Department in these areas is limited only in administering the harvesting and utilization of timber and other forest products therein as provided for under existing forest laws, rules and regulations.

FOREST LAWS AND RELEVANT LEGISLATIONS

Apart from the Forest Laws and relevant legislations, there are also other regulations and guidelines that regulate the forestry harvesting operation. Brunei Darussalam has its own harvesting guideline in order to achieve the following objectives; to maintain and increase the timber production capacity in the designated production forest areas; and to balance the timber harvesting and biodiversity conservation *i.e.* extracting commercial timbers while meeting the need for forest conservation. The guideline covers every aspect that is considered crucial in maintaining the sustainability of the forest, *i.e.* from the construction of man forest roads to logging procedures and post-harvest reporting.

The guideline is primarily aimed in facilitating the day-to-day supervision and logging activities monitored by field officers, and ensuring that all such activities are in accordance with the Brunei Selection Felling System (BSFS). BSFS involves pre and post-logging assessment of the timber stand. Trees to be cut and harvested as well as trees to be left behind to constitute the next timber crop are marked. Felling includes commercial species and sized which are governed by a set of diameter limits.

To safeguard these forest resources from any illegal activities, the Department conducts regular forest patrol via air, land and water. This forest patrol activities are being carried out either as part of the departmental enforcement unit routine forest patrol activities or as part of the joint forest patrol operations with the relevant government enforcement agencies such as through joint operations under the "*Jawatankuasa Protap Salimbada*" - members of which comprises of Royal Brunei Police Force, Royal Brunei Armed Forces, Survey Department, Land Department and Forestry Departments.

The Management of the Forest Reserve

Out of 438,000 ha of the forested area, only about 235,520 ha [40% of the country's total land area] has been gazetted as Forest Reserve. In line with the national forest policy, the Department aims to gazette an additional 15% of the country's total land area as Forest Reserve, making the total Forest Reserve in the country to 55% of the country's total land area. For management purposes, this Forest Reserve is classified according to its functional categories namely protection, production, conservation, recreation and national park. The functional description, detail breakdown of this classification and the location of gazetted & proposed Forest Reserve throughout the country.

Forest Plantation Area in the Inter Riverine Zone

The Forestry Department aims to develop at least 30,000 hectares of timber plantation to support the demand for wood raw materials of the country's forest-based industry on a sustainable basis. This initiative also intends to ease the pressure of natural forest utilization and confine timber production in plantation forests. The large-scale timber plantation establishment was based on several trial plantings of different timber species conducted since the late 1960's. A comprehensive study conducted in 1984 recommended the development of the sawtimber plantation development within the Inter-Riverine Zone (IRZ) located between Tutong and Belait River

The private sector plays an important role in forest plantation development. They are being tapped by the Government as partners in this endeavor, being a project implementor on the ground. During the process, the private counterpart shares the responsibility of managing the resource and at the same time learns and develops their technical skills in timber plantation establishment under the guidance of the Forestry Department.

CHALLENGES IN FOREST LAND MANAGEMENT

The Forest Department also faces a great challenge in managing the forest resources of the country. Although 40% of the country's land resources has been protected by legislation and gazetted as forest reserves, huge tract of forested areas (35% of the country's land resource) remains outside of these reserves and have been the subject of different land development. The 1989 National Forest Policy calls for the gazettement of additional 15% of the country's land area under Forest Reserve – making 55% of the country's land area under sustainable forest management. To be specific, the Forestry Department has been continuously facilitating the gazettement of additional land area within the Inter-Riverine Zone to meet the National Forest Policy target and to protect the huge investment made by the Government in timber plantation.

It is an accepted fact that land is a scarce-resource in Brunei Darussalam just like the rest of the world and the modernization/development of other sectors such as agriculture, fisheries, industrial, and housing sector have continuously posed for the conversion of these forested areas into other land-uses.

FOREST CONSERVATION & PROTECTION INITIATIVES

The Tropical Rainforests are home to 50% of all the animal and plant species on Earth. They are not only providing shelters and habitats for wild fauna and flora but also giving direct benefit to human being by acting as a giant lung for the Earth - absorbing large amounts of carbon dioxide and producing oxygen in exchange.

In addition, the rainforests also provide us with many foods, raw materials, and medicines. Despite their importance, the rainforests, which are known to cover only 6% of the Earth's surface, are always facing threats of being destroyed. Every year about 20 million hectares of rainforest are being cleared worldwide due to logging activities and clearing of land for agricultural purposes. Most of the world's remaining rainforests are in developing nations and the financial return on timber and farming is crucial for survival of their people and thus to a certain extent takes precedence over rainforest conservation. As a result, destruction of the tropical rainforests continues at an alarming rate.

In Brunei Darussalam, the pressure on the natural forest is not that great as compared to other countries within the region, as economically, Brunei Darussalam is still heavily dependent on the production of oil and gas as the main source of income for the country. However, the pristine forest of the country should be conserved and managed in a proper and sound manner, within the context of sustainable forest management, for its social, economic, and environmental conservation values. Over the years, the Forestry Department has been conducting various initiatives, projects and programmes in trying to ensure that our forest resources will continue to be developed and managed in a sustainable manner.

The Heart of Borneo [HoB] Initiative

One of the major on-going initiatives implemented by the Forestry Department is the Heart of Borneo Initiative. The Heart of Borneo initiative aims at conserving a large tract of the forest resources - mainly upland rainforest, spanning the central highlands of Borneo, and extending through the foothills into adjacent lowlands where possible to retain ecological connectivity - involving the voluntary participation of three countries namely Brunei Darussalam, Indonesia & Malaysia. This initiative is considered as one of the largest tropical rainforest conservation initiative in this region. The implementation of HoB initiative in this country actually complements and strengthens the already existing efforts and initiatives of the Department in ensuring its forest resources are managed and developed in a sustainable manner. As part of the country's commitment and support to the implementation of this HoB initiative, Brunei Darussalam has committed to allocate 58% of its land area to be under HoB management. The conservation vision as envisaged by the HoB initiative is actually nothing new. However, with the international recognition and accolade received by this initiative, it had given the already existing conservation efforts implemented by the Forestry Department an added value under the branding of HoB initiative.

The impact of HoB initiative in the implementation of forest management and conservation in the country has been significant. In the past, forest planting activities were organized mainly by the Forestry Department as part of its annual activities, and the private sectors and other relevant stakeholder involvement in the implementation of such activities came in the form of passive reaction as a gesture of support to such planting activities. But now, under the HoB branding, a new dimension in the implementation of forest rehabilitation and replanting activities emerges. The private sectors and relevant stakeholders, now no longer become a passive partners in the implementation of such forest replanting activities, but instead some have voluntarily and proactively come forward to champion the implementation of some forest rehabilitation and replanting projects.

The implementation of Brunei HoB initiative can also be seen as an opportunity for Brunei Darussalam, especially the relevant government agencies, to promote a holistic approach in forest land use management. The establishment of HoB National Council of the 21 April 2008, with permanent members comprises of representative from Ministries which have relevancy over land use management, provides an important avenue for the relevant policy and decision makers to discuss and share ideas; exchange views and information; and streamlines projects and initiatives in order to reduce land use management conflicts and issues, hence ensuring the country to develop and grow in a sustainable manner, within the context of sustainable development. An important point to note about this HoB initiative is that, while the initiative put its emphasis on forest conservation, however, this initiative does not intent to lock away development. Apart from the HoB initiative, there are also other policies, strategies, programmes and initiatives that are being implemented by the Department over the years, as explained in the following sub-headings.

Reduced-cut Policy

The reduced-cut policy, which was enforced since 1990, is part of the conservation effort that is in line with the 1989 National Forest Policy. With the reduced-cut policy, the annual logging rate was reduced by 50%, that is, from 200,000 m^3 to 100,000 ^{3}. The shortage of supply to cater for domestic demand is offset by imported sawntimbers. This annual logging quota is still enforced until today.

Increase size of Forest Reserve

The 1989 National Forest Policy outlines the principles and strategies to, among other things, promote the conservation of biodiversity. One specific provision is the commitment of dedicating at least 55% of the country's land area as forest reserves, that is, an increase of about 15% from the 40% that has already been gazetted.

Limit the issuance of Logging Permits/Licenses

The Forestry Department has also stopped issuing new logging permits or licenses since 1980's. This measure was taken as part of our conservation strategy in order to take into account on the limited areas of the production forest. To date, there are only 24 sawmills cum loggers still actively involve with the sawmilling and logging activities in the country.

Ban on the Export of Raw Logs

The law bans the export of timber. This is part of the country's forest conservation strategies in order to ensure sufficient supply of forest products for future domestic requirements.

Enhancement on Public Awareness

The Forestry Department has been and will continue to organize programmes that could promote and generate awareness among the public on the importance of conserving and protecting the forests as well as instilling the sense of love and appreciation for nature. Among the programmes that are being carried out annually by the Department is the celebration to commemorate World Forestry Day, which we normally follow-up with several forestry-related activities throughout the year. The activities include the mass tree planting for all walks of life, nature camp and nature excursion for the students as well as scientific research project for the secondary school students, that is, the prestigious Princess Rashidah Young Nature Scientist Award [PRYNSA].

Increase Forest Productivity

The Forestry Department will continue to intensify its silvicultural treatment activities in the logged-over forests in order to increase the productivity of the designated production forests. Well-planned forest harvesting system will continue to be practiced to ensure the silvicultural objectives of the natural production forest reserves are met. The Department will also continue to develop 30,000 ha of forest plantation by planting fast-growing species with crop rotation ranging from 15 – 40 years. These forest plantations are aimed at reducing dependence on the natural production forests. This was prompted by the projection that the natural production forests would be exhausted by the year 2015 if the logging rate were allowed to continue at 200,000 m^3 per year. The implementation of this forest resource development activity *i.e.* plantation establishment and silviculture treatment is one of the programmes under our 5 year series National Development Programme.

Establishment of Conservation Areas

The Forestry Department has also conducted programmes, which are in line with conservation activities, such as the establishment of *ex-situ* and *in-situ* conservation areas; delineation of genetic resource areas; as well as germplasm collection, the later of which is conducted by the Agriculture Department. Excellent examples of *ex-situ* conservation collection in Brunei are those of selected tree species, palms, bamboo and rattans located at the Brunei Forestry Centre in Sungai Liang; and plant collections located at Forestry branch in Sungai Lumut. On the other hand, the Ulu Temburong National Park can be described as one of the largest in situ conservation area in Brunei. Situated mainly in the Batu Apoi Forest Reserve in Temburong District, the National Park is mostly a virgin jungle comprising various forest types classified according to altitudes and soil types. Apart from conservation, research, and education, the Park also caters to ecotourism.

Development of Brunei Tropical Biodiversity Centre [TBC]

One of the programme that has have been included under the current National Development Programme [NDP] is the development of the Brunei Tropical Biodiversity Centre. The main objective of the establishment is to ensure that conservation and utilization of biodiversity resources on a sustainable basis can be achieved. The main thrust of the centre focuses on providing opportunities for research, education, and eco-tourism. The master plan for the development of the Brunei Tropical Centre has been completed and now, under the current 2007-2012 NDP, actual implementation of the 1st phase of the project *i.e.* the construction of the TBC main building, is expected to be implemented by May 2010.

Strengthening International & Regional Cooperation

The Forestry Department has been very active in participating regional and international meetings, conferences, symposiums and workshops as well as strengthening cooperation with other research institutions, organizations etc. This effort is part of its commitment towards enhancing capacity building as well as facilitating the transfer of technology and sharing and exchanging information in forestry related fields, one of which includes the agenda on conservation of the forest resources.

9

Environmental Role of Forests

The forest can deliver the functions of protection or conservation expected from it only if it is either in its natural state and under good natural ecological conditions or, when in use, it is managed in a sustainable manner. Under such conditions, health and vitality are very important. It is the vitality of forests that allows them to grow with sufficient strength and vigour in a way that will counter physical forces affecting soils through water erosion. It is this same vigour that allows a well-structured architecture and rich foliage that can counter wind erosion. The health of forests is fundamental to many of their environmental functions. However, forests are often affected by insects and other pests. They may be affected by a number of physiological alterations depending on climate changes, especially droughts. With health, vitality and a proper state of conservation, management and development secured. the forest intervenes especially in the following major environmental and protective functions.

Protection of water resources. Through their foliage. craggy bark and abundant litter, trees and forests decrease the speed of water dispersion and favour slow but total infiltration of rainwater; particularly in dry areas, the capacity of trees to retain other precipitations such as mist that then can be collected and stored for use is also important.

Soil protection. The forest canopy slows down the wind while its dense network of roots holds the soil in place; added to the buffering function of the water flow, these characteristics protect against wind and water erosion, land movement (mass slides and falling rocks) and, under cold climates, the risk of avalanches. With the combination of slower water dispersion and percolation to phreatic and intermediary water tables, the forest exerts an important buffering effect that protects against flooding or severe river bank erosion.

Influence on the local climate and reduction of gas emission impacts. Through the control of wind velocity and air flows, the forest influences local air circulation and may thus retain solid suspensions and gaseous elements; it can filter air masses and retain contaminants. The forest exerts a definite protective effect on neighbouring human settlements and crops in particular. This capacity is useful in the protection of inhabited areas that adjoin industrial zones and in urban forestry in general.

Conservation of the natural habitat and biological diversity. The forest offers a habitat to flora and fauna and, depending on its health, vitality and ultimately the way it is managed or protected, secures its own perpetuation through the functioning of the forest ecological processes. In Europe, almost half of the ferns and flowering plants grow in the forest. Owing to its size and structural diversity, more animal species are found in the forest than in any other ecosystem.

Recreational and other social functions of forests. Apart from direct physical and biological protective functions, forests in general have gained increasingly important recreational functions during the past five decades. In the vicinity of cities, tourism and health resorts have flourished, benefiting from the forest environment; in the forested areas of developed and developing countries alike, secondary residences are getting people back closer to the forests.

Protecting the cultural dimension of forests. While urban communities, particularly in the industrialized countries, are striving to be closer to nature, at the same time the evolution of the global and local forest economies may threaten other protective functions of natural forests in the developing world where forests have still maintained their cultural and religious functions. It is a challenge to twenty-first century forestry to cater also for these needs and maintain the cultural dimension of the protective functions of the forests. A number of innovative management options and many social and community forestry initiatives have addressed these needs.

PROTECTIVE AND ENVIRONMENTAL FUNCTIONS OF FORESTS IN SELECTED FRAGILE ENVIRONMENTS

Mountain ecosystems are among the fragile ecosystems targeted by UNCED Agenda 21, "Sustainable mountain development". The many activities developed intensively during the last five years since the Earth Summit have heightened awareness of the many functions of mountains. They are a repository of rare and rich plant and animal biological diversity. They contain unique gene resources underpinning agriculture and animal husbandry in the particular circumstances and farming systems prevailing in high valleys and plateaus. They provide a steady flow of water resources with related renewable energy potential. They are host to and protectors of a diversity of races and cultures.

In high mountains the forests protect settlements and communications systems against avalanches, falling rocks and landslides. In the European Alps, parts of the protection forests have been subjected to long-established management practices that have tended to maintain a correct mix of evergreen and broad-leaved species. But trends are strongly pushing towards unsatisfactory changes.

Mixed stands are giving way to monospecific forests and many protection forests have been weakened and have become overaged. They grow more and more sparsely. In many cases the presence of game prevents natural regeneration through browsing and increases the degradation of the forest ecosystem and reduces its protective capacity. Measures to counter this negative trend or to restore the forest ecosystem include biological measures such as reforestation, engineering work and active silvicultural practices that promote and assist natural regeneration.

In other parts of the world the same decline may be observed but most of the degradation occurs as the result of the attempts of the rural poor to eke out a living, especially in marginal areas. In North Africa, in the African plateaus, in the Andes and in the Himalayas, the search for fuelwood, the grazing of animals, particularly well-adapted, ever-coping goats and marginal agriculture have been mentioned as being among the major factors of mountain forest fragmentation, degradation and loss. In many parts of the world mining also disrupts mountain ecology and leads to the degradation or destruction of unique biological diversity, in some cases affecting endemic species and the processes in which they are involved. Inappropriate mining in watershed areas may also often be the sources of contamination of major water courses.

Many ecological groups have highlighted the possible negative influences of communication lines and roads as their establishment entails the deforestation of large stretches which affect the habitat of species of high biological significance. However, the most potent root causes are highlighted by Hernandez (1997) in Latin America as being: i) the state of inherent fragility of young mountains; ii) the extreme poverty of the population and the loss of stamina and resolve in the face of drastically poor and yet rapidly degrading social conditions; iii) the segmented approach to development, the lack of institutional cohesion and coherence and the lack of suitable and participatory extension.

The protective functions of mountain forests and their relation to climate change require special attention. This subject has been largely studied worldwide. Gottle (1997, personal communication) reports the results of the research of the Bavarian State Office for Water Management on the possible consequences of climate change on the way mountain forests in Europe may then provide buffering functions:

- *Higher temperatures* will lead to a rise of the snow and ice line resulting in more favourable conditions for vegetation but with higher "erodibility", the shifting of the permafrost line creating more instability in areas yet to be colonized by vegetation;
- *Frequent changes between thawing and freezing* will release more weathering material and increase the related risks of rock fall and landslides while making the consolidation of vegetation more difficult;

- *More water precipitation with the predominance of rains* will lead to water-saturated soils which will eventually reduce shear resistance and loss of stability, and lead to conditions unfavourable to the establishment of vegetation. More surface runoff may also occur throughout the year leading to more erosion.

Similar changes in intensity but not in the direction of change may be experienced more strongly in mountain areas than on plains, meaning that more spectacular changes in mountain forest ecosystems, owing to the concentration on short horizontal distances of phenomena that otherwise happen in a differentiated manner and over greater distances on flat land.

Special attention to mountain forests for the future and greater efforts in watershed management. The Mountain Agenda which has been very active in the framework, and the creation of the worldwide Mountain Forum and its regional chapters have stressed the need to balance the unidirectional flow of resources including forest resources and services from the mountain downwards. The need to find innovative funding mechanisms and fresh policy options that restore the overall balance between mountain economies and societies in lowland areas was strongly advocated. The efforts to conserve mountain forests in this case cannot be supported only by mountain communities.

Hernandez (1997) sends the same message regarding tropical mountain forests; he stresses that interested societies are more and more aware of the importance of the cloud forests in the production of high-quality water in tropical mountain watersheds. Hence, a better comprehension of the needs of sustainable forest management and development must be shared throughout the watershed, if mountain forests and beyond, mountain natural systems and socio-economic systems are to remain prosperous for mountain and national communities, and also for regional solidarity.

In this context, many communities have called for more efforts for the management of watersheds that aim not only at the physical restoration of benign processes in the flow of water resources and solid material, but which aim at the sustainable development of mountain systems. This will be achieved through, inter alia; i) the stabilization of livelihood systems; ii) the improvement of living conditions in mountain areas; iii) the identification and promotion of innovative income-generating activities and alternative employment; and iv) the restoration of equity and solidarity between upstream and lowland communities.

ENHANCING NATURAL SEQUESTRATION

FORESTS

Forests are carbon stores, and they are carbon dioxide sinks when they are increasing in density or area. In Canada's boreal forests as much as 80% of the total carbon is stored in the soils as dead organic matter. A 40-year study of African, Asian, and South American tropical forests by the University of Leeds, shows tropical forests absorb about 18% of all carbon dioxide added by fossil fuels. Tropical reforestation can mitigate global warming until all available land has been reforested with mature forests. However, the global cooling effect of carbon sequestration by forests is partially counterbalanced in that reforestation can decrease the reflection of sunlight. Mid-to-high latitude forests have a much lower albedo during snow seasons than flat ground, thus contributing to warming. Modeling that compares the effects of albedo differences between forests and grasslands suggests that expanding the land area of forests in temperate zones offers only a temporary cooling benefit. In the United States in 2004, forests sequestered 10.6% of the carbon dioxide released in the United States by the combustion of fossil fuels. Urban trees sequestered another 1.5%. To further reduce U.S. carbon dioxide emissions by 7%, as stipulated by the Kyoto Protocol, would require the planting of "an area the size of Texas every 30 years". Carbon offset programmes are planting millions of fast-growing trees per year to reforest tropical lands, for as little as $0.10 per tree; over their typical 40-year lifetime, one million of these trees will fix 0.9 teragrams of carbon dioxide.

In Canada, reducing timber harvesting would have very little impact on carbon dioxide emissions because of the combination of harvest and stored carbon in manufactured wood products along with the regrowth of the harvested forests. Additionally, the amount of carbon released from harvesting is small compared to the amount of carbon lost each year to forest fires and other natural disturbances. The Intergovernmental Panel on Climate Change concluded that "a sustainable forest management strategy aimed at maintaining or increasing forest carbon stocks, while producing an annual sustained yield of timber fibre or energy from the forest, will generate the largest sustained mitigation benefit". Sustainable management practices keep forests growing at a higher rate over a potentially longer period of time, thus providing net sequestration benefits in addition to those of unmanaged forests.

Life expectancy of forests varies throughout the world, influenced by tree species, site conditions and natural disturbance patterns. In some forests carbon may be stored for centuries, while in other forests carbon is released with frequent stand replacing fires. Forests that are harvested prior to stand replacing events allow for the retention of carbon in manufactured forest products such as lumber. However, only a portion of the carbon removed from logged forests ends up as durable goods and buildings. The remainder ends up as sawmill by-products such as pulp, paper and pallets, which often end with incineration at the end of their lifecycle. For instance, of the 1,692 teragrams of carbon harvested from forests in Oregon and Washington from 1900 to 1992, only 23% is in long-term storage in forest products.

OCEANS

One way to increase the carbon sequestration efficiency of the oceans is to add micrometre-sized iron particles in the form of either hematite or melanterite to certain regions of the ocean. This has the effect of stimulating growth of plankton. Iron is an important nutrient for phytoplankton, usually made available via upwelling along the continental shelves, inflows from rivers and streams, as well as deposition of dust suspended in the atmosphere. Natural sources of ocean iron have been declining in recent decades, contributing to an overall decline in ocean productivity. Yet in the presence of iron nutrients plankton populations quickly grow, or 'bloom', expanding the base of biomass productivity throughout the region and removing significant quantities of CO_2 from the atmosphere via photosynthesis. A test in 2002 in the Southern Ocean around Antarctica suggests that between 10,000 and 100,000 carbon atoms are sunk for each iron atom added to the water. More recent work in Germany suggests that any biomass carbon in the oceans, whether exported to depth or recycled in the euphotic zone, represents long-term storage of carbon.

This means that application of iron nutrients in select parts of the oceans, at appropriate scales, could have the combined effect of restoring ocean productivity while at the same time mitigating the effects of human caused emissions of carbon dioxide to the atmosphere. Because the effect of periodic small scale phytoplankton blooms on ocean ecosystems is unclear, more studies would be helpful. Phytoplanktons have a complex effect on cloud formation via the release of substances such as dimethyl sulfide that are converted to sulfate aerosols in the atmosphere, providing cloud condensation nuclei, or CCN. But the effect of small scale plankton blooms on overall DMS production is unknown. Other nutrients such as nitrates, phosphates, and silica as well as iron may cause ocean fertilization. There has been some speculation that using pulses of fertilization may be more effective at getting carbon to ocean floor than sustained fertilization. There is some controversy over seeding the oceans with iron however, due to the potential for increased toxic phytoplankton growth, declining water quality due to overgrowth, and increasing anoxia in areas harming other sea-life such as zooplankton, fish, coral, etc.

SOILS

Since the 1850s, a large proportion of the world's grasslands have been tilled and converted to croplands, allowing the rapid oxidation of large quantities of soil organic carbon. However, in the United States in 2004, agricultural soils including pasture land sequestered 0.8% as much carbon as was released in the United States by the combustion of fossil fuels. The annual amount of this sequestration has been gradually increasing since 1998. Methods that significantly enhance carbon sequestration in soil include no-till farming, residue mulching, cover cropping, and crop rotation, all of which are more widely used in organic farming than in conventional farming. Because only 5% of US farmland currently uses no-till and residue mulching, there is a large potential for carbon sequestration. Conversion to pastureland, particularly with good management of grazing, can sequester even more carbon in the soil. Terra preta, an anthropogenic, high-carbon soil, is also being investigated as a sequestration mechanism. By pyrolysing biomass, about half of its carbon can be reduced to charcoal, which can persist in the soil for centuries, and makes a useful soil amendment, especially in tropical soils.

SAVANNA

Controlled burns on far north Australian savannas can result in an overall carbon sink. One working example is the West Arnhem Fire Management Agreement, started to bring "strategic fire management across 28,000 km^2 of Western Arnhem Land". Deliberately starting controlled burns early in the dry season results in a mosaic of burnt and unburnt country which reduces the area of burning compared with stronger, late dry season fires. In the early dry season there are higher moisture levels, cooler temperatures, and lighter wind than later in the dry season; fires tend to go out overnight. Early controlled burns also results in a smaller proportion of the grass and tree biomass being burnt. Emission reductions of 256,000 tonnes of CO_2 have been made as of 2007.

FRESHWATER ENVIRONMENTS

Only 2.58 per cent of the water on Earth is fresh. The largest share of that—1.97 per cent—occurs as ice in the world's glaciers and ice caps. The remaining freshwater—0.61 per cent—is stored in lakes, rivers, and groundwater. The Earth's climate—the processes of precipitation, evaporation, and water vapour transport—determines the amount and distribution of freshwater at any given time on Earth. Without freshwater, life could not exist. Freshwater is the most essential resource. Wetlands and river systems provide food and water for life. Wetlands are nature's natural filters of harmful substances and serve to purify water. It is the availability of water that partly determines the distribution of major biome types. Humans depend directly on freshwater for drinking and cooking, irrigation (agriculture), industry, transportation, fisheries, and recreation.

IPC ASSESSMENT

The IPCC, while growing populations are already overtaxing freshwater supplies in many parts of the world, global warming will make freshwater supply even scarcer in certain areas. Warming temperatures will also have consequences. Winter ice cover of streams and lakes will decline, and there will be a trend towards later freeze and earlier ice breakup, as is already happening in Europe. The timing and duration of freeze and the breakup of ice are important because they affect both biological and ecological processes. If precipitation decreases, the water flow rates will drop, and this could affect lakes and streams and lead to changes in habitat and breeding locations of aquatic flora and fauna.

The IPCC has also determined that hydrologically isolated systems, such as wetlands and topographical depressions, would be the most vulnerable areas to global climate change. Areas along larger rivers and lakeshores would feel the least impact. As freshwater supplies diminish with increasing temperatures and lack of precipitation, competition for diminishing freshwater resources will increase. Even if precipitation increases in some areas during the winter, if it cannot be stored as groundwater and runs off the surface when it melts in the spring, it will be unusable. Summers are predicted to become warmer and drier, which will also lead to a deterioration of freshwater ecosystems.

IMPACTS TO FRESHWATER AND WETLAND ECOSYSTEMS

Global warming has been determined to be devastating to freshwater aquatic ecosystems based on the results of a study conducted by scientists from Colourado State University (CSU), East Carolina University (ECU), and Louisiana State University (LSU). In their report, titled "Aquatic Ecosystems and Global Climate Change," the researchers believe that global warming will be devastating to trout, salmon, and several species of aquatic plants and animals. CSU researchers forecast significant shifts in fish habitats, decreasing water quality, and disappearing wetlands. Rivers, lakes, and wetlands will all be seriously affected. N. LeRoy Poff, a freshwater biologist at CSU, reports, "Wetlands and aquatic ecosystems are quite vulnerable to climate change. Projected increases in the Earth's surface temperature due to global warming are expected to significantly disrupt current patterns of aquatic plant and animal distribution and to alter fundamental ecosystem processes, resulting in major ecological changes." Two of the cold-water fish that are expected to suffer most under increased global warming conditions are trout and salmon.

Both species will likely disappear from their current geographic ranges. Species that attempt to migrate either northward or higher may not be able to because of urbanization blocking the way. They may become extinct because there is nowhere else to go. It will be a different scenario for those freshwater fish that thrive in warm-water environments, however. Biologists expect that fish, such as largemouth bass and carp, will actually expand their geographic range throughout the United States and Canada. Rising water temperatures—even of just a few degrees—will have a significant effect on wildlife in terms of their adaptation requirement. For instance, a 5.7°F (4°C) rise in surface temperature would require aquatic species to migrate 422 miles (680 km) northward to maintain the same thermal habitat conditions that they initially enjoyed before the rise in temperature. The CSU, water quality will also become an issue. It will decline because there will be a reduction in springtime melt and run-off, as well as higher temperatures.

Higher temperatures encourage the growth of algae. The more prolific the algae, the lower the amount of dissolved oxygen there is in the water. In the study, Poff also stressed that "reduced oxygen in lake waters due to increased production of blue-green algae can lead to the loss of large predatory fish and have negative effects on the food chain." In addition, he says that "climate change could have other subtle effects on aquatic ecosystems as precipitation patterns change. In the coming century, areas accustomed to snow may instead get rain in the winter, causing floods that destroy fish eggs left in streams and leaving little snowpack to sustain rivers during the dry summer months." The authors of the report—N. LeRoy Poff (freshwater biologist) of CSU, Mark Brinson (wetlands biologist) of ECU, and John Day, Jr. (estuarine biologist) of LSU—agree that the effects of global warming are very difficult to predict because the exact temperature changes at this point are based only on the most accurate climate models developed to date. They also agree that "in the face of inevitable climate change, humans can take actions to minimize the risk of ecosystem disruption."

ECOSYSTEM AND HUMAN ADAPTATION

The Pew Centre, the ability of aquatic ecosystems to adapt to climate change is very limited. Where animals and plants will need to migrate northward or higher in elevation, aquatic species differ in their dispersal abilities. If they are not able to migrate, they will be doomed. Many aquatic species are spatially isolated due to human activities and interference, making it more difficult for them to adapt. The extinction of many local species is to be expected across all aquatic ecosystems. In addition, it is unknown at this point what will happen when species do try to migrate and come into contact with new, nonnative, more resilient, introduced species for the first time.

The Pew Centre stresses that one of the critical uncertainties in projecting future aquatic ecosystem responses to global warming is how humans will interact with these ecosystems during the changes. Human activities have already severely interfered with aquatic ecosystems through the building of dikes, levees, and reservoirs, as well as pumping significant amounts of groundwater. Over the years, these have all modified natural processes and increased the vulnerability of aquatic systems. One major challenge for humans in the future will be to adapt in a way to global warming that will minimize the impact on aquatic ecosystems by minimizing pollution, introduction of exotic species into the ecosystem, habitat destruction, and further fragmentation of existing freshwater marine habitats. The Pew Centre stresses that humans greatly depend on inland freshwater and coastal wetland ecosystems to provide critical goods and services every day. If populations want that to continue, then scientists and policy-makers must focus more on minimizing negative effects to freshwater ecosystems and educating people about the best ways to maintain the health and longevity of these precious natural resources in a changing world.

INVERTEBRATE SPECIES IN FOREST ECOSYSTEMS

The diversity and abundance of invertebrate species in forest ecosystems are determined by the interaction of complex ecological and environmental factors. Research into these factors is gradually providing data that can be incorporated into models, both descriptive and mathematical, of the processes involved. At the core of such models is the description of the availability of resources that enables a particular invertebrate species to reproduce, whereas the rate of reproduction is dependent on multiple factors, including temperature, competition, etc., that determine utilization of available resources. In relation to exploitation of resources within a forest ecosystem, the primary focus for forest managers is the potential of invertebrates to cause damage and to reduce yields of the trees themselves. This is certainly the case in managed forests where the primary aim is to grow trees for commercial gain, often with associated value from recreational, landscape or environmental considerations. Any invertebrates that feed directly on the trees, or have an influence of how the trees grow, may be classified as pests and hence may require some form of pest management to reduce the threat. But it is pertinent to consider how ecological theory and practice can be used both to understand why a pest problem has arisen and to provide guidance on how to combat that problem. The same principles can be applied to understanding and enhancing the biodiversity of forests, including conservation of rare and beneficial invertebrates. Indeed, the two extremes of pest outbreak and invertebrate rarity are part of the same continuum of interaction between invertebrates and their host trees.

In general, considering that trees are known to harbour the greatest numbers of insect species (diversity), it is surprising that serious pest status is reached only relatively infrequently. For example, although oak *(Quercus* spp.) is known to support at least 450 species of insect in Britain, Bevan (1987) lists only 35 species as pests, of which five were classified as important or severe (XXX-XXXXX on Bevan's scale). Wider listing of pest species has been noted for forest insects in continental Europe, although even here the number of consistently serious pests is small relative to the total diversity of insects present (Klimetzek 1993).

Naturally, the severity of a pest outbreak will be classified differently depending on the purposes of the forest manager, and thus it is not always possible to classify insects consistently as pests across geographical ranges or, indeed, between years. Further, an insect may normally be classified as uncommon or rare but, given an abundance of resources, may be described later as a pest. This was certainly the case for oak pinhole borer, *Platypus cylindrus* (Coleoptera: Platypodidae), which prior to the 1987 great storm in southern England was listed in the British Red Data Book as rare.

Winter (1988) warned that the great increase in availability of freshly dead oak following the storm could result in an increase in numbers of *P. cylindrus* and, subsequently, it has been noted that the insect is now causing damage to high-value oak logs as a result of its habit of boring directly into the heartwood. Therefore, consideration of why invertebrate abundance might increase above a given threshold level can provide a more consistent appraisal of status and also potential approaches for reducing populations below that threshold or indicating measures that can be taken to encourage rare or endangered species.

Among the many interacting processes that determine insect diversity and abundance, a few attributes that are particularly well linked to the environmental status of trees can be distinguished. These provide a means to describe intrinsic processes but also allow comparison with other plants, thus offering the potential for a more consistent approach to descriptive population ecology of invertebrates irrespective of the host plant being considered.

Longevity

Trees are the longest-lived components of an ecosystem and, in the majority of natural situations, approach or represent the climax stage of vegetation succession. This provides opportunities for continuity of insect population growth and over time this will give a particular set of insect associates for a given tree species. Continuing the theme of resource availability, the fact that trees may be present in the same location for tens or hundreds of years ensures that both migrant and, particularly, relatively sedentary invertebrates can exploit them at some stage during their growth. However, longevity *per se* is not a quantifiable attribute but does provide the means for the other components, to influence invertebrate diversity and abundance. Increasing stability in the environment, as represented under forest conditions, leads to stability in both microclimate and resource availability, particularly favouring exploitation by specialist insect feeders and their associates.

Biogeography

Tree species richness in a given region or locality varies considerably with geographical location and thus the climax community structure and longevity of individual species may vary accordingly. Although subject to uncertainty and lack of unequivocal evidence of the underlying processes, there is certainly a gradient of species richness of plant cover, especially trees, from the tropics to the poles. Not surprisingly, the same applies to a wide range of animals and plants; as an extreme example, the species richness of orchids ranges from 15 to 2500 in a latitudinal range from 0 to 66° S. Further examples are provided by Brown (1988). More specifically, in an analysis of species richness of deciduous forests both in the tropics and, latitudinally, within North America, Brown and Gibson (1983) contrasted the high diversity of tree species in the tropics (up to 100 ha^{-1}) with the low diversity from north to south within North America (1–30ha^{-1}). However, even within such a generalization there are exceptions, so that for example conifers are more diverse in temperate latitudes.

Geological Time

The ecological structure of forests and of their associated invertebrate faunas is determined by both present and recent characteristics, particularly area, and by their history in geological time. The latter characteristic is driven as much by *time per se* as by the history of major disturbances (glaciations, volcanic eruptions, land shifts, etc.) within geological time. Thus, present flora and fauna may well be determined by recolonization since the last major disturbance, the diversity of species being a reflection of both colonization and speciation relative to more recent ecological conditions. The gradient of species from polar to tropical regions could therefore be explained by the high level of disturbance closer to the polar masses compared with the relative stability of climate in the tropics. Evidence from habitats that have become isolated since the last major disturbance indicates that current species diversity may reflect the historical link to the original major habitat. For example, islands that were once joined by land bridges to the larger land masses of New Guinea have species diversities of birds greater than expected compared with other islands that were never connected. However, although local ecological communities can be explained by historical patterns, the gradients over wide geographical scales are not so easily attributed to time as the major determinant.

The concept that major disturbances drive change undoubtedly carries weight, but closer analysis of tropical rain forests reveals that they too have been subject to disturbances of various sorts and are, in many cases, much changed over geological time. Based on this knowledge Haffer proposed a theory that disturbances in the tropics during the Pleistocene (alternating wet and dry periods giving rise to forest blocks surrounded by grassland) gave rise to refugia in which high levels of speciation took place, ultimately leading to the current high diversity characteristic of the tropics. While this theory is plausible and has been invoked by other authors, tests of the hypothesis have cast doubt over its wide applicability.

Irrespective of the mechanistic bases for variation in species richness over wide latitudinal ranges, time is a significant determinant at a more local scale. Birks (1980) analysed the insect faunas of trees in Britain during the period since the last glaciation 13000 years ago and concluded that there was a significant correlation with the time that a given tree species had been continuously present in Britain during that period. Taking this further, Kennedy and Southwood (1984) included Birks' radiocarbon estimates of time in Britain as a variable in a multiple regression analysis of insects on British trees. They showed that time was a significant predictor, both in its own right and as a co-predictor with log abundance (area). This is logical considering that over time there will be a tendency for well-established tree genera to increase their abundance, despite the recent proliferation in areas arising from commercial afforestation.

Size

A consequence of basic structure and longevity is that trees tend to be the largest components of the phylloplane, thus providing a much wider number and range of ecological niches compared with herb and shrub layer plants. Such attributes enable more insect species to colonize the plants without facing competition, both interspecific and intraspecific. Size has many interrelated characteristics that help to explain the presence of both herbivores and their natural enemies.

- A larger physical presence is more easily detected and colonized by invertebrates. This concept, termed 'apparency', reflects a higher visual profile for colonizing adults or passively migrating immature life stages (*e.g.* mean height was a good predictor in models of the distribution of Scolytidae in Finland; Heliovaara & Vaisanen 1995) and a greater production of olfactory cues.
- Complex plant architecture is a consequence of greater size and for trees is particularly linked to the provision of many niches capable of supporting invertebrates with a wide range of feeding strategies (guilds). Although the concept of the guild is not always clearly defined, it is nevertheless a useful means of describing the trophic interactions of invertebrates and trees without having to consider their precise taxonomic status. Thus, seed, bud, leaf, bark and wood feeders can be distinguished, each occupying a different part of the tree and therefore potentially avoiding competition.
- Increased biomass provides greater food resources per individual, thus encouraging growth and breeding success whilst reducing potential intraspecific competition. Price (1992) has reviewed the role of plant resources in insect population dynamics and concludes that, in part, the carrying capacity of an ecosystem depends on plant succession in time and space. Carrying capacity is therefore greater for trees, which generally appear late in succession, compared with herbs or shrubs. Thus biomass is available for longer and in greater quantities on trees compared with other plants.
- The concept of enemy-free space applies to the likelihood of

herbivores being overlooked by searching parasitoids or predators. One aspect of size is that there is more opportunity for herbivores to occupy a part of the plant that offers protection against natural enemies (and competition from other herbivores). The complexity and number of niches present on a plant (plant architecture; Lawton 1983) therefore has an effect on enemy-free space. The greater the architectural complexity, the greater the time that natural enemies may have to spend in searching for prey, thus reducing their overall effectiveness and enhancing the survival probability of herbivores.

Area

Forests tend to occupy greater areas than other plants. There is evidence to support the view that the greater the area occupied by the host plant, the greater the number of insect species associated with that host. This is called the species-area relationship and was originally linked to the island biogeography theories of species isolation and colonization.

Species-area relationships have been demonstrated for many plants and animals, including both highly mobile and sedentary species. With regard to the numbers of insect species on trees, data on the British insect fauna, one of the best characterized in the world, have been analysed by a number of authors.

Arising from earlier tentative analyses, Kennedy and Southwood (1984) analysed the relationships between insect species and characteristics of tree species in Britain. They concluded that there was a significant relationship between the area occupied by British tree species and the numbers of insect species associated with them. The area occupancy data were based on county and subcounty information of plant records and were thus potentially overestimates of actual areas occupied.

Claridge and Evans (1990) reanalysed the species-area relationships using the same insect data as Kennedy and Southwood but employing actual area information derived from Forestry Commission census data. They concluded that area was not a good descriptor for insect species on deciduous trees but was just significant for conifers.

However, further unpublished analysis of the data broken down into numbers and areas of habitat patches occupied by the tree genera indicated that whereas total area was not a good descriptor, the degree of fragmentation of that area provided a better explanation for the observed insect species numbers. A similar conclusion, concentrating on Lepidoptera that have moved hosts to include conifers, was reached by Fraser and Lawton (1994) in an analysis of new associations on British trees. The overall finding that area provides a broad descriptor of insect diversity remains valid, although this does not explain the mechanisms that lead to these associations.

Taxonomic Relatedness

In a given ecosystem, the closer the taxonomic relatedness of host plants, the more similar the numbers of insect species associated with those plants. This has been documented for tree species that have a long history in a particular location and thus provide rich sources of invertebrate species that may subsequently colonize other tree species, especially if they are taxonomically related. A particularly good example of the process in action is the colonization of *Nothofagus,* a member of the Fagaceae newly introduced to Britain, which has been colonized rapidly by insects associated with closely related members of the native Fagaceae.

Geographical isolation as a critical factor in determining invertebrate diversity and implications for international movement of pest organisms. Geographic isolation drives the acceleration of speciation and provides the means to derive principles on the processes involved. In addition, the fact of isolation means that some invertebrates abundant in one region will be absent elsewhere, despite conditions being suitable for their establishment. In such situations, it is pertinent to examine the international plant health implications of geographical isolation and how modern patterns of trade influence the degree of isolation and later colonization by new insect species.

Geographical Isolation as a Factor in Determining Insect Diversity

In general, the basic principles apply across ecosystem boundaries and geographical zones extending from the poles to the tropics. In addition, the isolation afforded by relatively small islands provides further evidence of the principles of species colonization over space and time. Great Britain provides a particularly good example, because of its island position in the Atlantic Offshore climatic zone and the fact that it has been subjected to glaciation in recent geological time. Among the ecological characteristics that arise from this set of circumstances are a relatively depauperate mix of both animal and plant species that have arisen from the combination of losses resulting from the last ice age and the slow rate of recolonization arising from island status. British flora and fauna are therefore impoverished relative to continental Europe. Further influences arise from the preponderance of exotic tree species in commercial forestry in Britain, which although dominated by exotic conifer species also has increasing numbers of exotic broadleaved species.

Expanding on the principle that colonization of tree species is determined by the features, the ranges of numbers of insect species associated with trees in Britain reflect both the area (with provisos) and length of time that trees have been continuously present in Britain. However, there are exceptions, so that Sitka spruce in particular would appear to support far fewer species than predicted on the basis of area, although this probably also reflects the relatively short history of the species and recent expansions in areas planted in Britain.

Implications for International Movement of Pest Insects

Trading patterns between countries have had a considerable influence on both the flora and fauna of most of the countries in the world. This reflects deliberate introductions of plants and animals for agricultural, forestry and amenity purposes and also accidental introductions by association with other commodities. In the majority of cases, the introduced species have had a neutral to beneficial effect, reflecting their 'domesticity' within managed crop or animal husbandry systems. However, there are many exceptions to this generalization and in most cases the results of introduction of new species were not foreseen, either through lack of knowledge of the potential impacts or because the introduction was accidental and undetected. The literature on international biological control is dominated by examples where attempts have been made to rectify the effects of new pest introductions (both animal and plant). A classic example is the control of the woodwasp *Sirex noctilio* (Hymenoptera: Siricidae) in exotic pine forests in Australia with the introduced nematode *Deladenus siricidicola* (Nematoda: Neotylenchidae) isolated from *Sirex* in New Zealand. This nematode has a life cycle that includes a free-living form that feeds on the fungus *Amylostereum areolatum* and a parasitic form that invades *S. noctilio* larvae and subsequently migrates to the reproductive organs of the female wasp, causing sterility. In this situation, all elements of the interaction are exotic to the new location, with little prospect of natural control being realized.

HEMIPTERA

Aphids (Aphididae)

With short generation times and low developmental threshold temperatures, aphids are a group of insects that can be expected to be strongly influenced by environmental and climatic changes. In general, it has been predicted that aphids will appear at least eight days earlier in the spring within 50 years, though the rate of advance will vary depending on location and species. This could potentially result in greater damage to host plants depending on the phenology of host plants and natural enemies. Zhou *et al.* (1995), for example, investigated the timing of migration in Great Britain for five aphid species (*Brachycaudus helichrysi, Elatobium abietinum, Metopolophium dirhodum, Myzus persicae, Sitobion avenae*) over a period of almost 30 years and concluded that temperature, especially winter temperature, is the dominant factor affecting aphid phenology for all species. They found that a one degree Celsius increase in average winter temperature advanced the migration phenology by 4-19 days depending on species.

Elatobium abietinum (Walker) (Aphididae)

The green spruce aphid (*Elatobium abietinum*) is also believed likely to benefit from the increase in winter survival, leading to more intense and frequent defoliation of host spruce trees (*Picea* spp.). This aphid is native to Europe but has also been reported in both North and South America.

Infestations in Great Britain have resulted in large losses of spruce foliage and height both during the active infestation and in subsequent years. Westgarth-Smith *et al.* (2007) showed that warm weather associated with a positive North Atlantic Oscillation (NAO) index caused spring migration of *E. abietinum* to start earlier, last longer and contain more aphids. Positive NAO values correspond to warmer atmospheric conditions over Great Britain. Since global warming is believed to increase NAO variability, shifting the system to more positive values, this will most likely lead to further increases in aphid activity and more damage to spruce trees and forests in the area.

VALUE OF ENVIRONMENTAL ASSETS

While man-made and human capital may be valued with relative ease by observing existing market sys-tems, where available; the existence value of clean wa-ter and air, tropical forests, wetlands, coral reefs and other environmental assets and their functions is much more difficult since not even market prices can reflect their full contribution to other economic ac-tivity and to human welfare. In particular, the mar-ket price of water does not reflect the various services it provides nor do market values can accurately re-flect what happens when irreversible loss or damage of natural resources occurs as environmental degrada-tion exceeds a critical threshold level.

COMPLEXITY OF ENVIRONMENTAL VALUATION

Complexity of environmental valuation also arises due to the multiple functions of being a source of raw materials and energy, being a sink for assimilating man-made wastes and providing other services such as recre-ational/tourism services, storage of genetic diversity and scientific and educational benefits.

The following classification is useful:

- Direct use values are derived from the economic uses made of the natural system's resources and services. Examples of these are outputs such as timber, game and recreation from forests or fish and scuba tourism from coral reefs.
- Indirect use values are the indirect support and protection provided to economic activities and property by the resource system's natural func-tions or environmental services. Examples of these are watershed protection and soil erosion prevention provided by forests, and beach sand and mooring facilities protection provided by coral reefs.
- Non-use values lie in the special attributes of the natural system as a whole, its cultural and heri-tage uniqueness; it includes both

existence and option values. Existence values reflect public goods which can be enjoyed by more than one consumer without decreasing the amounts en-joyed by others such as clean air, beaches and forests. The existence value is the utility that consumers derive from just knowing the public good exists. A way of measuring that utility would be to measure the willingness to pay or the contingent value assigned if the public good were to disappear. Option values reflect what current generations wish to bequeath for future generations to inherit. They imply both an ethi-cal commitment to sustainabilty for the children of our children and in a shorter time-frame the maintenance of options to solve current prob-lems. Examples of option values have been devel-oped for forests relating to biodiversity and the search for cures of cancer and AIDS-"if forest were to disappear then the options to find such cures will vanish" or "as long as the forest is pro-tected there is the option to find the cure".

Direct Effects Valued on Conventional Markets Some methods are directly based on market prices or productivity. This is possible where a change in envi-ronmental quality affects actual production or pro-ductive capability.

Change-in-Productivity: Development projects can affect production and productivity positively or neg-atively. The incremental output can be valued by us-ing standard economic prices where available. Loss-of-Earnings. Environmental impacts can signifi-cantly affect human health. In theory, the value of health impacts should be determined by the willing-ness to pay of individuals to maintain their health. In practice, one uses earnings lost upon early death, dis-ease or job absence. This approach is used in highway and industrial safety, and in air pollution studies.

The "implicit value of human life" approach is reject-ed by many as dehumanizing since human life can be said to have infinite value. However, society, govern-ment regulations, insurance companies and judicial courts implicitly and explicitly place finite values on human life and health. This is a necessity reflecting limited resources to be allocated for health expendi-tures. The relatively high level of health expenditures in Cuba would indicate a high implicit value of hu-man life and health. However, one can also discern a political motivation and its attendant benefits behind it.

Preventive Expenditures: Individuals and governments invest in prevention measures to avoid or reduce un-wanted environmental effects. Environmental dam-ages, are often difficult to assess, but historical infor-mation on preventive measures and their costs may be interpreted as a minimum value for the expected benefits that the preventive measures seek. If it is found, for example, that there is a historical pattern of under investment in Cuba for natural disaster planning and prevention; then, it can be inferred that the benefits expected from such preventive measures have been very low. This conclusion would in turn imply that the loss of human life would be assigned a relatively low value. While this possible conclusion may appear to conflict with the high political priority for health expenditures, it can be observed that the level of damages or loss of lives caused by natural di-sasters cannot be easily attributed to government in-action and thus the political cost may be easily ex-plained away.

Potential Expenditure

Valued on Conventional Markets

Replacement Cost: Simply, the costs that would have to be incurred in order to replace a damaged asset. The estimate is not a measure of benefit of avoiding damage since the damage costs may be higher or low-er than the replacement cost.

Valuation Using Implicit (or Surrogate) Markets

Sometimes one must use market information indi-rectly. Approaches to be considered are the travel cost method, the property value approach, the wage dif-ferential approach, and uses of marketed goods as surrogates for non-marketed goods. Each technique has its particular advantages and disadvantages, as well as requirements for data and resources. One must determine which techniques might be applica-ble to a particular situation.

Travel Cost: This approach measures the travel cost which reflects the willingness of consumers or users can serve to measure the benefits produced by recreation sites (parks, lakes, forests, wilderness). It can also be used to value "travel time" in projects dealing with fuelwood and water collec-tion.

Property Value: This valuation method is based on the general land value approach and can determine the implicit prices of certain land areas. The property value approach can help analyze willingness to pay for properties with different pollution levels and infer the implicit cost of pollution. The method compares prices of houses in affected areas with equal size and similar neighborhood characteristics elsewhere in the same metropolitan area.

Valuation Using Constructed Markets

Contingent Valuation: When society's preferences as revealed in market prices are not available, the con-tingent valuation method tries to obtain information on individual preferences by posing direct questions about willingness to pay. It basically asks people what they are willing to pay for a benefit, and/or what they are willing to accept by way of alternate compensation to tolerate an environmental cost. This process may be achieved through a direct questionnaire/sur-vey. Willingness to pay is difficult to measure and de-pends on the income level of the sample subjects, and involves problems of designing, implementing and interpreting questionnaires. While its applicability may be limited, there is now considerable experience in evaluating the quality of supply of potable water and electricity services. Artificial Market: Such markets can be constructed for experimental purposes, to determine consumer willingness to pay for a good or service. For example, a home water purification kit might be marketed at various price levels, or access to a game reserve may be offered on the basis of different admission fees, thereby facilitating the estimation of values placed by individuals on water purity or on recreation facilities.

THE DISCOUNT RATE

Discounting is the process by which costs and benefits occurring in different time periods may be com-pared. The discount rate to be used has been a general problem in cost-benefit analysis, but it is particularly important with regard to environmental issues, since some of the associated costs and benefits are very long-term or irreversible in nature. In standard analysis, past costs and benefits are treated as "sunk" and are ignored in decisions about the present and future.

Future costs and benefits are discounted to their equivalent present value and then compared. In theory, in a perfect market, the interest rate reflects both the subjective rate of time preference (of private individuals) and the rate of productivity of capital.

Higher discount rates may discriminate against future generations. This is because projects with social costs occurring in the long term and net social benefits occurring in the near term, will be favored by higher discount rates. It is often argued that discount rates should be lowered to reflect long-term environmental concerns and issues of intergenerational equity. However, this would have the drawback that not only would ecologically sound activities pass the cost-benefit test more frequently, but also a larger number of projects would generally pass the test and the resulting in-crease in investment would lead to additional envi-ronmental stress.

Many environmentalists believe that a zero discount rate should be employed to protect future generations. However, employing a zero discount rate is inequitable, since it would imply a policy of total current sacrifice, which runs counter to the proposed aim of eliminating discrimination between time periods-especially when the present contains wide-spread poverty.

In the case of projects leading to irreversible damage (*e.g.*, destruction of natural habitats, etc.), the benefits of preservation may be incorporated into standard cost-benefit methodology using the Krutilla-Fisher approach. Benefits of preservation will grow over time as the supply of scarce environmental resources decreases, demand (fueled by population growth) increases, and possibly, existence value increases. The Krutilla-Fisher approach incorporates these increasing benefits of preservation by including preservation benefits foregone within project costs. The benefits are shown to increase through time by the use of a rate of annual growth. While this ap-proach has the same effect on the overall CBA as lowering discount rates, it avoids the problem of distorted resource allocation caused by arbitrarily manipulating discount rates.

CONCLUSIONS

In order to achieve economically sustainable manage-ment of natural resources and environmental protec-tion, one must effectively incorporate environmental concerns into decision making through the EIA pro-cess.

This presentation has reviewed concepts and tech-niques for economic valuation of environmental im-pacts within EIA procedures. The process of internal-izing these environmental externalities can be facilitated by making rough qualitative assessments early on in the project evaluation cycle-the advantages of which would include:

- early exclusion of alternatives that are not sound from an environmental point of view;
- more effective in-depth consideration of those al-ternatives that are preferable from the environ-mental viewpoint; and
- better opportunities for redesigning projects and policies to achieve sustainable development goals.

In order to fully reflect society's values and preferenc-es about environmental values, non-market methods of estimation can be used and will enhance commu-nity participation through well designed and admin-istered questionnaires and surveys. Research and training about EIA and embedded economic analysis methods is needed in Cuba. As developing countries learn to successfully apply EIA methodologies the goals of sustainable development will become more attainable.

FOREST ECOSYSTEM SERVICES BY LOCAL PEOPLE

Climate change is expected to increase the frequency and intensity of extreme weather events, such as hurricanes, torrential rains and droughts. Rural people often depend on emergency supplies during or just after such events. Forests, in many cases in the past, have provided such emergency supplies or safety nets *e.g.* wood for construction and repair of houses, woodfuel for cooking and fruits and other food to replace the lost crops. The need for these safety nets will further increase when climate change increases the loss of crops. Indigenous groups are often vulnerable to extreme events, especially those events that restrict access to the outside world and markets. However, in such cases, they can usually find sufficient emergency supplies from within the forest until access is restored. In addition, more people have become aware of the different ecosystem services and want to use such services even under non-extreme weather conditions.

Forests as regulators of water quality and quantity have become ever more important, in particular, in areas with frequent droughts and/or frequent torrential rains that may cause erosion, sedimentation and flooding. In Central America, this function may be one of the main reasons for forest protection or restoration by private landowners even though it is possibly based on an erroneous perception of the benefits of the forest, since such functions may not be beneficial in some climate and soil conditions. The impact of climate change on this ecosystem service, however, is still not very well understood, since different species, different environmental and geological settings and different socioeconomic conditions may affect the response of this service to climate change.

LAND TENURE AND OTHER FOREST RIGHT ISSUES

Deforestation and forest degradation in tropical and some of boreal forests are serious problems that contribute to the emission of greenhouse gases as well as to the fragmentation of forests. Deforestation and degradation have a series of direct and underlying causes, but none of these can be resolved if land and forest tenure are not clear or are not enforced. State land is more frequently subject to conversion into agricultural land than privately owned land.

Privately owned and concession forests, however, are increasingly coming under pressure, especially in countries with policies that recognize traditional rights or favour the rights of community inhabitants to their surrounding forests. In the Amazon region, community lands also receive increased pressure, possibly due to the regional infrastructural plans (IIRSA), speculation of future forest values under new international agreements on 8 climate change (REDD+), investment in bioenergy, the relatively large size of many community lands in relation to their population and the lack of financial and human resources to secure their borders.

Since many of the areas with land and forest rights concerns are in remote areas and refer to areas where people may have conflicting interests, regularizing these rights has been a major challenge in the past. Some progress has nevertheless been made in Latin America.

CHANGES IN POLICY ENVIRONMENT

REDD+ Expectations

Probably one of the more notable short-term changes in the policy arena is the discussion of GHG emissions reduction through REDD+ and management, conservation and restoration of forest carbon stocks. Large sums of money have been pledged against the demonstrable reduction of GHG emissions through REDD+, but so far, no international agreement has been reached on emissions reduction targets for developing countries. Further, in many pilot projects, measurable results have been interesting but financial benefits limited.

REDD+ expectations are manifold, depending on the interest group. Some of these expectations are justified, others not, and most are probably too ambitious.

Implementation of REDD+ strategies will have to deal with most, if not all, of the challenges mentioned in this chapter. At the same time it will require the implementation of a monitoring system, the extent and detail of which has not yet been agreed upon. While this has serious implications, the current (international and national) political environment is set to enable projects and countries alike, to meet at least some of these challenges. For the forest manager much of the challenge lies in adjusting management practices in favour of carbon accumulation, while at the same time maintaining biodiversity, recognizing the rights of indigenous people and contributing to local economic development.

Changes in Legislation

In Latin America, many countries implemented new forest legislation in the period between 1995 and 2000. While in some countries this was based on a thorough analysis of the forest sector, in others it was more in response to different pressure groups and based on changes in neighbouring countries.

In some countries (for example Costa Rica), new legislation was relatively successful in achieving the objective of forest conservation, although reducing forest use for timber production considerably (Louman, in print). In others, it has been difficult to implement new legislation if unaccompanied by other measures and if the process was not participatory and consultative.

More recently, countries have realized that they have better results when their new legislation is developed using more participative processes (for example in the DRC and Honduras). However, these processes are too young to be able to assess the true success in terms of increased implementation of legislative requirements.

Climate change will increase the challenge of designing and implementing new legislation that considers new international agreements, conflicts of interest in forest areas, as well as the need for coordination with other sectors. This may involve legislation on land and forest tenure, indigenous rights, the production of fuels and land use planning including restricting the access and use of certain areas or of some species, due to the risk of climate change impacts or the need of soil and water protection or maintenance of biological corridors. In revising forest legislation, it is important to consider all related legislation, so that, for example, legislation or policies oriented at increasing forest area on private land is not nullified by policies or legislation that define forest land as 'un-used' or 'luxury possessions', taxing them relatively heavily or even threatening to expropriate the owners.

Social Responsibility Requirements

Concerns for sustainable development, for the deterioration of the environment and of social relations, as well as for the negative effects of climate change at different scales are influencing market decisions. This can above all be noticed in agricultural product markets, where buyers are looking for products that meet specific environmental and/or social standards. Some banana plantation owners that export to the European market, for example, have started to invest in forest land for conservation and carbon emissions compensation.

New standards have just recently been developed to monitor and evaluate carbon dioxide equivalent emissions from livestock farms in Costa Rica, while the COOPEDOTA coffee cooperative was recently declared carbon neutral. These new developments pose interesting opportunities, more research is required to determine how these mechanisms can be used to improve the maintenance of other ecosystem services (such as water regulation and biodiversity maintenance), strengthen the adaptive capacity of natural and human systems and complement conservation and sustainable use of the existing forest areas within the agricultural landscapes.

Opportunity Costs of Land Use

Meeting REDD+ expectations has much to do with being able to identify the opportunity costs of local actors when they choose forest conservation and management rather than other land uses. Many of the REDD+ cost analyses are based on compensation for lost opportunities, although it has been found that forest conservation on private lands does not only occur for financial reasons.

If lands surrounding forests have high opportunity costs, there is the likelihood of increased pressure to convert those forests to the adjacent land use in order to make them more profitable. Opportunity costs may vary due to variations in market prices of the crops cultivated, government policies that subsidize agricultural inputs or the exportation of the outputs, or policies favouring the production of biofuel. The forest user or owner does not easily influence these factors.

As a group, in particular, if acting within the framework of REDD+, it may be possible to influence legislation, reduce the unequal treatment of forests as compared to agricultural crops, thus making forest management more competitive with other forms of land use.

UNCERTAINTY AND RISK MANAGEMENT IN FOREST CLIMATE

Climate change projections for the future involve a series of uncertainties. It is still not sure what emission scenario will best reflect reality, how these emissions change climate, in particular in relation to the distribution of precipitation or what other factors may play a role in influencing local vegetation and how local vegetation will react to climate and other factors. Thus, forest management for climate change has to deal with a range of uncertainties.

The challenge is to reduce those uncertainties and to design management systems that can deal with unexpected changes. Uncertainty and risk management options may involve monitoring systems (*e.g.* climate, biodiversity, production, and social impacts), early warning systems, working groups that analyze the implications of data obtained through monitoring, mechanisms dealing with risk of income loss, appeal systems for unpopular decisions as well as free prior and informed consent of indigenous and local communities. Flexible adaptive management approaches need to be a part of any management strategy that involves risk and uncertainty. Such strategies will need to include a set of tools, rather than one specific approach, to be able to switch from one to another tool, depending on local conditions, changes in those conditions, and success of already applied tools.

Bibliography

A.K. Shrivastava: *Global Warming*, APH Publication, Delhi, 2007.

Anup Chatterjee: *Global Warming and Climate Change*, Global Publications, Delhi, 2010.

Chanchal Singh: *Global Warming and Climatology*, Akansha Publication, Delhi, 2007.

Chander Bhan Yadav: *Global Warming : India's Response and Strategy*, Rajat Publication, Delhi, 2005.

Ernesto Zedillo: *Global Warming : Looking Beyond Kyoto*, Pentagon Press, New York, 2009.

Gopal Bhargava: *Global Warming and Climate Changes : Transparency and Accountability (3 Vols-Set)*, Isha Books, Delhi, 2004.

Ishita Haldar: *Global Warming : The Causes and Consequences*, Mind Melodies, Delhi, 2011.

Ishita Haldar: *Global Warming*, Mind Melodies, Delhi, 2011.

Karan Pal Singh: *Global Warming and Grassland Ecosystem*, Axis Publication, Delhi, 2010.

M.G. Chitkara: *Global Warming and Climate Change*, APH Publishing, Delhi, 2012.

M.H. Syed: *Global Warming and Gases*, Shubhi Publication, Delhi, 2009.

M.K. Mishra: *Global Warming*, Ritu Publication, Delhi, 2012.

M.L. Dewan and Devendra Sahai: *Global And Regional Perspectives On Global Warming*, Jnanada Prakashan, Delhi, 2010.

M.P. Singh, Bijay S. Singh and Ranveer Kumar: *Global Warming and Climate Change*, APH Publication, Delhi, 2012.

M.P. Singh: *Global Warming and Forest*, Daya Publication, Delhi, 2011.

Mahendra Pandey: *Global Warming and Climate Change*, Dominant Publication, Delhi, 2005.

P. Brown: *Global Warming : Can Civilization Survive?*, Universities Press, Delhi, 1998.

Parikshit Ballabh: *Global Warming : An International Emerging Issue*, Cyber Tech Publication, Delhi, 2009.

Pawan Sikka: *Global Warming : India's Response to Climate Change, Disaster Mitigation and Adaptation*, Uppal Publication, Delhi, 2010.

R K Sharma: *Global Warming : Technology, Nature and Society*, Arise Publication, Delhi, 2006.

R. Chattopadhyay and Mandira Chatterjee: *Global Warming : Origin Significance and Management*, Global Vision Publishing House, Delhi, 2012.

Rajeev Gupta: *Global Warming and Climate Change*, Sonali Publications, Delhi, 2012.

Rajendra Chauhan: *Global Warming : An International Issue*, Saurabh Publication, Delhi, 2010.

S K Agarwal: *Global Warming and Climate Change (Past, Present and Future)*, APH Publication, Delhi, 2010.

S.K. Paneer Selvam: *Global Warming and Climate Change*, APH Publication, Delhi, 2012.

Shagoon Tabin: *Global Warming : The Effects of "Ozone" Depletion*, APH Publication, Delhi, 2011.

Vichitra Kumar Sharma: *Global Warming : A Challenge*, Ancient Publication, Delhi, 2011.

Vijay Mittal: *Global Warming and Climate Change*, Oxford Book Company, Delhi, 2012.

Index

I

J

L

M

N

O

P

R

S

T

U

V